An Introduction
to Experimental Design
in Psychology

An Introduction to Experimental Design in Psychology

A Case Approach

FOURTH EDITION

ROBERT L. SOLSO
University of Nevada–Reno

HOMER H. JOHNSON
Loyola University of Chicago

HarperCollins*Publishers*, New York
Grand Rapids, Philadelphia, St. Louis, San Francisco,
London, Singapore, Sydney, Tokyo

Editor in Chief: Judith Rothman
Sponsoring Editor: Laura Pearson
Project Editor: Jo-Ann Goldfarb
Text Design: Lucy Krikorian
Cover Design: Brand X Studio
Text Art: Accurate Art Inc.
Production Manager: Willie Lane
Compositor: David E. Seham Associates Inc., Typographers
Printer and Binder: R. R. Donnelley & Sons Company
Cover Printer: Phoenix Color

AN INTRODUCTION TO EXPERIMENTAL DESIGN IN PSYCHOLOGY:
A CASE APPROACH, Fourth Edition

Copyright © 1989 by HarperCollins*Publishers*, Inc.

Library of Congress Cataloging in Publication Data
Solso, Robert L.
    An introduction to experimental design in psychology : a case
approach / Robert L. Solso, Homer H. Johnson. — 4th ed.
        p.   cm.
    Bibliography: p.
    Includes index.
    ISBN 0-06-046436-4
    1. Psychology, Experimental. 2. Experimental design.
3. Psychology—Case studies. I. Johnson, Homer H.
II. Title.
    [DNLM: 1. Psychology, Experimental. BF 181 J67i]
BF191.S65   1989
150'.724—dc19
DNLM/DLC
for Library of Congress   88-38859
                                                                CIP

90 91 92 9 8 7 6 5 4 3

Contents

To the Instructor

The purpose of this book is to assist you in teaching students the basic principles of experimental design. Too often the teaching design is obscured by lengthy discussions of statistics, by the philosophy of science, or by a concentration on a highly specialized area of research. After several years of experimenting, we have developed a method of teaching design by which the student learns the principles rather easily and quickly and is able to generalize and apply these principles in a variety of content areas. This method uses cases of actual experiments, and through the analysis of these experiments, the student learns how design principles are applied in research. In this new edition the student will read, critique, or analyze some 80 cases or experiments that exemplify various design principles and problems. In addition to understanding design, the student should become comfortable with the research literature as well as learn much of the content material of psychology. This edition is based on the feedback and suggestions of numerous instructors who used the first editions in a wide variety of courses, ranging from introductory psychology through graduate psychology, as well as courses in other disciplines. Their recommended improvements, as well as our class testing, have strengthened the text considerably.

Teaching by means of examples, or cases as they are sometimes called, has been a traditional means of instruction and, in the best educational environments, is still widely practiced—whether the subject be high-energy physics, carpentry, accounting, computer programing, psychotherapy, creative writing, or cellular biology. And yet, in courses involving experimental design in the behavioral sciences, or more plainly stated, "how to do to research in . . . ," the common practice is to meander through a series of philosophic/theoretical issues that are essentially important to any scholar's education, with little or no effort to establish a link between these principles and the real world of research. We believe this link is necessary to the development of critical thought and the practice of research, and hope that this book will help you be instructive in teaching experimental design. In general we wrote the book from a tutorial standpoint: as if we were a private tutor instructing a student as he or she reads the material. First we present a principle or a problem of experimental design. Then we show how the principle or problem has been dealt with in the psychological literature. In the second section, we provide an annotated review of actual articles, much as a master teacher might do if he or she sat down with a student and critically read an article with him or her. From the comments received from students and instructors alike, the technique is remarkably successful. We hope the present edition will be even more useful and instructive than the previous editions.

Part One of this new edition, which deals with the basic principles of experimental design, has been expanded and is more inclusive. We have long advocated the view that experimental psychology is a laboratory science, but in the current edition we have expanded the view expressed in previous editions—that a researcher needs to be prepared to ply his or her trade in a variety of settings, in or outside the laboratory. New sections on quasi-experimental studies and naturalistic observations have been added. In addition, we have expanded the section on ethics in experimental research to reflect changes in that area over the past few years. A new chapter entitled "The Psychological Literature: Reading for Understanding and as a Source for Research Ideas" has been added to Part One. We trust that this chapter will be helpful to your students in using the library and electronic data banks for accessing information. Also, we discuss the development of research ideas.

In Part Two the format has been changed and several articles have been added and deleted. The scope of articles in Part Two was carefully selected to sample the major areas of psychology, from industrial and practical issues to cognitive psychology, to social psychology, to animal and ethological studies, to practical problems, to cross-cultural studies, to psychotherapy, to single subject designs, to educational psychology, to behavioral modification, to child psychology, and so on. We find that students learn a great deal about the field of psychology in addition to learning about the vast diversity of experimental designs and techniques currently being employed in the behavioral sciences. In this edition, the case analysis notes on articles appear on (or near) the same page as the referenced section. Thus, the student does not have to flip back and forth between the case analysis and the article. Judging from students' remarks to us, this change will be greatly appreciated!

Finally, in Appendixes A and B we have responded to frequent requests to include a section on basic statistics. This section is modest, but will allow the book to be used in a much wider range of courses where basic statistics is necessary and will allow instructors to demonstrate the computational procedures for a large number of the statistical tests demonstrated in this book.

It would be inappropriate for us to tell the instructor how the text should be used; however, we would like to note briefly how we have used these materials with students having only a minimal knowledge of psychology. Our procedure was to make daily reading assignments of a small amount of material and to discuss this material during the next class session. Chapter 1 is introductory material and was usually handled in one or two class sessions. Chapter 2 is expanded for discussion in seven or eight class sessions. The first three sections of this chapter were usually understood by the students without much help from the instructor, but some students needed help with the later sections. In Chapter 3 the four control-problem examples were discussed at length in class to insure that the students understood the design and procedures involved.

The design critiques in Chapter 4 were handled in two ways. One way was to have the students redesign the experiments outside of class and then select students to present their designs in class. The rest of the class was asked to comment on and criticize the new design. The other way was similar to the first, except that the students were put in groups of three or four, and the group presented the new designs. This generated considerable discussion and debate.

The exercise at the end of Chapter 4 asks the students to develop a series of questions to be asked of experiments with respect to their design. Having students do this as a homework assignment and comparing lists in class seems to work quite well.

In the class discussion of Chapter 5, initial emphasis was placed on the problems (through examples) that could arise if a biased assignment of subjects occurs. This emphasizes why the various assignment procedures are used, and the emphasis on the problem seems to generate more appreciation of possible solutions. The procedure for Chapter 6 was similar to that for Chapter 4.

Chapter 7, on the ethics of experimental research, promises to create a lively class discussion. Although the examples are intended to illustrate clearly an ethical principle, there is, in many cases, another side to the issue and it may be possible to have different students debate the issue. Chapter 8 is on practical matters of using the library and developing research ideas. Chapter 9 provides a model experimental write-up and we suggest that students try their hand at preparing a similar paper, using a hypothetical experiment and results.

Part Two of the book contains 15 reprinted articles. These articles were carefully selected to illustrate the design problems raised in the first section. In this addition we have identified 10 "special issues" of experimental design that are embodied in some articles. These issues range from control problems to field-based experiments, to small n experiments, to research with animals, to clinical research. The 15 articles are arranged in four groups with different types of experiments in each group. In general, the articles in each group are ordered from less complex to more complex. Some instructors may choose to assign all articles in the sequence presented, while others may select only those articles that reflect the nature of the course. We hope we have presented a bountiful enough smorgasbord of articles to satisfy the most selective diet.

For beginning students of pyschology, it is important that some of the experiments be discussed thoroughly, even to the point of reading them aloud in class and discussing the article point by point. Reading the psychological literature is a new and somewhat anxiety-provoking experience for most students. A thorough discussion of the articles seems to relieve much of this anxiety as well as acquaint the student with design and technique in a specific content area. Chapters 13, 17, 21, and 25 are to be analyzed by the students. This can be done individually or in groups and makes for good class discussion. Our goal in this book is to make psychological research more understandable, more interesting, and perhaps even exciting to the student. We invite communication from instructors as to how these goals might be better achieved within the context of this book.

An Instructor's Manual, with answers to questions and comments on the ethical section (Chapter 7), is available. The manual also contains test questions.

Finally, we would like to acknowledge the assistance of the following reviewers: Mary Gauvin, Oregon State University; Larry Grimm, University of Illinois–Chicago; and Hal Mansfield, Fort Lewis College. Each read the entire new manuscript and provided us with cogent remarks and valued advice. Brian Oppy and Curt Mearns, who are graduate assistants and graduate students in experimental design at the University of Nevada–Reno, provided thoughtful remarks on this project. To all, we offer our sincere thanks.

Robert L. Solso
Homer H. Johnson

To the Student

You have chosen an excellent time to study experimental design in psychology. Exciting new developments are emerging in nearly every domain of the behavioral sciences, and a strong background in the methodology of psychology will form a basis for understanding these new developments and provide you with the skills necessary for conducting research on your own.

This book is about the methods of psychology. A great deal of the book is devoted to the topic of controlled psychological experiments and the collection of reliable data based on observations. We use a "case approach," which means that each of the principles of experimental design is illustrated by an example, or case, drawn from the professional literature in psychology. Study these cases with great care as they represent excellent examples of skillfully conducted experiments from every major area in psychology—from animal studies to clinical observations, to child psychology, to social psychology, to cognitive psychology, to applied psychology. We also emphasize a section on ethics in experimental research and give some guidelines on how you might go about developing research ideas and writing a research paper.

We believe that a powerful lesson can be learned by studying the techniques of experts. In addition to helping you learn proper experimental design, we hope the book will be instructive in the various topics of psychology and inspire some of you to continue your studies and the work in psychology some of us have started. We would be interested to hear your reaction to the material in this book and how it affected your study and work in the field. Good reading!

RLS
HHJ

An Introduction
to Experimental Design
in Psychology

PART ONE

BASIC PRINCIPLES IN EXPERIMENTAL DESIGN

This book is divided into two distinct sections. Part One focuses on the basic principles of experimental design in psychology while Part Two deals with the analysis of experiments that have appeared in the psychological literature.

Part One begins with a brief introduction to scientific inquiry and methodology in psychology. The remainder of this part spells out the fundamentals of research design as they apply to experimental psychology. A large portion of the material in Part One is devoted to the issue of experimental control, or the means that have been developed to insure the integrity of a psychological experiment.

Two chapters in Part One are called "Design Critiques," which are brief descriptions of experiments that contain at least one conceptual or technical flaw. As you read these critiques, try to discover the error. Practice with these problems may strengthen your own ability to design experiments devoid of error.

Part One ends with three important and practical chapters: "Ethics of Experimental Research," "The Psychological Literature," and "Conducting Research and Writing a Research Paper." These final chapters are intended to guide you through the experimental process from the conceptualization of an experiment, through running an experiment, to submitting a manuscript for publication.

The whole of Part One is intended to give you the rudiments of good experimental design to the completion of a well-controlled and valid experimental paper.

CHAPTER
1

An Introduction to Scientific Inquiry

WHAT IS SCIENCE?

If an undergraduate student is asked "What is science?" a common answer is that science is physics, chemistry, and biology. This definition suggests that science is expressed in certain specific subject areas. There is some support for this definition of science. For example, college students have to take a number of required "science courses," and these courses usually are drawn from physics, chemistry, and biology. However, this definition does not appear to be adequate. If you ask why chemistry is a science but history is not, or why physics is a science but music is not, the definition of science becomes more complex and usually confusing. People usually argue that science deals with facts (yet, so does history), or that science deals with theories (so does music), or that science involves laboratory experiments (but what about astronomy or the classification of plants in botany). This is usually where the discussion ends, with the layman saying that he or she does not know exactly how to answer the question.

The layman's difficulty with an adequate definition of science is also shared by the scientist. There are various definitions of science. Many of these specify the collection of facts, the use of experimentation as a method of "proof," the use of theories as tentative explanations, and so on. Some definitions emphasize the ongoing or dynamic nature of science—the search for and discovery of new "facts," and new theories arising to replace the old, much as the theories of Einstein replaced those of Newton. James Conant (1951, p. 25) expresses this quality of science when he defines science as "an interconnected series of concepts and conceptual schemes that have developed as a result of experimentation and observation and are fruitful of further experimentation and observation." By observation and experimentation, the scientist attempts to ascertain what is related to what, or what "cause" is related to what "effect." These "facts" are put into conceptual schemes. These schemes (or theories, or models) are tentative; they are the best way we can explain or interrelate the information we have on hand. New information, and then new conceptual schemes, will arise to replace the old.

Modern scientists—be they physicists, geologists, astronomers, anthropologists, or psychologists—approach their specialized field with a few basic assumptions about the structure of the universe. Common to scientific thought is the assumption that nature is somehow structured by laws that govern its operation. Objects, such as a lead ball, fall toward the earth if dropped from a high structure, such as the Leaning Tower of Pisa—a simple, but fundamental law of physics. Furthermore, this observation can be repeated with essentially identical results. Natural laws are reliable.

The speed at which objects fall is not consistent. An object, such as a lead ball, falls slowly at first and then accelerates. First, through observation, the general "lawful" properties of acceleration are discovered. From these observations are sometimes derived general principles or models that incorporate more and more features of the universe in ever-widening theories.

Behavioral scientists, including psychologists, go about their scientific work with the same assumptions about the lawful nature of nature. Only the subject matter and technical equipment, not the scientific method, is different. For the most part behavioral scientists study human and animal behavior, not accelerating lead balls, geological pressure plates, the evolution of supernovas, or simian development. On the surface, the subject matter and the specialized tools used to study the subject matter seem to separate disciplines, but the scientific method and the assumption of a lawful universe whose secrets may be revealed through observation and experimentation unites all scientists in a common adventure. Join us as we further examine this theme.

WHY A SCIENCE OF PSYCHOLOGY?

You may ask, "Why are psychologists so concerned with developing a science of psychology?" The simple answer is: Psychologists attempt to understand the lawfulness of behavior; and to understand, to predict, or to control behavior with any precision is a very difficult task. Scientific inquiry seems to offer the possibility of handling such a task, and psychologists adopt it because it is the best approach we have that may give us the precision we want.

Laypersons also understand behavior; at least, they understand behavior to the extent that they can coexist with other people. They have principles of behavior; for example, "absence makes the heart grow fonder." But there is a contrary principle that goes "out of sight, out of mind." To *predict* precisely which of these principles will hold in a given case is difficult. Furthermore, their understanding of behavior is highly colored by their own perspectives. Campus disturbances, for example, are called everything from left-wing plots to right-wing plots to anarchist plots, depending on one's perspectives (and anxieties). Nonscientists do not normally engage in systematic studies of behavior; they do not formulate hypotheses about behavior and then test these hypotheses; they do not test their conceptual schemes in any organized manner. But psychologists do, and the techniques that psychologists use to test their hypotheses are described as scientific inquiry.

ON THE SCIENTIFIC METHOD

From the earlier discussion an exact definition of science is somewhat elusive. The same problem arises when one attempts to define the scientific method. Some authors seem to imply that *the* scientific method consists of a few simple steps that, if followed, will inevitably lead to amazing and dramatic discoveries about nature. If there is one method of science, why does science progress in such a slow, fumbling, bumbling manner? A well-publicized example is the fact that we have attempted to find a cure for cancer for at least half a century, and in spite of the use of some highly talented researchers and millions of dollars, the progress has been slow.

Scientific inquiry consists of a variety of techniques, approaches, strategies, designs, and rules of logic. These vary from problem to problem and discipline to discipline. In this book an attempt is made to give the student an indication of the experimental techniques used in psychology.

Psychological Experiment

We use the term *psychological experiment* to refer to investigations in which at least one variable is manipulated in order to study cause-and-effect relationships. Emphasis will be placed on experimental research in which the researcher manipulates some factors (variables), controls others, and ascertains the effects of the manipulated variable on another variable. This type of research best illustrates the researcher's attempts to control relevant variables and to find the cause-and-effect relationships between them. It is the search for such relationships that is so characteristic of scientific research.

Other types of scientific inquiry are commonly made in psychology. These are valid techniques and have led to important discoveries.

Naturalistic Observation

When Charles Darwin planned an expedition to the Galapagos Islands during the early part of the nineteenth century, he did so with the intention of making observations of the natural life he might find there. He did not plan a series of "experiments" in which factors would be manipulated and their effects measured. Given Darwin's interest and the subject matter, *naturalistic observations* (i.e., observations made in natural or native settings) led to a theory that has been one of the most influential developments in science.

Naturalistic observation involves the systematic recording of perceived information. The setting may be as undefiled by humans as parts of the Galapagos Islands in Darwin's time, or they may be as "civilized" as the interior of a health club in the heart of Los Angeles. Scrutiny of the minute details of the mating rituals of giant tortoises, iguanid lizards, craneated tastos, and geckos on a south Pacific island or the subtle precopulatory maneuvers of yuppies in a health club are both examples of naturalistic observations.

For a period, naturalistic observations in American psychology seemed as taboo as some of the things observed. Recently, naturalistic observations have gained greater popularity and are once again considered an important method of gathering data. When making naturalistic observations it is important to make *objective* and systematic observations to guard against the distortion of information through personal prejudices, feelings, and biases.

SCIENTIFIC METHODOLOGY

Our approach to the use of scientific methodology in the study of psychology is established on two principles. The first of these principles is that scientific observations are based on sensory experiences. We see, hear, touch, taste, and smell the sensations of the world. Observations made under certain defined circumstances, or *controlled conditions* as they are called in experimental psychology, should correspond to the observations made by another scientist under comparable conditions.

This feature of replicating the results of an experiment is called *reliability of results* and is a major requisite of scientific credibility.

However, because our sensory systems are limited in capability as well as scope, many signals outside the range of normal sensitivity remain unnoticed, so that those things that are detected take on a disproportionately greater significance. We call this the *tyranny of the senses* as it is difficult to consider the importance of some of the real phenomena in the universe that lie outside the range of unassisted human perception.

Consider the electromagnetic spectrum whose presence is ubiquitous. At one end of the spectrum are cosmic rays, gamma rays, and X rays, and on the other end are radio and TV waves. In the middle, between about 400 and 700 nanometers, are waves that are detectable by the human eye. For the vast majority of the life of the human species, the energy that fell within the visual spectrum was considered to be "reality." The same constricted view of reality applies to all the other senses. Even though we have become aware of the presence of other forms of energy in our world, we continue to emphasize the sensations that we can detect through the ordinary senses.

Some augmentation of the senses has been achieved through the development of technology. Instruments and techniques in science are designed primarily to make "visible" those things that are "invisible" to the unaided sensory system. These instruments—the microscope, radio telescope, and spectroscope, for example—translate energy outside the normal range of human detection into signals that can be understood by humans. In psychology many sophisticated instruments have been developed that allow us to see deep within the psyche of a species and reveal secrets of human and animal life that were left to conjecture and speculation only a few years ago. We will encounter many of these techniques in this book, and we remind the reader that if a technique does not yet exist for his or her topic of interest, there is no prohibition against the invention of new techniques.

The second principle upon which the science of psychology is based is that observations from our senses are organized logically into a structure of knowledge. Frequently in experimental psychology these systems of knowledge are called *models*, or statements about observations and their relationships. Cognitive psychologists may, for example, develop a model of memory that may be based on their observations of two types of memory and the laws that govern their relationship and storage of information. The structural web that ties observations into models and models into theories is based on the principles of logic, which in the current context has developed out of the rich history of Western empirical thought. It is our purpose here not to deal in detail with these ideas, but rather to discuss some topics in the philosophy of science in relation to experimental psychology. The interested student will find extensive writings on this topic.

DEVELOPMENT OF THOUGHTS AND HYPOTHESES IN EXPERIMENTAL PSYCHOLOGY

One of the most difficult tasks confronting beginning students in experimental psychology is to organize their thoughts and develop a testable hypothesis on a given topic. For many reasons this is difficult not only for the novice but also for the seasoned researcher, but one major reason is the lack of knowledge. New research ideas rarely, if ever, erupt spontaneously out of an intellectual void; rather new ideas and hypotheses usually are built on existing knowledge and past research. Therefore, our best advice on how to develop new thoughts and hypotheses in experimental psychology is to immerse yourself in the literature in a branch of psychology that holds some real interest for you.[1] Read, discuss, investigate, and otherwise become well versed in the subject matter. But passive knowledge is not enough. As you acquire knowledge about a topic, question the premise, the conclusions, and the technique and relate it to your knowledge of other matters. The development of new ideas in psychology, as well as other disciplines, rests on the acquisition of the fundamental elements of a subject *and* flexibility in thinking that allows one to combine and recombine the elements of thought in increasingly novel and meaningful patterns. It is essential that once the elements are combined in a meaningful pattern or a new hypothesis, it is recognized for its originality and significance.

The final ingredient in the development of creative and meaningful research in psychology is dedication. As we review the lives of great scientists—Pavlov, Bartlett, Curie, Tinbergen, Darwin, Wundt, James— we are impressed with their ability to devote endless hours of their lives in the pursuit of truth. An inquiring, creative mind can be developed, given knowledge, flexibility in thinking, the availability of resources, and dedication.

New ideas are based on old ideas, new inventions are based on old inventions, and new hypotheses are based on old hypotheses. Contrary to popular lore and media fiction, scientific advancements are made through a series of experiments in which small increments of progress are reported, rather than in a single brilliant experiment in which a major

---

[1] Recently, the growth of knowledge in nearly every branch of scientific inquiry— including psychology—has expanded so rapidly that it is difficult for students and professional scientists to keep up with current facts and theories in their field. More and more we are seeing the use of data banks, which store vast amounts of information in computer memories. (See Solso, 1987, for a more complete discussion of this topic.) As a consequence of the explosion of scientific information a first step in the experimental process is frequently a computer-assisted search of the literature.

discovery is disclosed. Of course, we all aspire to make scientific break-throughs and your budding enthusiasm for achieving scientific eminence should not be discouraged, but such profound achievements are few and far between, and very significant research investigations that fall short of seminal programs can nonetheless contribute mightily to the overall growth of scientific knowledge.

To illustrate the point of the accumulation of knowledge, consider an innocent question asked by one of the authors'[2] sons a few years ago: "Who invented the automobile?" Trying to be instructive, the author told him that in about 1886 Karl Benz invented the automobile. "Wow, he must have been a real genius to figure out the engine, the brakes, the sparkplugs, the wheels, and how everything worked together!" "Well, there were others, such as Henry Ford and Olds, and Daimler and even the person who invented the wheel . . . Oh!" the author remarked in a moment of self-knowledge, "I think I may have misled you. Not one person invented all of the components of the automobile any more than a single person invented the television, or theories of learning, or the symphony. Many people made many significant discoveries which led up to the invention of the automobile and other scientific discoveries. It is almost as if the inventors of the automobile *combined* the knowledge of others in a unique way and then 'invented' only those additional things that were necessary."

It seems to us that the development of knowledge in psychology progresses along similar lines. Given an inquiring mind and determined heart, many important scientific truths lie waiting to be uncovered by new scientists. Past discoveries beget future discoveries, past knowledge begets future knowledge and even past wisdom may beget future wisdom.

In this chapter, an attempt was made to answer several questions frequently raised by students: What is science? What is the scientific method? Why are psychologists so insistent on being scientific? What is a psychological experiment? What is naturalistic observation? What is the source of hypotheses in psychology? Although the answers have been brief, it is hoped that they will provide some general understanding and justification for what follows in this book. The next chapter will offer a more specific attempt to examine some of the basic concepts of experimental design.

DEFINITIONS

The following terms and concepts were used in this chapter. We have found it instructive to go through the material a second time and define each of the following:

[2]RLS.

scientific method

psychological experiment

naturalistic observation

controlled conditions

tyranny of the senses

models

dedication of scientists

accumulation of knowledge

experimental design

reliability of results

new ideas in science

development of thoughts

CHAPTER

2

Anatomy of Experimental Design: Design Strategies

THE LOGIC OF EXPERIMENTAL DESIGN

An Italian scientist named Spallanzani attempted to determine what part of the semen stimulates the egg cell to develop into a fetus and then into a child. On the basis of some earlier studies, Spallanzani hypothesized that it was the sperm cell. To test this hypothesis, Spallanzani artificially inseminated dogs with either the normal semen or the seminal fluid with the sperm cells filtered out. The bitches inseminated with the normal fluid became pregnant, whereas the bitches inseminated with the sperm-free filtrate did not become pregnant. Thus, Spallanzani demonstrated that it was the sperm cell that stimulated the egg cell to develop.

This experiment was conducted around 1785, and it was conducted by a biologist, not a psychologist. However, in spite of its age and despite the fact that it was a biological experiment, Spallanzani's work illustrates basic principles of experimental design that are frequently used in psychology today. Spallanzani started with a hypothesis that was based on previous research. In order to test this hypothesis, he used an artificial insemination technique (actually, he invented the technique). By using this technique, he could *control* the insemination process and *manipulate* the content of the fluid prior to its injection into the dogs. Having this control, he set up a two-group experiment. One group of bitches (the experimental group) was inseminated with the sperm-free filtrate, and another group of bitches (the control group) was inseminated with the normal semen. The two groups were treated alike in all other ways. Since the *only* difference between the two groups was whether or not sperm cells were present in the seminal fluid, then any difference in the pregnancy rates between the two groups must have been due to this manipulation. It was this type of logic and this type of design that enabled Spallanzani to arrive at his (valid) conclusion.

Most psychological research uses the very same type of logic. In the simplest case, one factor *(variable)* is manipulated by the experimenter, and all other factors are held constant. The manipulated variable is called the *independent* variable, and the effects of this manipulation upon another variable, called the *dependent* variable, is observed. (In Spallanzani's experiment the dependent variable was the number of dogs that became pregnant.) Usually, one group of subjects receives one *level* (type or amount) of the independent variable, and another group of subjects receives another level of the independent variable. Since both groups of subjects are treated exactly alike *except* for the independent variable, then any difference observed in the dependent variable is due in all probability to the independent variable. In this manner, the psychologist hopes to ascertain precisely the effects of one variable on another and to build knowledge about cause-and-effect relationships *(functional relationships)* in behavior. Some writers have talked about this whole process in terms of a "theory of control." This label is a good expression of the idea behind research. An attempt is made to control variables either by manipulation or by holding them constant. Once this control is obtained, the determinants of behavior can more likely be discovered.

With this overview of experimental logic, the following material will illustrate some of the concepts and designs in psychological research.

INDEPENDENT AND DEPENDENT VARIABLES

It has already been noted that in the simplest of experimental situations we manipulate one variable and observe the effect of this manipulation on another variable. The manipulated variable is called the *independent* variable. The variable being observed is called the *dependent* variable. The experimenter simply measures the subjects' responses, which are called *dependent responses*.

EXAMPLE

Lorge (1930) investigated the problem of whether performance on a task is better when a person practices the task continuously without interruption (massed practice) or when a person distributes his or her practice sessions with rest intervals in between them (distributed practice). Lorge chose for his task a mirror-tracing problem in which the subject traces a pattern (e.g., a star) but can only see the pattern (and his or her hand) in a mirror. Lorge had three groups of subjects, each group tracing the pattern 20 times. For one group of subjects the 20 trials were completed one after another with no rest between trials. For a second group of subjects each trial was followed by a 1-minute rest period. A third group of subjects practiced one trial a day for 20 days; thus there was a 24-hour interval between trials. Lorge's measure of performance was the length of time it took the subjects to trace the pattern—the shorter the time, the better the performance. His results indicated that except for the first trial, the average performance of the 24-hour interval group was better than that of the 1-minute interval group, and that of the 1-minute interval group was better than that of the continuous-practice group. The conclusion was that on this type of task, spaced practice on the average leads to better performance than massed practice.

In the Lorge experiment the independent variable was quantitative: the length of time interval between trials—none, 1 minute, or 24 hours. Note that the experimenter actively manipulated this variable. The dependent variable was the time it took the subjects to trace the pattern, and the experimenter merely measured (or recorded) this variable. Lorge attempted to keep all other factors constant. For example, the subjects all performed the same task. They all had the same number of trials on the task. The task was unusual and required special equipment. Had the task involved learning lists of words, the subjects might have received additional (uncontrolled) practice during rest periods.

EXAMPLE

Asch (1952) conducted an experiment to determine if the first information you hear about another person is more important in forming an impression of that person than later information (primacy effect), or if later information is more important (recency effect). Asch used two groups of subjects. A series of adjectives that were said to describe a certain person were read to both groups; however, one group received positive information first and negative information last, while the second group received negative information first and positive information last. The adjectives (and order) read to the group receiving positive information first were intelligent, industrious, impulsive, critical, stubborn, and envious. The group receiving the negative information first were read the same list but in reverse order. Asch then asked the subjects to write down their general impression of the person. The group receiving the positive information first described him as an able person who had certain shortcomings. The group receiving the negative information first described him as a "problem" whose abilities were hampered by serious difficulties. Since the group that received the positive information first tended to have a positive evaluation of the stimulus person, and the group that received the negative information first tended to have a negative evaluation, Asch concluded that there is a primary effect in impression formation.

The independent variable in the above example was the order of presentation of the information—either positive to negative or negative to positive. The dependent variable was the subjects' descriptions of the person. All other variables were held constant; for example, both groups received exactly the same adjectives. Since the only difference between the treatment groups was the order of presentation, then the different impressions must be due to this manipulation. Thus, Asch was able to demonstrate a "law of primacy" in the formation of impressions.

Frequently, a study will be undertaken in which a subject variable acts as an independent variable. A subject variable is a variable such as IQ, authoritarianism, sex, or some other trait or characteristic that a person "carries with him." For example, an experimenter may want to ascertain the influence of the personality trait of authoritarianism on concept learning, so he or she *selects* two groups of subjects. One group consists of people who receive high scores on a standard test for authoritarianism and the other group consists of people who receive low scores on the same test. Both groups perform the same concept learning task, and the amount or speed of learning is the dependent variable. Note here that the experimenter has not actively *manipulated* authoritarianism but has *selected* for it. Some experimental psychologists would view this type of study as being "nonexperimental" as the independent variable (in this case a subject variable) is *selected*. Strictly speaking, an experiment in psychology is a piece of research that measures the effects of a carefully *manipulated* independent variable.

EXAMPLE

A study by Camilla Benbow and Julian Stanley (1980) used the subject variable of sex in trying to differentiate mathematical ability between boys and girls. The researchers gathered test scores on 9927 seventh- and eighth-graders who were *matched* on (i.e., had equal amounts of) mathematics courses. The male and female students were given the Scholastic Aptitude Test. On the mathematics portion of this test, the boys' average scores were significantly higher than the girls' average scores. Also, more than 50 percent of the boys scored above 600 (out of a possible 800), while not one of the girls scored above 600. The researchers also reported extreme scores. The highest score for a boy was 190 points higher than the highest score for a girl. As might be expected, the results were controversial (see *Science*, April 10, 1982, p. 114, for a sample of the reaction), but the researchers stand by their data and suggest that one of the purposes of scientific inquiry is to search for the cause of such data.

There are numerous subject variables that have been "selected" as independent variables. For example, children of wealthy parents and children of poor parents have been asked to draw pictures of dimes or quarters in order to examine the relationship of economic background to estimates of size of money. There are many experiments that compare the responses of males versus females on a variety of tasks. A comparison of the incidence of lung cancer for people who smoke cigarettes versus people who do not smoke has been made. Typically, such investigations make use of correlational analysis in which the incidence of lung cancer is compared with cigarette use.

EXPERIMENTAL AND CONTROL GROUPS

Whereas in many experiments the various treatment groups consist of different levels of the independent variable (see the examples on the preceding pages), on other occasions an experimental group and a control group are used. Although these experiments can be described using our definition of independent variable, these concepts are discussed here in a separate section because they present some unique problems of experimental design.

The *experimental group* of subjects is the group that receives the experimental treatment—that is, some manipulation by the experimenter. The *control group* of subjects is treated exactly like the experimental group except that they do not receive the experimental treatment. The Spallanzani experiment is a good example of this. The group of bitches receiving the sperm-free filtrate was the experimental group, and the group receiving the normal semen was the control group.

EXAMPLE

Blind persons are very adept at avoiding obstacles; however, little was known about how they do this. One hypothesis was that blind people have developed a "facial vision"; that is, they react to air pressure on exposed surfaces of the skin. A second theory was that avoidance of obstacles comes through the use of auditory (hearing) cues. Supa, Cotzin, and Dallenbach (1944) set out to test these theories. They had blind people walk around in a large room in which obstacles (screens) had been set up. Two experimental treatments were used. In the first treatment, blind subjects wore a felt veil over their face and gloves on their hands (thus eliminating "skin perception"). In the second treatment, blind subjects wore earplugs (thus eliminating auditory cues). A third treatment was the control treatment, in which blind subjects walked around the room as they would normally. The results indicated that subjects in the control and in the felt-veil treatment avoided the obstacles every time, but the subjects in the earplug treatment bumped into the obstacles every time. Based on these results, the authors concluded that the adeptness of the blind in avoiding obstacles is due primarily to their use of auditory cues and not to any "facial vision."

The above experiment is an abbreviated version of a series of experiments on the perception of normal and blind subjects. In this example, it is difficult to specify an independent variable. The experiment is most easily described as having two treatment groups—one in which "facial vision" is eliminated and one in which auditory cues are eliminated. The control group is treated the same as the other treatment groups except they do not receive the precise treatment. The control group provides a "normal" baseline in order to determine whether the treatments improve or hamper the avoidance of obstacles. The dependent variable in this study—that is, the ability to respond to sensory-deprived cues—was measured by the number of times the subjects walked into the obstacles.

Sometimes more than one control group is needed. For example, in pharmacology a *placebo* control group is frequently used. A placebo group is best described as a group of subjects who are told that they are getting a treatment that will improve their performance or cure some symptom but actually are not. This type of control group is also used in testing the effectiveness of therapy.

EXAMPLE

Paul (1966) conducted an experiment to test the effectiveness of two types of therapy in treating "speech phobia." His subjects were students enrolled in public speaking classes at a large university. Paul took 67 students who had serious performance problems in the course (speech phobia) and assigned them to one of four conditions. One group of 15 subjects received a form of behavior therapy. A second group of 15 subjects received an insight therapy. A third group of 15 subjects received a placebo condition in which they were given harmless and ineffective pills and were told that this would "cure" them of their problems. A fourth group of 22 subjects was not given any treatment but simply answered questionnaires given to the other three groups. All subjects had to give one speech before the treatment began and one after the treatment had been completed. One dependent variable was the amount of improvement shown by the subjects from the first to the second speech based on ratings made by four clinical psychologists. The four psychologists were not involved in the therapy to the subjects nor did they know which subjects were in which treatment group. The results indicated that 100 percent of the behavior therapy subjects improved, 60 percent of the insight therapy subjects improved, 73 percent of the placebo subjects improved, and 32 percent of the no-treatment control subjects improved.

The Paul experiment illustrates the need (in some experiments) for different types of control groups. The interpretation of the results of the above experiment would have been quite different if Paul had not used a placebo control group. Without the placebo control group it would have appeared that insight therapy was actually effective (as a therapy) in improving speech difficulties. With the placebo group placed in the design it now appears that the insight therapy was not really effective as a therapy, but may have only acted as a placebo. In fact, there was some tendency for the placebo group to improve more than the insight therapy group. The experiment also points out the need for a no-treatment control group. Over 30 percent of the subjects in this treatment improved in spite of the fact that they received no treatment, and this may form a baseline to measure the improvement rate if subjects are just left alone. Different types of control groups are used in different areas of research. Experiments in which animals are given operations and a part of the brain is removed sometimes use a control group that undergoes all of the surgical procedures except that the brain is not tampered with. This would control for such factors as postoperative shock causing the effect found in the experimental group. The point to be remembered is that the control group is treated exactly like the experimental group except for the *specific* experimental treatment.

This experiment also illustrates an important control procedure that is used to avoid *experimenter bias*. The psychologists who rated the subjects' speaking performances were not the same people who treated the subjects in therapy, nor did they know which subjects were in which experimental group. It is reasonable to assume that therapists might be biased (or defensive!) when it comes to judging the improvement of their own patients. Furthermore, the four judges might have their favorite therapy, and if they knew which subjects had received this therapy, they might be prone to see more improvement for these subjects than for subjects in the other experimental conditions. Or perhaps the judges would have assumed that the subjects in the no-treatment control group could not have improved (because they received no treatment) and would therefore rate the performance of these subjects as very poor. Paul controlled for these potential biasing effects by using independent judges, as well as by keeping the judges blind as to what experimental group a particular subject was in.

The term *blind* is used in a special sense in experimental research. *Single blind* usually means that the subjects of an experiment are not informed as to which treatment groups they are in and may not be informed as to the nature of the experiment. The *double blind* is frequently used in drug research or any research that involves observers who are judging the performance or progress of the subjects of the experiment. Using this latter procedure, both the judges and the subjects are kept blind as to the type of drug (or the experimental treatment) that is being used as well as the type of effect that might be expected.

EXAMPLE

An obvious demonstration of experimental bias was shown by Rosenthal and Fode (1963) in a study that seemingly allowed for little subjective judgment on the part of the experimenter. A group of student experimenters who had some background in experimental psychology was asked to evaluate maze performance of two groups of rats. One group, so the experimenters were told, was selected from a long strain of maze "bright" rats, while the second group was selected from a long strain of maze "dull" rats. The experimenters conducted a study of maze performance and, as expected, the maze "bright" rats did significantly better than the "dull" rats. The odd thing was that the rats were randomly divided from a standard sample of rats. Had the maze "bright" rats actually performed better than the "dull" rats? Probably not, but the experimenters who observed the "bright" rats expected them to perform better, and this expectation seemed to cloud their observations.

   Rosenthal has written several books on experimental bias that are recommended to the interested student (see Rosenthal, 1966, 1969; Rosenthal and Rosnow, 1969).

Within-Subject Control

Basically, control of subjects in psychological experiments can be divided into two simple types. In the first type, as illustrated in the previous examples, two (or more) sets of subjects are treated to different conditions, one of which may serve as a control condition. Contrasts are made between the results of treatments. This design is called a *between-subjects design* (see Chapter 3, Models 1 and 2) as the contrasts are made between two or more groups of subjects. We can state the relationship as:

Ss A	Experimental Condition A	Measure Effects
Ss B	Absence of Experimental Conditions	Measure Effects

In the second type of design each subject is "treated" to two or more experimental conditions. The experimental observations are made between the results obtained on one treatment as contrasted with the results obtained on another treatment or treatments. This type of design is called *within-subject design* (see Models 3 and 4 in Chapter 3), as the experimental observations are made within each subject when treated under different conditions. We can state the relationship of experimental results based on contrasts of XC_1 vs. XC_2 vs. XC_n as follows:

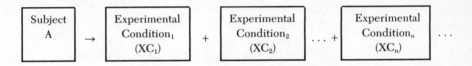

Within-subject designs offer several advantages over between-subject designs. They require fewer subjects, as each subject is treated to several conditions rather than, as in the between-subject design, requiring two sets of subjects for each experimental condition. Also, there is no need to match experimental and control subjects as each subject serves as his or her own control. Finally, from a statistical point of view, the "variability" (the amount scores vary around a center score) may be less for within-subject experiments than for between-subject experiments.

Nevertheless, some real problems may be encountered with a within-subject design that may render it inappropriate for some experiments. Generally, experiments in which subsequent results may be affected by previous experience should avoid within-subject designs (unless, of course, that is the focus of the study). As an obvious example, consider the ease of learning two computer word-processing programs: Starry Words and Perfect Words. In a typical within-subject design a subject might learn Starry Words and then Perfect Words. The contrast would be between the ease of learning Starry Words and Perfect Words. However, the experience with Starry Words might make it easier to learn Perfect Words (or, equally contaminating, there might be a negative effect). In either case, the experimenter does not know what effects a previous experimental experience might have on the subsequent experience. Thus, the results would be spurious.

One additional problem may exist with these designs. That problem is called the *demand characteristics* of experimental design, which means, plainly stated, that the subject figures out what the experimenter wants and then tries to comply (or not give the experimenter what is wanted!). For example in the within-subject design, because a subject has had previous experience with the independent variable he or she has a better idea as to what the experiment is about and can then bias his or her responses to satisfy (or frustrate) the experimenter.[1] Demand characteristics are a hassle in experimental psychology, especially in studies of social psychology, consumer psychology, child psychology,

---

[1]Good students often satisfy their professors by giving in to the "demand characteristics" set forth in a class. Sometimes they discriminate between what is wanted by the professor and what they think is true. One bright student wrote on his examination "The *correct* answer is . . . The answer the professor wants is . . ." The professor wrote back "Answer 1 = F, Answer 2 = A. Overall grade C."

cognitive psychology, and abnormal psychology, and they should be seriously considered a contaminating factor before running an experiment.

In Chapter 3 some sophisticated means for getting around some of these problems are discussed.

EXAMPLE

Psychologists have been interested in the way information is "coded" or recorded in memory after it is perceived. In a series of important experiments done by Posner and his associates (see Posner 1969; Posner, Boies, Eichelman, & Taylor, 1969; Posner & Keele, 1968), it was found that subjects formed a visual code initially and then a name code for letters. Following these experiments Solso and Short (1979) did an experiment on color codes. The Solso and Short experiment allows us to introduce several relevant features of contemporary experimental design, including within-subject design, reaction times, tachistoscopic procedures, and visual representation of data.

In the Solso and Short experiment subjects were shown a color square and either an associate to the color, the name of the color, or another color square. The second stimulus appeared either simultaneously with the first color, after 500 msec ($\frac{1}{2}$ sec), or after 1500 msec ($1\frac{1}{2}$ sec). Subjects were asked to respond as quickly as possible by pressing a reaction-time key if the two stimuli (the color square and the secondary stimulus) matched. An equal number of nonmatched secondary stimuli were shown to assure that the subjects were really reacting to the matching task. For example, in a given series of trials a subject might see a red square and then 500 msec later the word BLOOD, which, being a correct associate, should produce a match response. In a second trial the same subject might see a green square presented simultaneously with the word GREEN, which is also a match. In a third trial the same subject might see a blue square followed 1500 msec later by a green square, which is not a match, and so on. The design could be thought of as a 3 × 3 × 2 design (factorial designs are discussed in Chapter 3), with the first three representing the color associates, color–color words, and color–color conditions, and the second representing the "delay" between stimuli: 0, 500, or 1500 msec. The 2 represents either a match (RED–BLOOD) or a mismatch (RED–BLUE).

The design is a within-subject design. Each subject was treated with each of the experimental variables. The results are shown in Figure 2.1 and indicate that initially subjects respond fastest to color–color (RED–RED) pairings and slowest to color–association pairings (RED–BLOOD), but as the interval between the first stimulus and the second stimulus is lengthened to 1500 msec, the reaction time to matching the associate (e.g., BLOOD) increases significantly, so that the reaction times for all three groups are similar. The authors interpret the results in terms of the parallel development of codes to colors (see Figure 2.2).

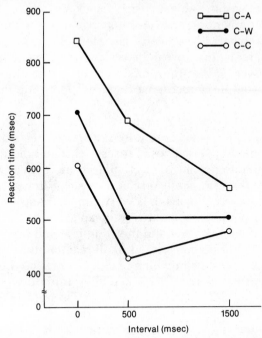

Figure 2.1   Reaction times of various match conditions as a function of priming interval: C–A = color to associate; C–W = color to word; and C–C = color to color. From Solso & Short, 1979.

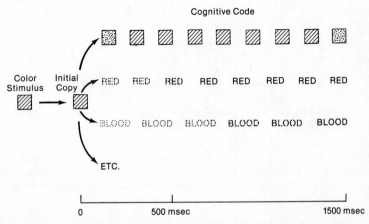

Figure 2.2   The development of color codes. From Solso & Short, 1979.

In addition to within-subject design issues, the above experiment also introduces the topic of *mental chronometry*, or measuring the speed of mental events. This topic has been a favorite tool of many recent cognitive psychologists, but its history can be traced to the last century. The basic principle involved is that the processing of information requires time and that more complex processing requires more time than simple processing.[2]

Reaction-time experiments are usually based on a *forced choice* procedure in which the subject is required to make a decision between two (or sometimes more) alternatives. Because reaction-time experiments are highly susceptible to practice effects, subjects are normally given a series of warm-up trials to, as some researchers say, "limber up their index fingers." Much more is involved, however, in warm-up, or practice, trials. Familiarization with the apparatus, the experimental environment, the experimenter, the procedure, and the stimuli are all important factors in warm-up and should be considered in all reaction-time experiments, as well as other experiments in which practice effects might be significant.

Finally, the experiment by Solso and Short introduces us to a standard piece of equipment: The *tachistoscope*, or T-scope. This device is not new to experimental psychology, but has recently become very sophisticated. It is primarily used in experiments that require great precision in the presentation of visual stimuli for brief time periods and accuracy in measuring reaction times. It is an enclosed box that may have several "channels," or viewing chambers. A typical T-scope has three channels and is capable of exposing three different stimuli. The exposure time is controlled by small fluorescent tubes that show the stimuli when illuminated. Computers have replaced many of the functions of tachistoscopes (see following box), but for very precise presentation of stimuli, the T-scope remains an essential tool.

COMPUTERS AND EXPERIMENTAL PSYCHOLOGY

In recent years a significant change in laboratory technology has taken place largely due to the use of small personal computers (PCs). These desktop versions of their ancestral behemoths now find a place in many experimental laboratories.

Originally computers were used in the statistical analysis of data, or "number crunching"; however, psychologists soon discovered that PCs were excellent devices for the presentation of stimuli, especially for the presentation of timed stimuli.

Programs can be written that make the PC one of the most useful and necessary pieces of equipment in the laboratory. Some corporations specialize in programs

---

[2]A more complete history and discussion of mental chronometry can be found in Snodgrass, Levy-Berger, and Haydon (1985).

for PCs. For example, Psychological Software Tools, in Pittsburgh, has a device called Micro Experimental Laboratory (MEL), in which a researcher may design an experiment; the program will translate the design into a computer program that will collect the data (with millisecond accuracy) and pass it to a data management system.

The Two Meanings of Control

The preceding discussion on control groups and on single and double blinds points out that there are two uses of the word *control*, and both are very important. In the first, and literal, sense the experimenter makes things happen when he or she wants them to happen. This is what was meant by control when it was stated that Spallanzani had control over the insemination process—that is, he could control the contents of the seminal fluid. The manipulation and/or selection of independent variables are prime examples of this type of control. Another example comes from the study of factors controlling nest building in canaries. Under natural conditions, these birds do small amounts of nest building scattered over rather long periods of time. In order to study this behavior experimentally, it is necessary to get the birds to build during those times when the experimenter is prepared to make his or her observations. This problem is solved by keeping canaries in cages and giving them the nest material only when the researcher is there to observe. Here the presentation of nest material has been controlled (manipulated), and precise observations of nest-building behavior can be made.

The second use of the word *control* is in arranging conditions so that the experimenter can attribute the result to the independent variable and not to some other variable. Paul controlled for the judges' bias by keeping the judges blind. The use of control groups is an attempt to insure that the results are not due to some other variable. In the Paul experiment the placebo control group "controls" for placebo effects, and the no-treatment control group controls for spontaneous remission. This second use of control will be discussed more fully in Chapter 3.

REPORTING DATA: CURVES, FIGURES, AND FUNCTIONS

The results of an experiment are reported verbally ("The racially prejudiced subjects saw more instances of aggression in the environment"), statistically ("The probability of obtaining these data by chance is less than 1 in 100, or $p < .01$), or graphically. In this book only the barest of statistically analyses are discussed[3]; not that they are unimportant—they

---

[3]In Appendix A there is a brief description of and mathematical procedures for some of the statistical tests covered in this book.

are—but many other books deal with this topic in far greater detail than can be done here. In this section we focus on graphic representation of experimental results.

In the beginning of this chapter independent and dependent variables were discussed, and in both of the examples cited we arrived at some general conclusions about the relationship between variables. In the first example it was concluded that spaced practice was superior to massed practice; in the second example it was concluded that the information given first about a person has more "weight" than information given later in forming an impression of that person. While these general conclusions are valuable, psychologists strive for greater precision in depicting the relationship between two variables. One way of illustrating data is by means of a figure that illustrates the relationship between two or more variables. A curve can be drawn any time the experimenter has values for the dependent variable for several levels of some other variable (usually the independent variable). The plotting of curves allows the researcher to "see" the relationships between two variables. For example, the shapes of "learning" curves can be compared to determine more precisely the effect of certain types of reinforcement on conditioning.

A researcher has, in addition to the use of statistics, numerous visual forms which display the procedure and/or data collected. In some instances, the use of these graphic forms is superior to a verbal presentation only. Consider the use of photographs and a drawing in this example.

EXAMPLE

A study of visual memory in the *Rhesus* monkey was done by Sands, Lincoln, and Wright (1982). Humans have an unusual ability to remember photographs, and some psychologists have argued that such an ability may be related to a verbal labeling of visual stimuli. This argument is: I am able to remember a picture of a flashlight because when I saw a flashlight I recognized it and labeled it "flashlight." Thus, the real code (the way the article is stored in memory) is verbal rather than purely visual.

Since monkeys are thought not to have verbal labels for objects, an answer to the question of how visual memory operates may be answered by looking at visual memory in nonverbal creatures. The experimenters constructed an apparatus that was best shown in the drawing in Figure 2.3. Next, the experimenters presented their subjects with several photographs, which were presented one above the other (a banana and an apple are illustrated). The monkey was trained to slide a lever to the right or left, indicating whether or not the objects were the same or different. Reinforcement is delivered by liquid in the bottle. Some of the faces used are shown in Figure 2.4.

The use of a line drawing and photographs helps the reader grasp the technique used, and these graphics greatly facilitate the reproduction of the experiment.

The researchers found that visual stimuli, such as a face, could be differentiated by monkeys, which suggested that primate visual memory can be represented by a visual code.

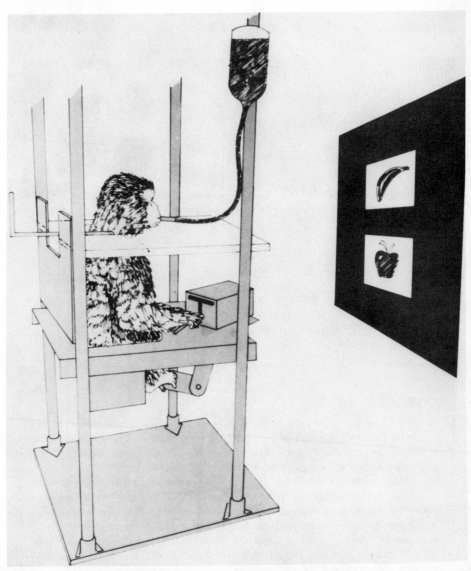

Figure 2.3 A *Rhesus* monkey operating an apparatus with a lever in a test of visual memory conducted by Sands, Lincoln, & Wright (1982).

Figure 2.4   Some of the faces shown to the *Rhesus* monkey in Sands, Lincoln, & Wright's study of visual memory (1982).

BAR GRAPHS

Another form in which data can be represented is a bar graph or histogram. This form of representation is a particularly vivid graphic form that allows the reader to form a quick impression of the results of an experiment.

EXAMPLE

An example of how a histogram effectively portrays information is shown in an article by Czeisler, Moore, Ede, and Coleman (1982) in which shift workers in a minerals and chemical corporation were given work schedules that either corresponded to their natural circadian (day/night biological schedule) schedules or not. That is, measures were made of worker preference after experience with two schedules: one in harmony with the circadian schedule and one out of harmony with the schedule. In this example the dependent variable was worker satisfaction, which was measured from questionnaires distributed three months after the introduction of the independent variable—schedule A (out of harmony with the natural circadian schedule) and schedule B (in harmony with the natural circadian schedule). As is clearly shown, the workers preferred the harmonious schedule (see Figure 2.5).

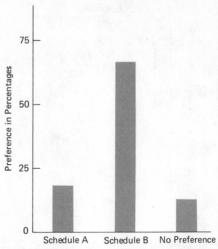

Figure 2.5   A bar graph showing the schedule preferences or lack of preference of shift workers in a minerals and chemical corporation. From Czeisler, Ede, & Coleman (1982).

Typically, the element manipulated (cause) is depicted on the horizontal plane while the consequence (effect) is shown on the vertical axis.

In this example the data along the *ordinate,* i.e., the horizontal axis (the vertical axis is called the *abscissa*) is discrete—that is, the categories represented on the horizontal plane are distinct and noncontinuous. In many experiments in psychology the variables may be more or less continuous, such as time, IQ, running speed, delay of reinforcer, or amount or frequency of drug administration.

Another example of the use of a histogram in which continuous data is represented is the dramatic portrayal of the incidence of various types of cancers among men who have discontinued the use of cigarettes. The length of time between the cessation of smoking and the risk of cancer is clearly shown in the three histograms in Figure 2.6. Also of interest in these figures is the addition of present smokers and nonsmokers on the extreme ends of the ordinate. Strictly speaking, these classes are noncontinuous and out of place with the continuous data (years of smoking cessation). Nevertheless, their inclusion as two types of *base* data (or in this case, a type of a control) gives a fuller picture of the results of stopping smoking on the risk of cancer. Note that the risk of lung and esophagus cancer is actually less for present smokers than for those smokers who have quit smoking within 1–3 years prior to the study. Why?

At a more sophisticated level, figures can be used to derive a mathematical equation that describes the relationship between two variables. This mathematical equation, or *function,* is simply a shorthand method

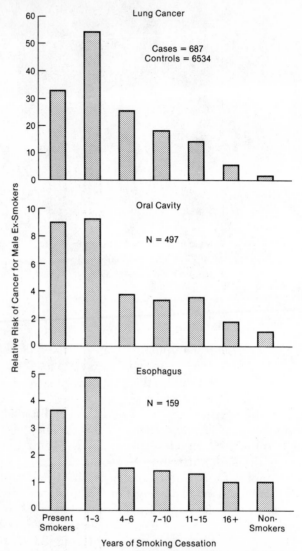

Figure 2.6   Male ex-smokers' relative risk of three types of cancer, by years since quitting. From the American Health Foundation's control study of 3,716 cancer cases and 18,000 controls (Wynder & Stellman, 1977). Copyright 1977 by Cancer Research, Inc. Reprinted by permission.

of describing the relationship. Once the function is derived, all one has to do is substitute the value for one variable, and with a few, usually simple, calculations one can find the value of the second variable.

Two examples are presented below. These are "classic" studies, one from the area of audition and one from the area of memory.

EXAMPLE

Several experiments in the area of audition have been conducted to determine the relationship between the frequency (pitch) of a tone (as measured in cycles per second) and a subject's sensitivity to the intensity of that tone. One procedure (Hirsch, 1966) for investigating this problem is to present a pure tone to the subject through earphones. The tone is presented at a clearly audible level, and then the intensity (loudness) of the tone (as measured in decibels) is gradually decreased until the subject signals that he or she can no longer hear the tone. The point of decibel level at which this signal is given is called the *descending threshold*. The procedure is then reversed, with the same tone being presented at an inaudible level and the intensity being gradually increased until the subject signals that he or she can hear the tone. The decibel level at which this signal is given is called the *ascending threshold*. This procedure is repeated several times for the same subject on the same tone.

The dependent variable is the *absolute threshold* for that particular tone, which is the decibel level above which the subject will hear the tone and below which he or she will not hear the tone. The absolute threshold for a given tone is determined by averaging the threshold values obtained in the descending and the ascending trials. For example, suppose a tone of 100 cycles per second were presented to a subject. The values of the descending threshold for three attempts were 40, 45, and 42 decibels, and the values of the ascending threshold for three attempts were 48, 49, and 52 decibels. To get the absolute threshold we would sum the six values (sum = 276) and divide by the number of values (6), which would yield an absolute threshold of 46 decibels for that particular tone and for that particular subject.

After the absolute threshold of a tone of a given frequency is determined for a subject, the above procedure is repeated for a tone of another frequency. Once the absolute threshold for six or eight tones of varying frequency has been ascertained, the experimenter can plot a curve. Figure 2.7 has been constructed from data presented in Stevens and Davis (1938). In this example, four subjects were tested at 10 levels of frequency (25, 50, etc.).

The points on the curve are the mean *absolute threshold* values for the four subjects at each level. For example, at the level of 50 cycles per second the absolute threshold values for the four subjects were 59, 60, 63, and 66 decibels. To get the mean, these four values are summed (sum = 248) and the sum is divided by the number of values (248 ÷ 4), which yields a mean of 62 decibels. This procedure is followed for each of the levels tested. The points are then connected to form a curve.

The results indicate that the subjects are rather insensitive to low-pitch or low-frequency tones. Sensitivity increases as frequency increases up to a point, with the subjects being most sensitive to tones that range between 1000 and 3000 cycles per second. Beyond that point, however, sensitivity decreases.

The independent variable in the above example is frequency as measured in cycles per second, and the dependent variable is the absolute

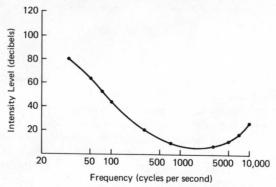

Figure 2.7   The relationship between frequency and intensity on absolute auditory thresholds. From Dempster (1981).

threshold as measured on a decibel scale. In plotting a curve it is customary to place the values of the dependent variable on the vertical axis *(ordinate)* and the values of the independent variable on the horizontal axis *(abscissa)*. The values of these two variables go from low to high as one goes out from the point at which the two axes intersect or meet. This procedure was followed in plotting the curve in the above example. The resulting curve quite clearly illustrates the relationship between the two variables.

The research strategy used here is sometimes referred to as *parameter estimation,* in contrast to *hypothesis testing.* The parameter testing research does not start out with some hypothesis to be tested but, rather, is an attempt to measure precisely the relationship between two variables. From a methodological viewpoint, however, there is little difference between this experiment and those described earlier in this chapter (see under *Independent and Dependent Variables*). The experimenter manipulates the independent variable, measures its effects on the dependent variable, and holds all other variables at a constant level. The only difference is that the experimenter can accurately manipulate several levels of the independent variable, and because of this he or she can determine fairly precisely the relationship between the independent and dependent variables.

In the above example, there were 10 levels of the independent variable (frequency), and each subject was tested at all 10 levels. The curve drawn in Figure 2.7 connects the performance of subjects at each of the 10 levels. Another type of curve that is frequently encountered in psychological research is one in which each subject is tested at only one level of the independent variable; however, he or she is repeatedly tested at this level over a series of trials or time periods.

In the second example of functional relationships we take a study from developmental psychology and memory research. The question posed

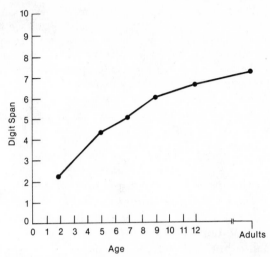

Figure 2.8   Memory for digits as a function of age. From Dempster, 1981.

is: "Does memory for digits increase as a function of age." The results of an experiment by Dempster (1981) shown in Figure 2.8 indicate that as age increases so too does the ability to remember digits increase.

EXAMPLE

It is hard to think of human behavior without memory. Recently the topic of memory has been a major focus of cognitive psychologists. Often the progress of a specific area of research, such as memory research, begins with the study of simple elements and then advances to more complex elements. In memory research, psychologists have for many decades examined memory span for immediate events, or "immediate memory span." Studies of this sort have become integrated into theories of memory, especially *short-term memory:* a hypothetical memory system that stores a limited amount of information for about 12 sec.

A commonly used task is memory for digits or the "digit span test." This test is so common that it is part of several popular individual intelligence tests.

Digit span tasks require that a subject repeat a series of numbers that have been presented. If the experimenter reads 4–8–3–6–9, the subject's task is to repeat the numbers 48369.

An intriguing question in memory research is: Does memory span for digits increase as a function of age? And, if so, how might that relationship appear?

In a study by Dempster (1981), children of various ages were given a digit span task. The results are shown in Figure 2.8, in which digit span increases as a function of age. The question of why this function occurs is an even more fascinating puzzle. The answer may be found in the ability of older subjects to group information into more meaningful "chunks." In the above example, an older subject may chunk the series into two well-known mathematical progressions (4–8 and 3–6–9), which would reduce the "information load" substantially. We all use these types of memory tricks whether or not we are aware of them.

Finally, this example also brings up the issue of stimulus control. If the purpose of the experiment is to study "pure" memory (i.e., not chunking of meaningful material), then the experimenter should try to avoid the use of any series that could be meaningfully chunked. (This is sometimes very difficult.)

FACTORIAL DESIGNS

The experiments discussed so far manipulated only one variable. In the first example presented in this chapter (massed vs. distributed practice), the single variable manipulated was the time between trials. While such experiments are important, it is evident that behavior is rarely a function of a single variable but, rather, is a function of several variables. For example, evidence from the field of developmental psychology indicates that if a child's parents are hostile and negative toward the child, as well as very controlling of his or her behavior, the child tends to become inactive and withdrawn. On the other hand, if the hostile and negative parental behavior is combined with a tendency to ignore the child and exert no control over his or her behavior, the child tends toward antisocial behaviors. These two variables—(1) degree of love-hostility and (2) degree of control—both seem to be quite important in determining the child's behavior.

In order to determine the influence of two or more independent variables on the dependent variable, researchers employ a *factorial design*. In a factorial design two or more variables may be manipulated at the same time. To illustrate a factorial design, consider the experiment by Solso and Short (1979) on color matching. Two independent variables were introduced: color-match and match delay. The dependent variable was reaction time. This experiment had three factors, each varied several ways. The factorial design was a 3 × 3 × 2 in which the first level corresponded to the type of match (e.g., RED–RED, RED–"RED", RED–BLOOD), the second level, the delay of second stimulus (0, 500 msec., 1500 msec.), and the third level, the match/non match condition (e.g., RED–RED, RED–BLUE).

EXAMPLE

Ehrenfreund and Badia (1962) examined the performance of rats under high and low food-deprivation conditions and high and low incentive conditions. The apparatus used in the experiment was a straight alley, 5 ft long. The alley contained a start box on one end and a goal box on the other end. The dependent measure was the speed at which the rats ran down the alley. Twenty rats were used in this experiment. During the experiment half of the 20 rats were maintained at 95 percent of their *ad lib,* or free-feeding, weight, and the other half were maintained at 85 percent of their ad lib weight. (Ad lib, or free-feeding, weight is the weight of the rat when it is allowed to eat as much food as it wants.) For the purposes of this experiment, the high food-deprivation treatment was defined as those rats maintained on 85 percent of ad lib weight, and low food-deprivation treatment was defined as those rats maintained on 95 percent of their ad lib weight.

One-half of the rats in the high-deprivation treatment received a 45-milligram (mg) food pellet (low incentive) in the goal box, and the other half of the rats received a 260-mg food pellet (high incentive) in the goal box. Combining the incentive treatments with the deprivation treatments yields the four experimental treatments of the experiment:

1. High deprivation—High incentive
2. High deprivation—Low incentive
3. Low deprivation—High incentive
4. Low deprivation—Low incentive

There were five rats in each of the four conditions.

Performance was measured in terms of the speed at which the rats traversed the middle 2-ft section of the runway. In the analysis presented here the authors took the median (the midpoint) running speed for each rat on the last 10 trials of the experiment. Each rat's median score was converted to its reciprocal by dividing each score into 10. The conversion to reciprocals simply changes the direction of the scoring such that the higher the score, the faster the rat is running. The means of the reciprocal scores for each of the four treatment groups are presented in Table 2.1.

It is evident from the means that both deprivation and incentive are influential in determining performance. The mean for the high-reward groups (260 mg) is higher than the mean for the low-reward groups within each deprivation level. Further, both of the means for the high-deprivation groups (85 percent body weight) are higher than the means for the low-deprivation groups.

The preceding example is called a 2 × 2 factorial design. Two levels of one variable (deprivation) are combined factorially with two levels of a second variable (incentive) to yield 2 × 2 or four separate treatments.

Table 2.1   MEAN RUNNING SCORES
FOR DEPRIVATION AND INCENTIVE GROUPS

		BODY	WEIGHT
		95%	85%
Reward	45 mg	10.26	13.92
	260 mg	13.86	15.15

Each level of the first variable occurs within each level of the second variable; that is, high and low incentive is tested both under high-deprivation and under low-deprivation conditions. The design can be extended to add other independent variables; for example, a 2 × 2 × 3 design would consist of two levels of the first variable, two levels of the second variable, and three levels of a third variable. In this latter design there would be 12 separate treatments. One of the advantages of the factorial design is that it allows the researcher to ascertain how independent variables combine with one another to determine the values of the dependent variable. The experiment presented below is a 2 × 2 factorial design that illustrates the power of such a design to pick out interactions between variables.

EXAMPLE

Several studies have shown that if people are somehow induced to argue for an attitude position to which they are opposed, the more they are paid for the task, the more they change their own attitude toward this attitude position. This result is consistent with a reinforcement explanation; that is, money is a reward, and the more money that is paid for arguing for an opposing attitude position, the more that position is "reinforced." On the other hand, several other studies have found the exact opposite effect. These studies support a dissonance theory (Festinger, 1957) explanation. This explanation assumes that dissonance (or a state of discomfort) is aroused when a person argues for a position he or she actually opposes. Furthermore, the smaller the amount of money given for doing this task, the greater should be the amount of dissonance aroused, since low sums of money provide inadequate justification for defending something to which a person is opposed. In order to relieve the dissonance, the person should "convince himself or herself" that he or she really is for the argued position. Therefore, more attitude change would occur in the high-dissonance (low-money) condition than in the low-dissonance (high-money) condition.

In an attempt to resolve the conflicting results reported above, Linder, Cooper, and Jones (1967) hypothesized that the reinforcement hypothesis would be verified when subjects have no choice in arguing for the position they opposed. These researchers further assumed that the dissonance hypothesis would be supported under conditions when the subject somehow chose to argue for the opposed position. The logic for these hypotheses was based on the assumption that dissonance can be created only when a person chooses to do something he or she opposes, but not when he or she is "forced" to do it.

The design used in this experiment was a 2 × 2 factorial design. Subjects were college students who wrote an essay supporting a speaker-ban law for colleges (a position they were against). The subjects either were simply told to write the essay (no-choice condition) or were given a choice as to whether or not they would write the essay (free-choice condition). In addition, half of the subjects in each of the above two conditions were paid $.50 for writing the essay, while the other half of the subjects were paid $2.50. Thus, the design consists of the four treatments, and 10 different subjects were randomly assigned (by chance) to be in each of the four treatments. The dependent variable was the amount of change in the subjects' attitudes toward the speaker-ban law. This was measured by having the subjects check the point on a scale that indicated the degree to which they approved of the speaker-ban law. The mean scale point attitude change for the four treatments is given in Table 2.2.

These data can be further shown in a graphic form in which the relationships are clearly presented (see Figure 2.9).

Positive scores indicate change toward the position argued in the essay (for a speaker-ban law), while negative scores indicate change against the argued position (a "boomerang" effect). The results support the authors' hypothesis. In the no-choice treatment, the $2.50 incentive elicited more attitude change than did the $.50 incentive. In the free-choice treatment, the opposite result was found.

The appropriate statistical test to analyze the above results is *analysis of variance*. The actual calculations of this statistic will not be explained here. However, some of the logic of this test will be described because it gives some insight as to the logic of a factorial design, as well as pointing out how the results are analyzed. The basic design was a 2 × 2 factorial design with two levels of one variable (no-choice or free-choice) and two levels of a second variable ($.50 incentive or $2.50 incentive). In the analysis of variance each factor (or variable) is analyzed separately, and then the interactions between these variables are analyzed. In the example experiment above there are two factors; therefore, the analysis of variance will contain (1) an analysis of the *main effect* of the first variable, (2) an analysis of the *main effect* of the second variable, and (3) an analysis of the *interaction* between the two variables.

One can conceptualize a 2 × 2 analysis in the following form:

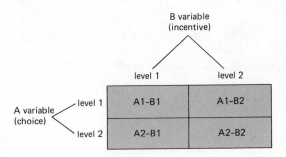

In this form we can see how information from each variable (A or B) and for each level (1 or 2) is combined in a single experiment.

Table 2.2 MEAN ATTITUDE CHANGE FOR CHOICE AND INCENTIVE CONDITIONS

		INCENTIVE	
		$.50 INCENTIVE	$2.50 INCENTIVE
Choice	No-choice treatment	−0.05	+0.63
	Free-choice treatment	+1.25	−0.07

The main effect of the choice variable is analyzed by comparing the mean attitude change of subjects in the no-choice condition with that of subjects in the free-choice condition, ignoring the incentive manipulations. The mean of the no-choice condition is calculated by adding the means for both incentive conditions and dividing by 2 (−0.05 plus +0.63 = +0.58 divided by 2 = +0.29). The same process is done in the free-choice condition with a resulting mean value of +0.59. These two means are then compared (+0.29 versus +0.59) in order to analyze for the main effect of choice. The same process is repeated in order to analyze the

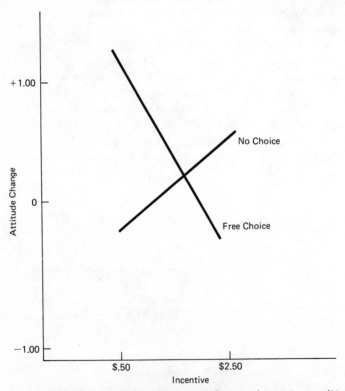

Figure 2.9   Mean attitude change as a function of choice and incentive conditions.

main effect of incentive. By doing the necessary addition and division, it is found that the mean attitude change for the $.50 condition is +0.60 and the mean for the $2.50 condition is +0.28. These two means are then compared to ascertain the main effect of incentive. The next step is to analyze the interaction effect. The test for an interaction effect is a test to determine if the two variables are independent of one another with respect to their influence on the dependent variable. The variables are nonindependent (and interact) if it is demonstrated that the one variable shows a different pattern of results on the DV under each of the two levels of the second variable.

Linder, Cooper, and Jones analyzed their results in the manner described above. Their computations indicated that although there were differences in the mean attitude change both for the main effect of choice and for the main effect of incentive, these differences were not *statistically significant* (see Box); that is to say, the magnitude of these differences was not greater than what would be expected by chance fluctuation in the means. *Chance* is the variation in the results that is due to uncontrolled factors such as guessing, experimental error, and failure to achieve a perfect matching of subjects in each treatment group. Because of a variety of uncontrolled factors, we would expect the means to differ even if the treatments were not effective. The interaction effect was found to be statistically significant in the direction hypothesized by the authors—high incentive facilitates attitude change in the no-choice condition, but low incentive facilitates attitude change in the free-choice condition. Thus, the two variables interact with each other to determine the amount of attitude change.

STATISTICALLY SIGNIFICANT

Prior to the collection and analysis of data, experimental psychologists commonly anticipate how the data will be analyzed statistically. In the analysis of data a researcher sets a *level of significance,* which is defined as the statistical point for inferring the operation of nonchance factors. Researchers set a level of significance in terms of a probabilistic statement. In many psychological experiments the level may be expressed as "the 0.05 level" or "the 0.01 level," which means that the results will occur by chance only 0.05 (or 1 in 20) or 0.01 (or 1 in 100) times by chance alone. Data that have been statistically analyzed and meet the pre-established criterion (be it 0.05 or 0.01 or even 0.001) are called *statistically significant.*

It should be pointed out that a result can be statistically nonsignificant but still be of interest. Often, especially during pilot studies or experiments with few subjects or observations, the results may fail to reach a level of statistical significance but do suggest that further investigation with perhaps better controls and/or a larger number of observations may be worthwhile. At the same time, one must be cautious not to "go on a fishing trip" for results, which means that you should not continue to run and fine-tune an experiment until you get the results you selectively report.

In yet another example of 2×2 design we find the use of subjects as an element whose effect is measured as well as type of task. Consider the following experiment by Chi (1978).

In a study of the importance of specialized knowledge on memory, Chi (1978) examined recall of digits (see previous discussion in this chapter) and the recall of chess pieces by children and by adults. In this experiment two types of subjects and two types of tasks were used, thus completing the 2×2 design. The children were 10-year-olds who were skilled chess players, while the adults were novice chess players. The task involved looking at chess pieces as they might appear in a normal game in process. In this arrangement the pieces would be located in conventional places; that is, the arrangement of pawns, kings, and other pieces would conform to a normal game. In the digit portion of the task, a standard series of digits, such as might appear on an IQ test, were used.

The design can be conceptualized as:

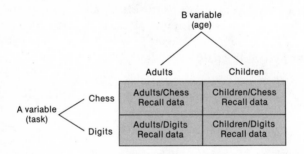

The results of this experiment are shown in Figure 2.10, in which a strong interaction is shown. It appears that specialized knowledge (chess knowledge) facilitates recall of information in that domain, but has little effect on digit memory. Adults, unsophisticated in chess, recall fewer pieces than children, but do better on digit recall.

In the example of chess and digit memory among children and adults we can show the strong interaction graphically (see Figure 2.10). In the text of the study, Chi presents a statistical analysis that confirms mathematically the results shown in the figure.

The 2×2 design is the simplest of factorial designs. These designs become increasingly complex as more factors are added, or when several levels of factor are used. The example below is illustrative of this point.

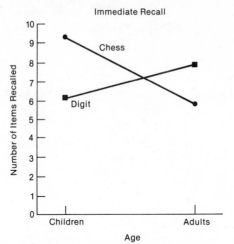

Figure 2.10   Children's and adults' recall of chess and digit stimuli. From Chi, 1978. Reprinted by permission.

Only two factors are used, with two levels of one factor and four levels of the second factor. This example was taken from research in educational psychology.

Much of the evaluation of learning in the classroom takes place through tests, quizzes, or papers. Quite frequently a student takes a test or turns in a paper on one day and it is returned several days later. In this situation there is *delayed information feedback;* that is, there is a time delay between the student's taking of the test and his or her receiving information as to where he or she was correct and where he or she made mistakes. This delayed feedback has been criticized as poor educational procedure. Some new methods, such as programmed textbooks and teaching machines, make use of an immediate feedback technique in that the student gives an answer to a particular question and finds out immediately whether he or she is right or wrong.

More (1969) designed an experiment to compare the effects of immediate and delayed feedback using verbal learning materials similar to those used in the school classroom. The subjects were eighth-grade students in four junior high schools. The material to be learned was the science information contained in a 1200-word article on glaciers. The student read the article and was immediately tested on its contents by a 20-question multiple-choice test.

To test for the effects of delay of information, a 4 × 2 factorial design was used. The first variable was the length of delay between the taking of the test and the feedback as to the correct answers. There were four time intervals. One group received this feedback immediately as they answered each question on the test. The other three groups received the feedback either 2½ hours, 1 day, or 4 days after taking the test. After the students received this feedback, they took the same test again. At this point the experimenter introduced the second variable; that is, the student took the test immediately after receiving the feedback (acquisition treatment) or 3 days after receiving the feedback (retention treatment). The dependent variable was the number of questions (out of a possible 20) answered correctly by the student on the second administration of the test.

Based on the symbolic description of factorial studies previously discussed, this procedure has two levels of an A variable (treatment) and four levels of a B variable (delay of feedback). This can be conceptualized as:

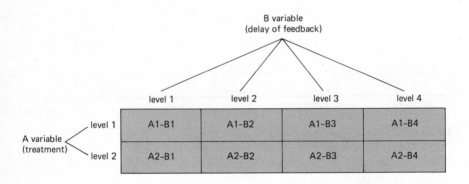

To summarize the procedure, the students read an article on glaciers and were immediately tested on the material. They then received feedback as to the correct answers after one of four time delays. They were then retested on the same test immediately after feedback or three days after feedback. There were three different classrooms of students in each of the eight treatment groups. The mean number of questions correct on the second test for each of the eight groups is given in Table 2.3. The results indicated that performance was higher in the acquisition group than in the retention group (16.4 vs. 14.8). This was expected as some forgetting would occur after three days. For the acquisition treatment it was found that the no-delay feedback group scored significantly lower on performance than the other three feedback groups (15.1 vs. 16.7, 16.9, 16.7). These latter three did not differ significantly from each other. For the retention treatment, the 2½-hour and the 1-day delay groups scored significantly higher than either the no-delay or the 4-day delay groups (15.4 and 16.0 vs. 13.0 and 14.6). No significant difference was found between the 2½-hour and the 1-day groups or between the no-delay and the 4-day groups.

Table 2.3   MEAN SECOND TEST PERFORMANCE OF TREATMENT GROUPS

		DELAY OF FEEDBACK		
	NO DELAY	2½-HOUR DELAY	1-DAY DELAY	4-DAY DELAY	TREATMENT MEANS
Acquisition treatment	15.1	16.7	16.9	16.7	16.4
Retention treatment	13.0	15.4	16.0	14.6	14.8
Feedback means	14.2	16.1	16.5	15.7	

The author concluded that the results offer no support for the assumption that immediate feedback maximizes learning of the specific materials used in his study. Quite the contrary—delayed feedback of 2½ hours or more produced superior learning in the acquisition treatment. For the retention treatment some, but not too much, delay maximizes the students' retention of the materials at a later date. The author felt that this latter finding was especially important since one primary objective of instruction is to maximize retention of material.

Most researchers find it valuable to draw a figure of the results of a factorial design since it enables them to "see" the relationship between the variables more clearly. Figure 2.11 illustrates the results of More's experiment.

The dependent variable is on the vertical axis, and one independent variable is on the horizontal axis. Separate curves are drawn for each level of the second independent variable. The effects of delay in the experiment are more clearly illustrated by this figure. This experiment

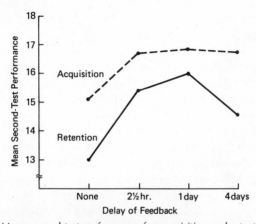

Figure 2.11   Mean second test performance for acquisition and retention conditions.

points out the possible complexity of the factorial design as well as the value of having several levels of the factors. If only two levels of the delay factor had been used, the interpretation of the results for the retention treatment would have been quite different, regardless of which two delay levels had been picked.

CORRELATIONAL STUDIES

The term "studies" rather than "experiments" is used in this section as correlational studies do not involve cause-and-effect relationships (even though they may infer such relationships). *Correlational studies* are studies of the relationship between two variables. They are valuable procedures if the interpretation of results does not go beyond the limited scope of the measure. A *variable* in experimental psychology is a phenomenon with diversity that can be quantified.

An example of a psychological variable is intelligence. Some people have more, some less, and a quotient of intelligence can be expressed quantifiably. Running speed is also a human variable that can be quantified. A correlational study could be done between intelligence and running speed. Members of a sophomore class at Stanvard University could be set off running a 50-meter race. Some would run fast, some slow, and quite a lot somewhere in between. The same group of subjects could be given a standard intelligence test, and the results would also yield variability on this measure. The two sets of data (intelligence and running speed) could be correlated by means of a simple statistical test. The results would indicate the degree of relationship between the two variables, which in the case cited would probably yield a small positive correlation; that is, intelligence and running speed would tend (slightly) to go with each other.

Or consider a correlational study done in the San Francisco Bay area between 1970 and 1980, which examined liquor consumption and birthrates. Both liquor consumption and birthrates are variables—some people practice both forms of activities at varying levels. The study yielded a high positive correlation between birthrate and the consumption of alcoholic beverages. Did one cause the other? Well, you could make a pretty good case that drinking caused more children to be born, but then, another (somewhat cynical) interpretation could be that the birth of children caused an increased consumption of alcohol. Or it could be that the coincidental data was caused by other factors. Both causal statements are equally invalid (based solely on the correlational analysis).

Does intelligence *cause* running speed (or does running speed *cause* intelligence) or does drinking lead to children being born (or vice versa)? Perhaps, but from a scientific position merely establishing a relationship between two variables *does not mean causality*. We wish to stress this point as all too frequently among lay thinkers, the co-occurrence of two events is taken as a reason for causal relationships. Many times superstition follows from this thinking, and you can think of many examples of people carrying good luck charms because on one or more occasions they had good luck while carrying such amulets.

The following example provides a correlational analysis of humor and body type:

EXAMPLE   HUMOR AND BODY TYPE: A CORRELATIONAL ANALYSIS
Glenn D. Wilson

The first test of a hypothesis involving two sets of data may simply be correlational analysis. If it is hypothesized that A causes B, then failure to find a correlation between A and B may be reason to question the hypothesis.

Wilson, Nias, and Brazendale (1975) hypothesized that a woman's own body shape would influence her liking for chauvinistic cartoons focusing on female anatomy. Sixty-two female student teachers rated the "funniness" of a set of British seaside postcards which featured lecherous male interest in sexually attractive women. Afterwards, the subjects' vital statistics were measured. Results of a correlational analysis between "shapeliness" as measured by bust/waist ratio and ratings of sexist cartoons indicated a positive correlation. The results are shown below:

CARTOONS SIGNIFICANTLY PREFERRED BY "SHAPELY" GIRLS ($N = 62$)

DESCRIPTION OF CARTOON	CORRELATION WITH "SHAPELINESS"
1. Two men lecherously surveying bikini-clad females	.34**
2. Two male swimmers admiring female sunbather	.31*
3. Busty pet-shop girl displaying birds in cage labelled TITS	.27*
4. Man and woman viewing painting of a pirate standing behind a very phallic cannon	.26*

*$p < .05$.
**$p < .01$.

The authors' interpretation was that women who are themselves physically attractive are more accepting of physical-sexual humor.

Of course, correlations do not determine the direction of cause and effect but merely indicate that a relationship between two variables may occur with a probability greater than one would expect by chance. It may be more logical to search for cause and effect relationships between body types and appreciation of risqué postcards than to conclude that appreciation of risqué postcards causes body changes. The judicious use of correlational techniques is a valid technique for searching for causative relationships.

QUASI-EXPERIMENTAL DESIGNS

The concept of quasi-experimental designs was introduced by Campbell and Stanley (1963)[4] to overcome some of the problems faced by psychologists who wish to study behavior that occurs in less formally structured environments than a laboratory. *Quasi-experimental designs* are studies in which the independent variables (the element whose effect is to be measured) are selected from the (natural) environment. Sometimes quasi-experimental designs are termed *ex post facto*, or evaluation research, or "as if" designs, as the collection and analysis of data take place after an event has happened. One part of the logic of the design is that if the experimenter had been able to introduce an independent variable into a situation, then that variable may have been the same as the variable that was naturally introduced. It is *as if* the experimenter was responsible for the introduction of the experimental variable.

In typical experimental work in psychology, subjects are selected for participation. Selection of subjects is either random or on the basis of certain well-defined characteristics (e.g., seamen between the ages of 19 and 25). The subjects are then presented with an experimental variable, frequently in a laboratory setting, which is a controlled environment. The results of the experiment are then (frequently) generalized to a larger population or in some way applied to real life. But real "real life" occurs naturally, and sometimes it is impossible to select subjects for an experiment and bring them to a well-controlled laboratory setting for precise recording of data. How could you study a riot in the laboratory? And if you could, would the conclusions be generalizable to a real riot? These topics are of interest to experimental and applied psychologists and need to be addressed. Essentially, the argument boils down to the premise that observations based on microcosmic life (as might be observed in the laboratory) may not be valid for macrocosmic life.

In laboratory work in experimental psychology we are dealing with a highly controlled environment, which has been called a *closed system*. The laboratory has many virtues and many experiments in psychology require rigid control over stimuli. An *open system* is an environment over which we have no or little control, such as what is vaguely called the "real world."

---

[4]In order to give the reader a sense of the importance of these types of designs in psychology, surveys of graduate departments of psychology have been conducted that have asked which books are most frequently recommended to their graduate students who are preparing for departmental examinations. *Experimental and Quasi-Experimental Designs for Research* by Campbell and Stanley has been one of the most frequently recommended books throughout several surveys. (See Solso, 1987b, for details.)

EXAMPLE

After a record number of automobile fatalities in the state of Connecticut several years ago, very harsh action was taken against speeders. After stringent measures toward speeders were introduced, a decline in traffic fatalities was noted. Campbell (1969) makes a detailed study of this phenomenon within the context of a quasi-experimental design. The data for Connecticut traffic fatalities are shown in Figure 2.12. In this figure we see a decline in fatalities from 1955 to 1956, after the introduction of harsh treatment of speeders. However, you may argue (correctly) that many other causes may have entered into this picture (e.g., road conditions may have been improved, less inclement weather may have occurred in 1956 than in 1955, or better driver education may have been available). One might note however, that the decline in fatalities continued downward after 1955.

Another way to look at the data is to make comparisons of the automobile fatality rates among comparable states. Those data are shown in Figure 2.13. In this figure we see that although the four states do report a slightly lower fatality rate over the years 1951–1959, the rate of decline for Connecticut is much greater, especially after the 1955 crackdown on speeders. The results therefore suggest that the 1955 treatment did have an effect on fatality rates.

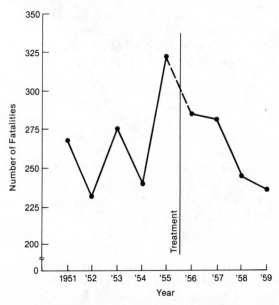

Figure 2.12   Connecticut traffic fatalities, 1951–1959. From Campbell, 1969. Copyright 1969 by the American Psychological Association. Reprinted by permission.

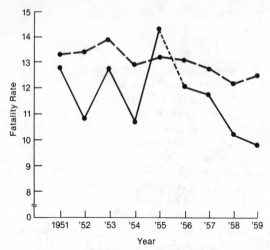

Figure 2.13   Connecticut traffic fatality rate (lower line) with the fatality rate of four comparable states. From Campbell, 1969. Copyright 1969 by the American Psychological Association. Reprinted by permission.

In the case of the automobile fatality rate among Connecticut drivers, no reasonable laboratory experiment can adequately resolve the effectiveness of the 1955 crackdown. Yet the issue is important, and many other similar "real world" problems cry out for solution. Given the circumstances, and the inherent lack of precision in these types of studies, quasi-experimental designs seem to render as close approximations to valid conclusions as one could expect.

Field-Based Studies

In yet another example of these types of studies we illustrate a field-based study, or research investigation conducted in a natural setting. In Chapter 12 we deal with this topic in detail (see "Distance and Rank," which describes the distance people of different military rank stand from each other). Here a field study was conducted using the "lost letter technique," which involves the distribution of bogus letters. The return rate (i.e., the number of "lost letters" mailed) is measured vis-à-vis neighborhoods, for example.

In the following example Bryson and Hamblin (1988) use the technique to evalute the return rate of postcards that contained either a neutral news or bad news. Note particularly the return rate by type of news and also the strong gender effect.

EXAMPLE   LOST LETTER TECHNIQUE AND THE MUM EFFECT
J. B. Bryson and K. Hamblin

A variant of the lost-letter technique, the lost postcard, was employed to examine attitudes toward informing people of their romantic partner's apparent infidelity. Stamped and addressed postcards were left on the windshields of 180 cars parked near mailboxes, with an accompanying handwritten note reading "Found this by your car—is it yours?".

One-third (N = 60) of the cards had a neutral/good news (control) message ("Glad to hear you've worked things out. We're getting along better too. Keep in touch . . ."); 30 of these were addressed to a male, 30 to a female.

The other 120 cards, equally divided by sex, informed the addressee of his (her) girlfriend's (boyfriend's) apparent infidelity, in the following message: Dear Bob (Judy), I hate to be the one to tell you this, but I think I saw your girlfriend Ann (boyfriend Bob) coming out of the TraveLodge off El Cajon Blvd. with another guy (woman) on Thursday. It might not be important, but I didn't know how to tell you in person—Barry (Beth).

Consistent with a MUM effect hypothesis, bad news was not transmitted as often as good news: 35 (58.3 percent) of the neutral/good news postcards were mailed versus 23 (19.2 percent) of the bad news postcards ($X^2 = 28.10\ p<.001$). There was also a substantial double standard effect caused by sex of the recipient of the bad news postcards: of the 23 cards returned, 19 were addressed to males and only 4 to females ($X^2 = 12.02$, $p<.01$). (See Figure 2.14 on page 48.)

These findings indicate that, while there is a general unwillingness to transmit bad news (the MUM effect), there is a definite double standard in the willingness to transmit information regarding infidelities—the partner is "always" the last to know, especially if she is female!

Why should people be less inclined to report a male's infidelities? Two possibilities seem likely: (1) it may be that female's infidelities are less accepted and hence, more likely to be reported, or (2) it may be that women are considered less capable of "dealing well" with their partner's infidelities, and need to be "protected" from learning of them.

Some field-based studies raise serious ethical questions because they may involve deceit and almost always the permission of the subject is not obtained. Bear this in mind and return to this problem when you read Case Study 7.9 in Chapter 7.

FUNCTIONAL DESIGNS

In the previously discussed factorial designs, the usual procedure was to assign several subjects to each experimental treatment. The mean (or percent) of the subjects' scores on the dependent variable were then

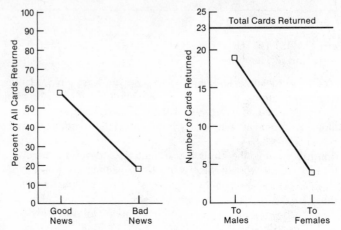

Figure 2.14   Percent of good/bad news cards mailed and number of bad news cards mailed to males and females.

compared with one another using the appropriate statistical test. While this procedure is common, a different research strategy is frequently used by those researchers who are interested in what they have called "the experimental analysis of behavior." This area of research originated with B. F. Skinner, and workers in this area are sometimes called "Skinnerians." The design has been called a *functional design* because of the use of functional definitions of terms and concepts. A functional definition of a concept (e.g., punishment) is accomplished by specifying the relationship between a set of determining conditions and their effects on behavior, both of which can be precisely measured. The requirement that functional definitions be used to some extent leads one to adopt the type of research strategy described here.

In contrasting this research strategy with those previously discussed in this book, several differences can be noted. First, researchers in this area tend to be *atheoretical* in the sense that they are more concerned with examining variables that control behavior than with testing some theory. Instead of viewing an experiment as a means of theory testing, variables that control behavior are systematically explored with the assumption that theory will emerge inductively from the data.

Small *n* Designs

A second difference is that researchers in this area will sometimes use only one (or two) subjects, called small *n* designs, rather than large groups of subjects in each experimental treatment. These researchers tend to report their data in the form of a "typical" response curve instead of comparing means (and variances) of several treatment groups. A typical response curve is a segment of the subject's behavior that is deemed typical of his performance under the particular experimental conditions. Another difference is that a statistical analysis of the data is sometimes not used, but the typical curve (or curves) is presented for "visual inspection" of the regularities in response that are representative of that particular stimulus condition.

A commonly used technique in small *n* studies is an ABA design in which a subject's untreated behavior (A) is first observed. This measure is sometimes called *baseline* data as it serves as a point of departure from which to contrast the effects of experimental treatment. In the second phase (B) the experimental variable is introduced and its effect is measured. In the final phase (A) the experimental variable is absent and behavior is observed. Behavior is observed throughout the sequence. Schematically, the design appears like this:

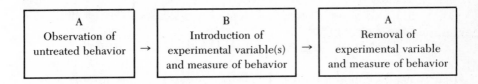

A behavioral therapist may, for example, be interested in treating a patient who overeats. In this case, the dependent variable could simply be the weight of the client (or, in more elaborate designs it could involve a whole range of physiological and psychological measures—metabolic rate, strength, feelings of well-being, absenteeism, and the like). Thus, the observation of untreated behavior in the A phase could be weight before treatment. In the B phase the treatment or dependent variable would be introduced, which could be a type of behavioral psychotherapy. Note that in some studies of this sort two or more independent variables may be used (e.g., the therapists may use positive reinforcement *and* exercise). It is critical to understand that the result of studies involving two or more independent variables do *not* allow for unequivocal specific cause-and-effect conclusions. It is possible, however, to make a valid concluding statement to the effect that treatment by means of conditions 1 and 2 has led to the following results. In the case mentioned, if weight loss occurred, then a statement to the effect that behavioral therapy and exercise led to (caused) weight loss would seem to be justified, but not that behavioral therapy *or* exercise (alone) led to weight loss. It *may* be that each of the variables alone would lead to weight loss, but the design does not permit such a conclusion. On the other hand, the combination of treatments (therapy *and* exercise) may effect weight loss in a *synergistic* way; that is, the cooperative action of treatments results in more effective reactions. Astute experimental psychologists are ever vigilant to possible synergistic effects, which may have profound behavioral consequences.

Many of the above designs are based on an internal validity in which the baseline data for a single subject serves as a control for subsequent observations. If an external control is used, it is of the following type, sometimes called AAA:

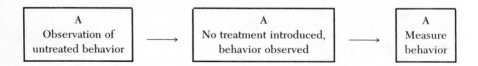

| A Observation of untreated behavior | ⟶ | A No treatment introduced, behavior observed | ⟶ | A Measure behavior |

The ABA design can be extended to an ABABA design, in which the experimental variable is reintroduced, or to an AB1AB2A or AB1AB2AB3A design, in which two or more different experimental variables are used. More elaborate designs are restricted only by the ingenuity of the psychologist.

Researchers in this area may use a highly controlled experimental situation. For example, an animal is placed in a "Skinner box"—a chamber that typically contains only a bar to be pressed by the animal, a food dispenser, and some signal lights. In this simple situation, numerous independent variables can be manipulated, such as the particular *schedule* of reinforcement under which the food is dispensed. For example, the reinforcement may be dispensed on a *fixed-ratio schedule,* in which the animal is reinforced with a food pellet after it presses the bar a fixed number of times (e.g., 1, 16, 47, or 100). In a *fixed-interval schedule* the animal receives a food pellet for pressing the bar at least once in every fixed time period (e.g., 30 seconds or four minutes). The dependent variable in the above cases is the frequency of bar-pressing responses and is usually presented in the form of a *cumulative frequency* curve. The example below is illustrative of research using this strategy and was taken from Ferster and Perrott (1968).

EXAMPLE

The apparatus used in this experiment was a Skinner box. This box is approximately 14 inches square. The only contents of the box are a small Plexiglas plate and a food magazine on the same wall. The plate and food magazine are connected such that if the plate is pushed, a food pellet will be released from the food magazine. Both the plate and the releaser mechanism are attached to a cumulative recorder. The recorder consists of a pen mounted on a sliding arm. The pen point rests on a strip of paper that passes slowly over a cylinder with the passage of time. If no responses are made by the animal, the pen point merely leaves a horizontal line as the paper passes over the cylinder. Each time the pigeon pecks the Plexiglas plate, the pen moves a small step in one direction on the paper and does not return to its original position. When the paper is examined, it is rather easy to see the animal's rate of response by noting the rate at which the pen moved (upward) in a given time period.

The subject used in this experiment was a pigeon that had had previous experience with this apparatus. The pigeon was kept at 80 percent of its free-feeding body weight throughout the experiment; that is, 80 percent of the weight it maintained when it was allowed to eat all that it wanted. The pigeon was placed in the Skinner box for one hour per day over a six-week time period. In the box the pigeon was reinforced for pecking the plate on a *fixed-ratio schedule;* that is, the pigeon received a food pellet after pecking the plate a fixed numer of times. Several fixed-ratio values were used. During the first week the pigeon received a food pellet after 70 pecks (FR 70); during the second week a food pellet was received after 185 pecks (FR 185); and during the third week the food pellet was received after 325 pecks (FR 325). The order was then reversed for the next three weeks.

Figure 2.15 shows the performance at each of the three fixed-ratio schedules. Each segment is an excerpt that is typical of the pigeon's daily performance on each schedule. The dots indicate the point at which reinforcement was delivered. The rate of performance in each segment can be estimated by comparing the overall slope of each segment with the slopes given in the grid in the lower right-hand corner of the figure. The slope of FR 70 indicates that when the pigeon was responding, it responded approximately three or four pecks per second. When 70 pecks are required for reinforcement, the bird's pecking is almost continuous, with a very slight pause after each reinforcement. For FR 185 there is a longer pause after each reinforcement; however, when the pigeon begins pecking again, it starts at a very rapid rate, which it maintains until the next reinforcement. The pause becomes much longer for FR 325; however, once the pigeon begins pecking again, it does so at the same rate that was noted for FR 70 and FR 185. The experiment indicates that the number of pecks necessary for reinforcement does not affect the rate of pecking but has influence on the length of pause between the dispensing of a reinforcement and the resumption of pecking.

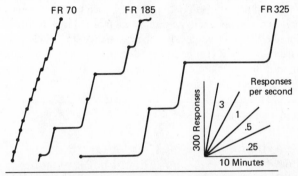

Figure 2.15   Rate of responding under three fixed-ratio schedules.

This example illustrates the design strategy used in this area. A single subject was used to participate in three experimental conditions (i.e., three different fixed-ratio schedules). The data are presented in terms of "typical" response curves that indicate the regularities of response under the three conditions. No high-level statistics are used. The experimental situation itself is highly controlled with precise measurement of the reinforcement conditions and the pecking responses.

This same design has been applied to a variety of subjects and experimental situations. The example below is an application of the same strategy in an experiment designed to control abnormal behavior in a chronic schizophrenic.

EXAMPLE

The strength of a response may decrease as a function of continued reinforcement. This phenomenon is called *satiation,* and it can be easily demonstrated in the laboratory situation. If the animal is given continuous reinforcement over a long period of time, the animal will stop emitting the reinforced response. Ayllon (1963) used the satiation procedure to control hoarding behavior in a psychiatric patient. The subject was a 47-year-old patient in a mental hospital who collected towels and stored them in her room. Although the nurses repeatedly retrieved the towels, the subject collected more towels and had an average of 20 towels in her room on any given day. Ayllon's procedure was first to establish a *baseline* as the average number of towels in the subject's room under these "normal" conditions. After a seven-week observation period, a satiation period began. In this satiation period the nurses no longer removed towels from the subject's room. Instead they began bringing towels into the room and simply handing them to the patient without comment. During this period the number of towels brought into the room by the nurses was increased from seven towels per day during the first week to 60 towels per day during the third week. The satiation period lasted for five weeks until the subject had accumulated 625 towels and had begun to remove the towels.

Figure 2.16 reports the mean number of towels per week in the patient's room over the period of the experiment. Note that this is *not* a cumulative record. After the satiation period the subject continued to remove towels from her room until at the twenty-second week there was an average of 1.5 towels in the room. This average continued through the twenty-sixth week. Ayllon made periodic observations throughout the next year and found that this average continued. The subject never returned to towel-hoarding behavior, and no other problem behavior replaced it.

While the Ayllon experiment is an interesting demonstration of a "cure" for hoarding behavior, one should not lose sight of the fact that it is also a nicely controlled experiment using a functional design. Ayllon

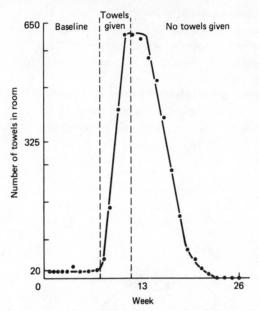

Figure 2.16   Number of towels in patient's room prior to, during, and after treatment.

(1) establishes a baseline for the frequency of this behavior, (2) institutes a well-designed experimental treatment, (3) terminates the experimental treatment, and (4) continues observation of the frequency of the behavior over an extended time period. A single subject is used, and no high-level statistics are used. A point to be emphasized here is that sound experimental design is necessary to ascertain the regularity of behavior, whether the subject be a pigeon or a mental patient and whether the behavior be pecking or towel hoarding.[5]

ADDITIONAL CONSIDERATIONS

This chapter has been concerned with some types of experimental designs used in psychology. While the basic logic of design is rather simple, the design and procedure may become fairly complex, as has been seen in the case of factorial designs and as will be seen in the following chapters.

Operational Definitions

There are two points that need to be introduced at this time. First, one decision that experimenters makes is how to operationally define their

[5]For additional experiments and theory, see Ayllon and Azrin (1968); see also Hayes (1981).

variables. Researchers usually have conceptual definitions of variables they wish to investigate. For example, psychologists talk about such things as anxiety, intelligence, ego involvement, drive, distributed practice, and reinforcement; their theories are based on the relationships between concepts such as these. However, to do research, psychologists must somehow operationally define these concepts by specifying precisely how the concept is manipulated or measured. An *operational definition* is a statement of the operations necessary to produce and measure the concept.

"... And then he raises the issue of 'how many angels can dance on the head of a pin,' and I say you haven't operationalized the question sufficiently — are you talking about classical ballet, jazz, the two-step, country swing. ..."

There is considerable variability as to the extent to which variables can be operationally defined in a manner that will be precise and that will retain the full meaning of concept that is being defined. On the one hand, variables such as the spacing of practice, as used in the Lorge experiment (see under *Independent and Dependent Variables*), or the delay of feedback, as used in More's experiment (see under *Factorial Designs*), are fairly easy to operationally define. On the other hand, psychologists also use abstract concepts such as intelligence or anxiety, which may be somewhat difficult to operationally define in a manner that includes the full complexity of that concept. Anxiety is a good example of such a variable. Almost everyone has some idea as to what anxiety is, and this concept is used by both psychologists and nonpsychologists. There are several dictionary definitions of this concept, most of which agree that it is a complex emotional state with apprehension as its most prominent component. In attempting to operationally define this variable, researchers have used pencil-and-paper tests, a palmar sweat technique, the galvanic skin response, heart rate, and eye movement. It seems probable that each of these operational definitions measures some part of this emotional state, although none of them measures its total complexity. A researcher somehow must pick or develop an operational definition that is suited for his or her specific situation.

The concern here is that the reader be aware of the necessity to operationally define the variables used in research. Below are a series of abstract concepts used in psychology, and it is suggested that the reader find operational definitions for these as part of a class exercise.

anxiety	memory	insight
creativity	learning	leadership
aggression	reinforcement	effort
intelligence	self-esteem	pornography
frustration	attitude	death
ego involvement	punishment	behavior

The second point is that the experimenter is always confronted with the question of how far he or she can *generalize* the results and conclusions of his or her experiment. In the Paul experiment reported earlier (showing the superiority of behavior therapy over insight therapy), there are several questions that can be raised concerning the generality of the findings:

1. How far can the results be generalized across "illnesses"? Is behavior therapy superior only for speech phobia? Only for phobias in general? Only for "mild" mental problems, including neurosis? Or is it superior for all mental problems?
2. How far can the results be generalized across subjects? Are the findings applicable only to college students? Would the same results be found for children or for middle-aged persons? Would the same results be found for lower-intelligence persons?
3. Would the same results be found if the therapy time had been longer? Is behavior therapy effective as a "quick" therapy, while insight therapy needs more time to produce its effects?
4. Did improvement generalize to other situations? Was the improvement only for speech class situations, or all classroom situations, or all speaking situations?

There are several additional questions that could be raised; for example, was the improvement temporary or permanent? However, the above four questions are sufficient to illustrate this point. The question of generalizability of results enters into all research and is not a criticism of the experiment. Rather it points out the limitations of a single experiment. In the single experiment, the researcher strives for tight control over his or her variables so that he or she can be certain as to the validity of the results. In many cases control is achieved by limiting an experiment to a specific behavior, to a specific sample of subjects, to a specific measurement technique, and to a specific time period. While these techniques of control are helpful (perhaps essential) in insuring the validity of the results, they also raise questions of the generalizability of the results. This is why one experiment leads to several other experiments, making research an ongoing process with new problems to investigate and new knowledge to acquire.

DEFINITIONS

The following is a list of experimental design or procedure-related concepts that were used in this chapter. Define each of these concepts:

independent variable
dependent variable
control group
placebo control group
subject variable
experimenter bias
experimenter blind
mean
operational definition
double blind
psychological experiment
ordinate
abscissa
base
ABA design
AAA design
baseline
warm-up
tachistoscope
correlational studies
variable

parameter estimation
generalization of results
ascending threshold
descending threshold
conceptual definitions
factorial design
statistical interaction
massed practice
functional design
level of significance
statistically significant
immediate memory span
short-term memory
synergistic effect
demand characteristics
mental chronometry
forced choice procedure
closed system
open system
quasi-experimental design

CHAPTER
3

Anatomy of Experimental Design: Control

CONTROLLED CONTRASTS

The nature of scientific inquiry is based on controlled observations of contrasts. This principle applies equally to physics, chemistry, botany, geology, astronomy, and psychology. In some scientific endeavors contrasts may be more obvious than in other endeavors. An example drawn from physics of a contrast might be criterion of objects of differing masses. Consider your reaction to hitting a regular tennis ball as contrasted to hitting a tennis ball made of lead. Most observations of contrasts are not nearly as obvious and far more subtle.

One example of a subtle contrast drawn from a nonpsychological science is the measurement by physicists of electromagnetic fields created by charged electrodes. The existence of these "invisible" fields is not immediately obvious; however, their influence can be measured by observing the distribution of finely machined iron particles in the absence of an electromagnetic field and under the influences of an electromagnetic field. In psychological terms, the electromagnetic field, created by a charged electrode, is an independent variable while the distribution of iron particles is a dependent measure from which a dependent variable (the characteristics of an electromagnetic field) is deduced. Nearly all high school students have seen this demonstration. Iron filings are sprinkled on a thin board and then a magnet is placed under the board. The distribution of iron particles immediately changes and a pattern emerges which clearly shows the presence of an electromagnetic field and its characteristics. From these observations theoretical characteristics are inferred and mathematical models that generalize to a wide class of similar physical situations may be developed. (See Figure 3.1 for a diagrammatic inference of such a field.)

The purpose of the above example is not to teach a principle of electrodynamics but to illustrate a basic tenet of experimentation: Science is based on observing contrasts. The observations may be in a controlled environment (as in a laboratory) or in nature. In the above example, the contrast was between the presence or absence of an electromagnetic field. In psychological research contrast is commonly made by observing the influence of an independent variable on one group of subjects as contrasted with another level of that variable in which a group of subjects is observed in the absence of an independent variable. The first group is called an experimental group, the second is a control group.

SCIENTIFIC INFERENCES

Another important scientific principle is also illustrated by the above example. Scientific conclusions are frequently based on inferences. The physicist infers that the distribution of iron filings changes as a result of an electromagnetic field; so too the psychologist infers that human behavior changes under the influence of a variety of conditions (e.g., drugs, social setting, motivation). Experimental psychologists observe behavior that is the consequence of some variable. Just as the physicist makes inferences about theoretical characteristics of electromagnetic fields,

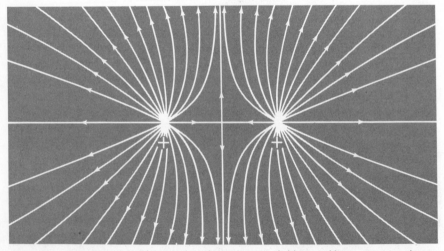

Figure 3.1   Hypothetical distribution of an electromagnetic field created by two positive charges.

psychologists sometimes develop models which generalize to a wide class of similar situations that lead to reliable laws about human behavior.

Because the validity of our observations of contrasting conditions is based on the characteristics of subjects (as well as the influence of the independent variable), it is extremely important that the experimental and control groups be as similar as possible. (Think of the erroneous conclusions about electromagnetic fields that would emerge if a scientist used iron filings in one electromagnetic experiment and aluminum filings in another!) Even though observations may be valid, if the experiment lacks control, the results may be meaningless or may misdirect future research. Those issues are important to the experimental psychologist, and careful procedures have evolved to safeguard against the errors that may be introduced by lack of proper control. We now turn to some error problems and procedures for their control.

TYPES OF CONTROL

Chapter 2 focused on design strategies. In the examples given there, the experimenter *manipulated* some variable and observed the effect of this manipulated variable (that is, the independent variable) on the dependent variable. The experimenter controlled the independent variable in the sense that he or she determined how much of the factor or what kind of factor was presented. This is one form of experimental control. It was also emphasized that good design requires that the only variable manipulated be the independent variable and that *all other conditions be held constant* for the various treatment groups. Holding all other conditions constant is a second type of experimental control. If the various treatment and control groups are treated exactly alike *except* for the independent variable, then any differences observed on the dependent variable must be due to the independent variable. If some other (unwanted) variable is affecting the results, this variable is usually called an *extraneous variable* or a *confounding variable.* The variable is extraneous in the sense that it is an "extra" variable that has entered into the experiment. It is a confounding variable because the experimenter cannot be sure if the results he or she obtained are due to the independent variable or are due to the extraneous variable. Thus, the results of the experiment are inconclusive, and the experiment should be repeated using a design that eliminates the influence of the extraneous variable.

A variable is extraneous in a given experiment *only* when it can be assumed to influence the dependent variable. In an experiment in the area of learning, such variables as color of the subject's eyes, height, or attitudes toward the USSR probably have no effect on learning, and the experimenter does not attempt to control these variables. On the other hand, there are variables, such as motivation or intelligence, that are known to affect learning, and these should be controlled for in a learning experiment. What variables are to be considered as extraneous (and therefore to be controlled) vary from experiment to experiment. Intelligence may be an extraneous variable in a learning experiment but may not be extraneous in an experiment to determine the absolute threshold of tones of varying frequencies. Thus the experimenter needs to control only a limited number of variables—those that can be assumed to influence the dependent variable.

This chapter is primarily concerned with the control of extraneous variables that occur when experimenters manipulate the independent variable. That is, when experimenters manipulate the independent variable, are they also introducing, or failing to control for, an extraneous variable? A second group of related problems deals with insuring that the various treatment groups are equal as to subject characteristics. This will be discussed separately in Chapter 5 because a set of rather specific techniques has evolved to handle this problem.

Up to this point considerable emphasis has been placed on holding all other conditions constant as a means of controlling for the effects of extraneous variables. Actually this is only one of two very general techniques of extraneous variable control. A second method is through the use of treatment or control groups. All experiments use the first technique of control in that the experimenters attempt to manipulate only the independent variable. Some experiments also add treatment or control groups to further control extraneous variables. This latter technique is frequently used when the experimental manipulation itself may contain an extraneous variable as well as the independent variable. By adding treatment or control groups it then may be possible to separate the effects of the extraneous variable from that of the independent variable. These two techniques are discussed in more detail below.

Holding Conditions Constant

In the Lorge experiment on massed versus distributed practice (Chapter 2), the independent variable was the length of time between practice sessions. This was the only variable that was manipulated, and all other variables appeared at the same level in all of the treatment groups. The same task was used for all treatment groups; all treatment groups had the same amount of practice; the task was such that it would be difficult for the treatment groups with spaced practice to rehearse between practice sessions; the subjects were assigned to treatments such that the abilities of the subjects in each treatment were equal; and so on. All these factors could operate as extraneous variables. For instance, if one treatment group received more practice than another group, the results could be explained as being due to the amount of practice rather than to the spacing of practice. If the different treatment groups had performed different tasks, then the results could be explained as being caused by a difference in task variables (such as task difficulty) rather than in the spacing of practice. If the subjects in one treatment were better in task-related abilities than those in another treatment, then the results could be explained as being due to differences in the abilities in the treatment groups rather than to the spacing of practice. By insuring that these factors or variables were the same for all treatment groups, Lorge eliminated these variables as explanations for his results. This is the logic behind holding all conditions constant except the independent variable.

Holding conditions constant is obviously essential to good experimental design, and this control strategy is easily understood by the beginning student of psychology. However, as will be seen in the following examples, even the most competent researchers may unknowingly violate this principle. While it is not possible to construct a checklist of extraneous variables (since they vary from situation to situation), there are some areas in which problems are especially prominent. For example, when the independent variable is a *subject variable,* that is, a variable such as intelligence, which the subject "carrries around" with him or her, there is always a danger that the subject variable is related in some systematic manner to another subject variable. If this is true, then the results might be due to the second subject variable introduced inadvertently rather than to the one the researcher introduced. This problem is called a *subject variable–subject variable confound.* To illustrate this problem, consider a study in which the researcher hypothesized that a person high in the personality trait of authoritarianism would have more difficulty learning complex material than a person low in this trait. This hypothesis was based on the assumption that high-authoritarian persons think in a rather simplistic manner and therefore would have difficulty learning complex material. To test this hypothesis the investigator had high-authoritarian subjects and low-authoritarian subjects learn some complex material. When tested for recall of the material, the low-authoritarian group showed considerably more learning than the high-authoritarian group. One criticism of several leveled against this study was that it is well known that there is an inverse relationship between authoritarianism and intelligence—persons high on a measure of authoritarianism tend to be low on a measure of intelligence, and vice versa. Therefore the fact that the high authoritarians learned less could be explained by the fact that they were less intelligent, and the trait of authoritarianism may not have had anything to do with the results. This criticism could be handled by matching the high- and the low-authoritarian groups on intelligence using a technique described in Chapter 5.

If a subject variable is not being manipulated, then a subject variable–subject variable confound is of little danger. However, the experimenter must be aware of other possible extraneous variables. Some of these problems will be discussed later in this chapter.

CONTROL AND EXPERIMENTAL PARADIGMS

A *paradigm* in experimental psychology is a model or pattern an investigator uses to organize his or her research.

In this section, we introduce several simple experimental paradigms and methods of control in each. Some of these paradigms and controls will be further discussed in later sections.

Suppose that you, as an experimenter, are interested in the influence of the color of a wine, as measured by ratings on a 5-point scale, on its enjoyment. You have developed a product that will change the color of wine without changing its taste. In one case, the natural color of the wine is a light ruby, while in the case of the altered wine, it is a deep burgundy. The experiment may be conducted in several ways.

Independent Subject Design: Model 1

In one paradigm (Model 1), called the *independent subject design* or *between-groups design*, subjects are given the artificially colored wine (the experimental group), while another group of subjects is given the naturally colored wine (the control group). A feature of the independent subject design is that subjects in one group are independent from the other group. The sample of subjects for this experiment is defined as students at a university between the ages of 21 and 30. From this sample, the subjects are *randomly* (i.e., by chance) assigned to either the experimental or the control group: A *randomized subjects design*. For convenience, we have labeled the subjects S_1, S_2, S_3, ... S_{10}. The arrangement of this procedure might look like the following:

MODEL 1.   INDEPENDENT SUBJECT
DESIGN OR RANDOMIZED GROUP DESIGN

Experimental Group (artificially colored wine) DATA		Control Group (naturally colored wine) DATA	
S_1		S_6	
S_2		S_7	
S_3		S_8	
S_4		S_9	
S_5		S_{10}	

Wine-tasting groups: Independent variable—colored wine
Dependent variable—evaluation of quality

This paradigm is frequently used in psychological experiments, and if the sample is large enough, you might assume that the subject variables that might influence the results (e.g., having a large number of expert wine tasters in one group) will be equally distributed between the two groups on the basis of random distribution.

Matched Subjects Design: Model 2

On the other hand, you may have reason to believe that subject variables (such as wine-tasting experience) may be so critical to the results that the subjects should be matched on that dimension. Such a design is called a *matched subjects design* (or matched pair) and might look like this:

MODEL 2.   MATCHED SUBJECTS DESIGN

Experimental Group (artificially colored wine) DATA		Control Group (naturally colored wine) DATA	
S_{1a}		S_{6a}	
S_{2b}		S_{7b}	
S_{3c}		S_{8c}	
S_{4d}		S_{9d}	
S_{5e}		S_{10e}	

Wine-tasting groups: Independent variable—colored wine
Dependent variable—evaluation of quality
Matching variable—wine-tasting experience

In this example we have labeled the subjects S_1, S_2, ... S_{10} but have added the subscript $_a$, $_b$, ... $_e$ to show that the S_{1a} subject and the S_{6a} subject are matched on some basis. In this case, we matched the subjects on the dimension of wine-tasting experience so that each pair of subjects have about the same level of experience. Other experiments may call for the matching of other attributes, e.g., sex, age, intelligence, running ability, training.

Repeated Measure Design: Model 3

A third design, called a *repeated measure design* or a *within-subject design* (Model 3), is characterized by each subject being exposed to two or more experimental conditions. In our example, each subject would be "treated" to both kinds of wine and, as such, each subject serves as his or her own control. We can illustrate this procedure as follows:

MODEL 3.   REPEATED MEASURE DESIGN
(WITHIN-SUBJECT DESIGN)

Experimental Condition (artificially colored wine) DATA		Control Condition (naturally colored wine) DATA	
S_1		S_1	
S_2		S_2	
S_3		S_3	
S_4		S_4	
S_5		S_5	
S_6		S_6	
S_7		S_7	
S_8		S_8	
S_9		S_9	
S_{10}		S_{10}	

Wine-tasting groups: Independent variable—colored wine
Dependent variable—evaluation of quality

As shown, S_1 tastes the experimental (colored) wine *and* the (natural) control wine. In addition to having each subject serve as his or her own control, it is possible to gather more data on the group as two measures are made for each subject rather than one. Sometimes it is possible to reduce the number of subjects in an experiment with the use of the repeated measured design, and this economy may be a practical solution to a limited subject sample.

Repeated Measure Design: Model 4

There are some problems that the previous type of design might create. One of these problems might be the sequence in which the substances are presented. It may be that in wine tasting the second taste of wine may seem more enjoyable than the first, not because the taste is better but because the taster has a slightly more rosy outlook on life in general. To safeguard against this eventuality, one could *balance* the sequence (Model 4). In this case, balancing the sequence is achieved by having half the subjects taste the colored wine first and then the natural wine, while the other half would taste the natural wine and then the colored wine. We can express this design in Model 4.

MODEL 4.   REPEATED MEASURE DESIGN
WITH SEQUENCE COUNTERBALANCED

Control Condition (naturally colored wine) DATA		Experimental Condition (artificially colored wine) DATA	
S_1		S_1	
S_2		S_2	
S_3		S_3	
S_4		S_4	
S_5		S_5	

Experimental Condition (artificially colored wine) DATA		Control Condition (naturally colored wine) DATA	
S_6		S_6	
S_7		S_7	
S_8		S_8	
S_9		S_9	
S_{10}		S_{10}	

Wine-tasting groups: Independent variable—colored wine
Dependent variable—evaluation of quality

In this case, S_1 through S_5 taste and evaluate the control wine first and then the experimental wine, while S_6 through S_{10} reverse the sequence. Data collected in this form can also be analyzed by statistical procedures that will identify the influence of sequencing. Further elaboration of the design is also possible. One could, for example, have a double or triple repeated design, but then the considerations of these complex designs might be intoxicating.

Factorial Design: Model 5

In the previous models we have been concerned with the effect of a single independent variable on a dependent variable. Frequently, however, psychologists are interested in studying the effects of more than a single independent variable as they may affect a dependent variable. Such designs are very useful in experimental psychology, and their utility has been pointed out in Chapter 2. Even though this design is not specifically designed to "control" for subject variables, it is reiterated here to show how these variables may be used in a factorial design. In the present context, suppose that the experimenter chose to measure the influence of the color of the wine *and* the type of grape used in producing the wine as two classes of independent variables on tasting enjoyment as the dependent variable. In this paradigm, the color of the wine would be either "natural" or "deep burgundy," as mentioned, but in addition, three types of wines (derived from three different grapes) would be used. For sake of illustration, a Pinot Noir, a Zinfandel, and a Chardonnay (a "white" wine) will be used.

A simple representation of this type of design can be shown in the following 2 × 3 matrix:

MODEL 5

		Factor 1 (color)	
		level 1 (natural)	level 2 (burgundy)
	level 1 (Pinot Noir)	S_1, S_2, S_3, S_4	S_1, S_2, S_3, S_4
Factor 2 (type of wine)	level 2 (Zinfandel)	S_1, S_2, S_3, S_4	S_1, S_2, S_3, S_4
	level 3 (Chardonnay)	S_1, S_2, S_3, S_4	S_1, S_2, S_3, S_4

In this design, the same subjects are treated to all conditions (repeated measure design: Model 3). Thus, in Model 5, it is possible to identify the integration of subject control techniques in a factorial design. It should be pointed out, however, that other subject control techniques may be used in a factorial experiment such as independent subject design Model 1. The latter design could be illustrated as follows:

		Factor 1 (color)	
		level 1 (natural)	level 2 (burgandy)
	level 1 (Pinot Noir)	S_1, S_2, S_3, S_4	$S_{13}, S_{14}, S_{15}, S_{16}$
Factor 2 (type of wine)	level 2 (Zinfandel)	S_5, S_6, S_7, S_8	$S_{17}, S_{18}, S_{19}, S_{20}$
	level 3 (Chardonnay)	$S_9, S_{10}, S_{11}, S_{12}$	$S_{21}, S_{22}, S_{23}, S_{24}$

USE OF TREATMENT AND CONTROL GROUPS

A *control group* is defined as a group of subjects similar to an experimental group and exposed to all the conditions of the investigation except the experimental variable (independent variable). In some cases the control group and the experimental group should be drawn at random from the entire population in order to make generalizations about the results of the experiment.

Some of the techniques for the control of extraneous variables have been briefly examined. Let us consider in more detail the Paul experiment (see Chapter 2, under *Experimental and Control Groups)* on the treatment of speech phobia. The experimenter wanted to compare the effectiveness of two types of therapy. In addition, he had to consider what extraneous variables might be varying along with the therapy. If an extraneous variable varied along with the specific therapy, then any improvement might be attributed to the extraneous variable rather than to the therapy itself. There are two well-known extraneous variables that are present when a subject receives some sort of therapy. First, it is well established that some persons showing symptoms of a behavior problem will improve over time without receiving any specific treatment for this problem. This phenomenon is called *spontaneous remission* because there is a disappearance of symptoms that takes place "spontaneously," that is, without any apparent treatment for the problem. In a therapy experiment the experimenter cannot be sure if the subject's improvement is due to the therapy itself or just to spontaneous remission.

Second, it is known that some people who think they are receiving some established treatment for their problems (but actually are not) may show considerable improvement. This is called the *placebo effect.* The term comes from the Latin word meaning "to please," and the effect was discovered by physicians who would give patients some medically inert substance (e.g., sugar water) that resembles an active medication in order to please the patient rather than to provide physical benefit. Interestingly enough, it was discovered that some patients improvd upon receiving placebo medication, especially those whose "illnesses" seemed to be psychosomatic. Since this placebo effect is assumed to occur in the treatment of psychological problems, any improvement in the therapy groups may be due to this effect rather than to the effect of the therapy itself.

To separate the amount of improvement due to spontaneous remission, due to the placebo effect, and due to the therapy itself, Paul used two control groups. Table 3.1 shows the four experimental groups as well as the variables influencing improvement in each of these groups.

As indicated in the table, the experimenter can now separate the effects of the various variables that are influencing the subject's improvement. For example, we can substract the amount of improvement shown in the placebo control group from the amount of improvement shown in each therapy group. This would give us an indication of the effectiveness of the therapy itself after we have eliminated the effect of the two extraneous variables.

Researchers will also frequently add treatment groups to the experimental design to insure that the results were not caused by an extraneous

Table 3.1   VARIABLES INFLUENCING IMPROVEMENT IN THE
FOUR EXPERIMENTAL GROUPS OF THE PAUL EXPERIMENT

| EXPERIMENTAL GROUPS | VARIABLES PRESENT INFLUENCING IMPROVEMENT | | | PERCENT IMPROVEMENT |
	THERAPY	PLACEBO	SPONTANEOUS REMISSION	
1. Behavior therapy	Yes	Yes	Yes	100
2. Insight therapy	Yes	Yes	Yes	60
3. Placebo	No	Yes	Yes	73
4. No treatment	No	No	Yes	32

variable. Suppose that in the Asch study on impression formation he had only used a single treatment group which received the positive adjectives first and the negative adjectives last (see Chapter 2, under *Independent and Dependent Variables*). With this single group the subject's evaluation of the person would be generally positive, and this would support the hypothesis of a primary effect in impression formation. If only this single treatment group were used in the experiment, several criticisms would arise. One criticism would be that Asch's negative adjectives were not really very negative, and thus the person would have been evaluated positively regardless of the order of presenting the adjectives. Another criticism would be that people tend to evaluate other people positively regardless of what information is given; that is, people "look" for good traits in others and tend to like others. If this hypothesis is valid, then Asch would have gotten a positive evaluation of the stimulus person regardless of the order of presentation or regardless of the type of adjectives used. The experimental design used by Asch eliminated these possible criticisms. One of his treatment groups received a positive-to-negative order of presentation of the adjectives. The second treatment group received a negative-to-positive presentation. The subject's evaluation of the person was positive in the first treatment and negative in the second treatment; therefore the above-mentioned criticisms or explanations of the results are not valid. However, note that they are demonstrated to be invalid because Asch used two treatment groups instead of one and obtained negative ratings from subjects.

It is difficult to formulate any specific principles about the control of extraneous variables. An experimenter usually begins with some problem to be solved or some hypothesis to be tested. In designing an experiment to test the hypothesis, he or she must keep in mind that any extraneous variables that could be used as an explanation of the results have to be eliminated. Certainly this would involve keeping all conditions constant except the independent variable, but it may also include the use of additional treatment or control groups. The best way for the student to learn what extraneous variables are to be controlled in any specific research area is to read experiments in this area. In this manner the student can become aware of the designs used by researchers in the area and also of what variables have to be controlled for. Some experiments that have appeared in the research literature involving control problems are presented below. The purpose of presenting these examples is not to criticize other researchers but to illustrate some problems that have arisen in the past. It is hoped that these examples will aid the student in analyzing a design for control problems.

CONTROL PROBLEM: SLEEP LEARNING

EXAMPLE

An experiment was conducted to determine if learning could take place during sleep. The material to be learned was the English equivalents of German words, and the subjects were 10 college students who reported that they had no knowledge of the German language. The subject slept in a comfortable bed in a soundproof, air-conditioned laboratory room. The subject retired about midnight, and at approximately 1:30 A.M. the experimenter entered the room and asked the subject if he or she was asleep. If there was no response, the experimenter turned on a recording that contained German words and their English equivalents: for example, "ohne means without." There were 60 different words on the recording, which was played continuously until 4:30 A.M. If the subject awoke during the night, he or she was to call out to the experimenter, and the recording would be stopped until the subject was asleep again. To test for learning, the 60 German words were played to the subjects in the morning, and after each word the subject reported what he or she thought was the English equivalent of the word. The number of German words correctly identified was the dependent measure. The results indicated that the mean number of words correctly identified was nine (out of a possible 60), and the highest number correctly identified by any subject was 20. The experimenters interpreted these results as supporting the hypothesis that learning can occur during sleep.

The experiment has several important implications. At a theoretical level it suggests that during sleep the brain is actively processing information received by the external senses. At a practical level it suggests that sleep learning may be an easy and effortless way to learn something. It should be a boon to college students who, instead of staying up all night to cram for an exam, can simply turn on a tape recorder and go to sleep. While these results are quite exciting, and one would like them to be valid, two major criticisms can be leveled against the experiment.

The first criticism was the failure to use a control group of subjects who had not been presented the learning material but who were given the recall test. While it is true that all subjects said that they had no knowledge of the German language, they may have been able to guess the meanings of some of the words. For example, *Mann* in German means man in English. Furthermore, some German words are frequently used in English, particularly in old war movies—for example, *Schwein* (pig), *nein* (no), and *ja* (yes). Thus, the apparent effects of sleep learning may actually be due to the subjects' ability to guess some words and some knowledge of other words, and a control group would have checked for this.

A second criticism dealt with the experimenter's operational definition of sleep. Sleep was defined as what the subject did between 1:30 A.M. and 4:30 A.M. unless the subject reported that he or she was awake. On the other hand, it is known that there are various levels of sleep, ranging from drowsiness to very deep sleep. It further is known that at drowsiness levels the subject has partial awareness of external stimuli; however, at the level at which sleep technically begins there is little or no awareness of external stimuli. In the experiment reported above there was no way of knowing what material was presented at what level of sleep. Therefore, it could be argued that any learning that occurred may have taken place at a drowsiness level rather than at a true sleep level.

Simon and Emmons (1956) designed an experiment to correct for the above problems. The materials to be learned consisted of 96 general information questions and their answers. They were presented in question form—for example, "In what kind of store did Ulysses S. Grant work before the war?" Then the answer was given: "Before the war, Ulysses S. Grant worked in a hardware store." Two groups of subjects were used: an experimental group, which was given the answers to the questions while sleeping, and a control group, which simply took the learning test without having the answers played to it. To begin the experiment both groups were given the questions and asked to guess the answers. Those questions that the subjects answered correctly on this test were eliminated. Next, the experimental group was presented with the questions and answers while asleep. During this period, recordings were made of their brain waves using an electroencephalograph (EEG). Because brain activity varies in a known manner at different stages of sleep, EEG records enable us to determine accurately the depth of sleep. As each answer was presented to the subject, the experimenter recorded the level of sleep of the subject. Thus, the experimenter had a record of the level of sleep at which each answer was given.

In the morning the experimental group was tested on the material that was presented to them during the night. This test was a multiple-choice test that consisted of the question and five alternative answers. The subject was to guess which of the answers was correct. A multiple-choice test was used since it probably is a more sensitive measure of sleep learning because the subject has only to recognize the correct answer rather than to recall it. The control group also took this test.

After the test scores were received, the experimenters separated the questions for each experimental subject into categories determined by the level of sleep at which the answer was played. The experimenters used eight levels of sleep, which have been condensed into two categories in Table 3.2, which reports the percentage of correct answers in these categories.

The data indicate that considerable learning took place when the experimental subjects were awake, and moderate learning appeared at a drowsy level; however, there was no apparent learning when the subjects were at a true sleep level. At this level the performance of the experimental group was the same as the control group subjects, who had no learning experience. The 23 percent correct for the control group represents the number of correct answers that could be expected if subjects guessed which of the five alternatives answers was correct.

In looking at the design of this experiment, it is important to note that the experimenters instituted several crucial control procedures to allow for a rather clear-cut test of sleep learning. First, to make sure that what appeared to be learned during sleep was not actually information that was previously known, the experimenters gave all subjects a pretreatment test on the material and eliminated the answers already known. Second, since a multiple-choice test was used as the test of learning, and since a certain proportion of the answers on such a test could be gotten correct by guessing, the experimenters used a control group to find out what the percentage correct by guessing would be. Third, the experimenters identified different levels of sleep and noted what answers were presented at each level. Using this technique it was possible to separate material presented when the subject was awake, was in a state of drowsiness, and was in a state of true sleep. Using these control procedures, the results suggest that no learning takes place at a true sleep level.

Table 3.2   PERCENTAGE OF ANSWERS CORRECT AT THREE SLEEP LEVELS

	LEVEL OF SLEEP		
	AWAKE	DROWSY	ASLEEP
Experimental group	92	65	23
Control group	24	23	23

CONTROL PROBLEM: SOCIAL DEPRIVATION AND SOCIAL REINFORCEMENT

EXAMPLE

It has been repeatedly demonstrated that for animals who have been deprived of food, the effectiveness of a food pellet as a reinforcer is considerably enhanced. An experiment was conducted to determine if the same results would occur with social "deprivation" and social reinforcement in children. Subjects were 6-year-olds in an elementary school.

The effectiveness of social reinforcement was measured by a marble game. The game consisted of a box with two holes in it, and the subject was to drop marbles one at a time into one of the two holes. For the first 4 minutes of the game the experimenter just watched the subject play the game. For the next 10 minutes the experimenter verbally reinforced the subject every time he or she put a marble in the hole that was least used in the initial 4-minute period. The verbal reinforcement consisted of the experimenter saying "Good" or "Fine" every time the subject put a marble in the "correct" hole. The dependent variable was the amount of increase in the subject's placing marbles in the "correct" hole from the 4-minute to the 10-minute reinforcement period.

In order to determine the effects of social deprivation the subjects were randomly assigned (i.e., on the basis of chance) to one of three treatments. In the social *deprivation* treatment, the subjects were left alone in a room for 20 minutes prior to playing the game. In the *nondeprivation* treatment, the subjects started playing the game immediately after leaving their classroom. In the social *satiation* treatment, the subjects spent 20 minutes talking with the experimenter while they were drawing and cutting out pictures. This took place just prior to the game.

The results indicated that the subjects showed more of an increase in putting marbles in the "correct" hole in the deprivation treatment than in either of the other two treatments. Further, there was a greater increase in the nondeprivation treatment than in the satiation treatment. The results were interpreted as supporting the hypothesis that the effectiveness of social reinforcement is influenced by conditions of social satiation or deprivation in a manner similar to that found for food or water deprivation.

The above experiment has wide theoretical implications in that it has produced evidence to suggest that social drives seem to be subject to the same "laws" that have been established using the primary appetitive drives, such as hunger. The authors also have developed a nice experimental situation to test this hypothesis since the experimental manipulations seem fairly clear-cut and the dependent measure (i.e., the number of marbles in the "correct" hole) is easy to record without ambiguity. However, soon after the experiment was published, research critical of the experiment began to appear, arguing that the results may have been caused by the failure to control extraneous variables.

One criticism of the experiment pointed out that when the experimenters were manipulating social deprivation ("social" being defined in terms of the amount of interaction with other people), they also manipulated general sensory deprivation. For example, in the deprivation treatment the child was not only isolated from other people but also had no toys with which to play. In the satiation treatment the child not only interacted with the experimenter but also drew and cut out pictures. Thus the experimenters comanipulated an extraneous variable (general sensory deprivation) along with social deprivation, and the results of the experiment might have been caused by this extraneous (potentially confounding) variable. Stevenson and Odom (1962) tested this alternative explanation of the results by comparing three groups of subjects. Before playing the marble game, one group of children was isolated and played with attractive toys for 15 minutes, one group was isolated for 15 minutes without toys, and a third group of children began playing the game immediately after being called out of the classroom. The results indicated no difference in the task performance of the two isolation groups, but both groups had higher levels of performance (i.e., more marbles dropped in the "correct" hole) than the no-isolation group. Since there was no difference in the task performance of the two isolation groups (both of which were socially deprived, but only one of which was deprived of toys), then the higher performance must have been due to the social deprivation. This supports the original interpretation of the experiment, and the influence of the extraneous variable was apparently very minor.

A second group of researchers raised a different criticism of the experiment. The crucial point of this criticism was that being placed in a strange environment by a strange adult should arouse anxiety in 6-year-olds. The greatest anxiety should occur in the deprivation treatment, in which the subjects were left alone in a strange room for 20 minutes. The next highest level of anxiety should occur in the nondeprivation situation, in which the subjects were led directly to the game situation. The least anxiety should occur in the satiation treatment, because after 20 minutes of friendly conversation with the experimenter the subject should be somewhat comfortable in his or her presence. Since there is evidence demonstrating that heightened anxiety improves performance in some learning tasks (especially simple learning tasks), then the results of the experiment could be explained by the difference in anxiety arousal in the three treatments, and there is no need to postulate some "social drive."

To test this hypothesis, Walter and Parke (1964) used a 2 × 2 factorial design in which they used two levels of isolation (either leaving the child alone for 10 minutes or starting the game immediately on coming from the classroom) and two levels of anxiety arousal. In the low-anxiety condition the experimenter treated the subject in a pleasant and friendly manner, and in the high-anxiety treatment the experimenter treated the subject in a rather cold and abrupt manner. Using these treatments, (1) no statistically significant difference was found in performance between the two levels of isolation, which is evidence against the social-drive interpretation; (2) subjects in the high-anxiety treatment performed better than subjects in the low-anxiety treatment, which supports the anxiety arousal interpretation; and (3) the interaction effect was not statistically significant. Thus it appears that an extraneous variable (arousal level) may have determined the results of the original experiment, and the social deprivation interpretation may be invalid.

CONTROL PROBLEM: PERCEPTUAL DEFENSE

The following experiment suggests that there is some process in the unconscious that determines whether a word is anxiety-provoking, and if this is so, conscious recognition of the word is prevented or at least delayed. This is the notion of perceptual defense, and it has far-reaching implications concerning human behavior.

EXAMPLE

It has long been suggested that the human organism has certain mechanisms that protect it from anxiety-provoking stimuli. One such mechanism is called *perceptual defense*. An experiment was designed to test the perceptual defense hypothesis by presenting to subjects neutral words and "taboo" words on a tachistoscope. This piece of equipment consists primarily of a shutter mechanism that can be manipulated so that a stimulus can be exposed to a subject for varying lengths of time. For example, the shutter can be adjusted so that the word is exposed for 0.01 of a second or for two minutes. The experimenter theorized that taboo words are anxiety-provoking, and while the subject may recognize them at an unconscious level, a perceptual defense mechanism would delay the subject's recognition of them at a conscious level. Based on this assumption, it was hypothesized that longer exposures would be necessary for the recognition of taboo words than for neutral words.

The subjects were eight male and eight female college students. Each subject was tested individually with both a male and a female experimenter present. Eleven neutral words (e.g., *apple, trade*) and seven taboo words (e.g., *whore, bitch*) were presented to each subject in a predetermined order. An ascending threshold method was used to determine the point at which the subject recognized the word. For each word, the shutter was set at a very fast exposure speed (0.01 of a second), and the exposure was gradually lengthened until the subject verbalized the word correctly. This process was repeated for each of the 18 words.

The mean threshold for the recognition of the neutral words was 0.053 seconds, and the mean threshold for the recognition of the taboo words was 0.098. The difference betweeen the two means was statistically significant. Since it took longer exposures (higher thresholds) for the subjects to recognize the taboo words than the neutral words, the experimenter concluded that the perceptual defense hypothesis was supported.

It did not take long for other researchers to criticize the design. Howes and Solomon (1950) raised two methodological points. First, they suggested that the results may be due to the subjects' reluctance to report a taboo word until they were absolutely positive of the identification of the word. The subjects might be particularly reluctant to verbalize these words in front of an experimenter of the opposite sex. A second methodological point was that neutral words appear much more frequently in print than the taboo words do, and therefore the subjects' quicker recognition of the neutral words was because they had seen these words more frequently. In a follow-up experiment Howes and Solomon (1951) demonstrated the validity of the "word-frequency" hypothesis. They obtained a listing of the frequency at which some 30,000 words appear in print. They chose 60 words of varying frequencies (all nontaboo words) and determined the recognition thresholds of each word using a procedure similar to that used in the above example. They found a high negative correlation (approximately -0.79) between the frequency at which the word appeared in print and its recognition threshold; that is, the more frequently the word appeared in print, the lower its recognition threshold. While this experiment demonstrates that word frequency is a plausible explanation for results of the original experiment, it still can be argued that taboo words show a higher threshold even if frequency is controlled for.

Postman, Bronson, and Gropper (1952) made a more direct test of the word-frequency explanation by determining how frequently the taboo words appeared in print and matching them with neutral words that appeared in print at the same frequency. The list of words was presented to the subjects using a procedure similar to that of the above example. The results of this experiment indicated no support for the perceptual defense hypothesis. In fact (surprisingly enough), it was found that the recognition threshold for the taboo words was significantly lower than that of the neutral words. This was probably due to an underestimation of the frequency of the taboo words. Although research in this area continues, it appears that the early demonstration of the perceptual defense phenomenon may have been due to the confounding variable of word frequency.

CONTROL PROBLEM: ONE-TRIAL LEARNING

EXAMPLE

When a child is learning to read the alphabet he or she is shown the letters A, B, C, and so on, while the teacher pronounces the names of the letters. This procedure is repeated until the child learns that association between the printed letter and the verbalized sound. There is a controversy among learning theorists as to whether this association is gradually built up (incremental process) or whether it occurs in an all-or-none fashion. The latter school of thought holds that if some learning trials have been given and the child is shown the letter and cannot verbalize its name, then no association between the letter and its name has taken place. The former school of thought would argue that some association has taken place, but this association is not yet of sufficient strength to allow the child to give the correct answer.

A rather ingenious experiment was performed to determine which of the above theories was correct. The task of the subjects was to learn eight nonsense syllable pairs. Each pair was presented to the subjects on a separate card. After the subject had seen all eight pairs he or she was then shown the first nonsense syllable of each of the pairs and asked to give the second nonsense syllable. This was the test of whether or not the pair had been learned. For example, one pair the subject was shown might have been POZ-LER. In the recall test the subject would be shown only the first syllable, POZ, and would have to supply the second syllable.

Two experimental groups were used. For the first experimental group the experimenter replaced every nonsense syllable pair that had not been learned on a single trial. For example, the subject was shown eight pairs and then given a learning test on these eight. Any pair that was not recalled was dropped out of the eight cards, and a new pair was substituted into the list of eight. The eight cards (which then included only pairs that were learned and new pairs) were again shown to the subject, and a recall test was given. The experimenter again eliminated those pairs that were not recalled and substituted new pairs. This process was repeated until the subject could recall all eight pairs given on a single trial. The second experimental group was treated like the first except that they were shown the same eight cards on each trial. Thus, eight pairs were shown, a recall test given, the same eight pairs shown, a recall test given, and so on. This process was repeated until the subject could recall all eight of the pairs on a single trial.

The experimenter then compared the number of trials it took to learn all eight pairs in each treatment group. The mean number of trials for perfect recall for both groups was exactly the same (8.1). Since *all* pairs learned in the first experimental group were learned on one trial (or else they were thrown out), and since there was no difference in the average number of trials necessary to learn the eight pairs perfectly in this condition from the one in which the same eight pairs were repeated, it was concluded that learning (i.e., associations) occurred in an all-or-none fashion. Stated a little differently, the experimenter argued that a gradual buildup of associations through repetition could have occurred in the second experimental group. If repetition is important in learning, then this second group should learn the eight pairs faster than the first group, who learned eight pairs on a single trial without any repetition. Since there was no difference between the two groups, it was concluded that repetition is unnecessary for learning.

As might be expected, this experiment caused some excitement, particularly among those who support an incremental view of associative learning, since the experiment suggests that this latter view of learning is invalid. It was not long after the experiment was published that research criticizing the experiment began to appear. One major criticism of the experiment was that the more difficult pairs were probably dropped out of the first treatment group, and thus the final list learned by each subject in this treatment consisted only of the easier pairs. No pairs were dropped out of the second treatment group, and thus the final list learned by these subjects consisted of both easy and difficult pairs. The failure of the experiment to find quicker learning in the treatment that involved repetition of the same pairs was due to the fact that the list learned by this group was more difficult than the final list learned by the subjects receiving the dropout procedure.

In one test of the "item selection" hypothesis, Underwood, Rehula, and Keppel (1962) repeated the above procedure except that they added a control group which received lists composed of the pairs that subjects in the dropout condition received on their last learning trial. Subjects in this condition were given the same word pairs on each trial. The results indicated that subjects in this latter condition learned all pairs on the list more rapidly than did those subjects given the list containing a random sample of all of the pairs used. Thus it appears that the subjects in the dropout condition of the original experiment were learning easier pairs, and this may be an explanation for the results.

SUMMARY

These four examples illustrate a variety of control problems. In the sleep learning experiment there was a failure to control for previous knowledge as well as level of sleep. In the social deprivation experiment there was a failure to separate social deprivation from sensory deprivation. Another criticism of this experiment was that results were due to arousal rather than to deprivation. In the perceptual defense experiment there was a failure to control for word frequency. In the one-trial learning experiment, the results were apparently due to the fact that the pairs learned in one treatment group were easier than those learned in the other treatment group.

The variety of control problems found in these examples illustrates why it is not possible to give the student a list of extraneous variables. Control problems vary so greatly from experiment to experiment, and from research area to research area, that no list is possible. To repeat an earlier statement, the student can best learn about experimental designs and control problems in a specific area of research by reading about experiments in that specific area. On the other hand, there is a variety of control problems that occasionally appears that is fairly easily recognized by the beginning student in psychology. Chapter 4 will allow students to practice their knowledge on some of these problems.

DEFINITIONS

The following is a list of terms and concepts that were used in this chapter. Define each of the following:

extraneous variable
holding conditions constant
spontaneous remission
electroencephalograph
confounding variable
placebo effect
tachistoscope
recognition threshold
paradigm

independent subject design
within-subject design
subject variable–subject
    variable confound
between-group design
matched subject design
matched pair design
repeated measure design
factorial design
control group

EXERCISE

For each of the four control problems described in this chapter, consider the experiment, the corrected problem, and the initial problem, and (1) identify the independent variable, (2) identify the dependent variable, and (3) explain how the design corrected for the initial control problem.

CHAPTER
4

Design Critiques I

The steps typically involved in psychological research start with searching the literature to become familiar with previous studies. The researcher may find an issue that needs further investigation, which is formalized in terms of an hypothesis. The next stage may involve designing an experiment that is both valid and practical. After all materials and subjects are lined up, the experimenter collects and analyzes data. Finally, conclusions are made in the form of a discussion. As simple as the process may appear, there are pitfalls lurking at every stage.

In this section (and other sections to follow) we will consider a special type of trap into which many seasoned researchers as well as neophyte researchers fall. It is basically a trap caused by imperfect logic that has been manifest in either faulty design and/or interpretation. Some students of experimental design identify these problems as presenting *alternative explanations,* as the results may be attributed to other causes than the one identified (see cartoon on page 80). Our experience leads us to believe that the more practice the student has on these problems, the better he or she will be in discovering design and/or interpretative problems in his or her own research. In addition to reading and discovering the flaw in the problems presented here, it is useful to make up some problems on your own.

On the following pages are a series of experiment "briefs." There is one or more design problems in each one of the briefs. The design problem occurs in the discrepancy between what the experimenter did in the experiment and the conclusion that he or she arrived at on the basis

"Well, shoot. I just can't figure it out. I'm movin' over 500 doughnuts a day, but I'm still just barely squeakin' by."

of the results. We previously illustrated basic principles of experimental design. Now our purpose is to expose you to a series of fictitious studies and let you apply your knowledge of experimental design in a critique of them. The use of the briefs is a quick way to expose you to a variety of problems in a variety of research areas. You do not need any expertise or technical knowledge in the research area being explored in the study. The problems can be recognized with a knowledge of design principles previously discussed in the text.

In criticizing the design of the experiment briefs, you should make use only of the information given in the brief. Do not criticize the design by inferring something that is not given. For example, if the experimenter used a pencil-and-paper anxiety test, you should assume that the test is valid and reliable unless information to the contrary is given. There is usually one major defect in each brief, and you should concentrate your criticism on this major problem. Be specific as to the defect. For example, do not just say that the experimenter should have used a control group, but point out exactly how this control group would be treated.

The following example illustrtes how the briefs should be criticized:

EXAMPLE

A certain investigator hypothesized that the hippocampus (a part of the brain) is related to complex thinking processes but not to simple thinking processes. He removed the hippocampus from a random sample of 20 rats. He had ten randomly selected rats learn a very simple maze and had ten randomly selected rats learn a very difficult and complex maze. The first group learned to run the maze without error within ten tries (or trials). It took the second group at least 30 trials to run the maze without error. Based on these results, he concluded that his hypothesis had been confirmed—rats without a hippocampus have more trouble learning a complex task than they do learning a simple task.

This experiment, in general, conforms to Model 1: the independent subject design or randomized group design described in Chapter 3. In that model, you may remember, subjects are randomly assigned to one or another condition.

In criticizing this design, it appears reasonable to assume that any rat would take more trials to learn a complex maze than it would to learn a simple maze. Thus, the results found by the experimenter may have nothing to do with the removal of the hippocampus—rats with the hippocampus intact might show the same results. In other words, although two independent variables were intended (task difficulty and presence/absence of hippocampus), only one independent variable was varied. This criticism would suggest that in redesigning the experiment a 2 × 2 factorial design should be used. One factor is hippocampus intact or hippocampus removed; the second factor is simple or complex maze. This is diagrammed below:

	SIMPLE MAZE	COMPLEX MAZE
Hippocampus intact	5 rats	5 rats
Hippocampus removed	5 rats	5 rats

This type of design conforms to Model 5, factorial designs mentioned in the previous chapter in which the effect of two independent variables is evaluated. This revised design would allow for a more reasonable test of the experimenter's hypothesis than the original design. Although the redesign might become more complex than that noted above, it does point out the major defect in the original experiment and indicates a way of correcting this defect. The student may first want to do the following exercise, which attempts to formulate questions that can be used to critique experiments.

EXERCISE

Formulate a series of questions that can be used to critique experiments. Each question should look at one aspect of the experimental design to test its adequacy. For example:

1. What is the independent variable? Are there at least two levels of it (or is there an experimental group and a control group)? If not, here is a design defect. In any experiment, one treatment has to be compared with another.
2. Is the independent variable a subject variable or a manipulated variable?
3. What is the dependent variable? How is it measured?
4. Assuming there are two levels of the independent variable, are all groups treated identically except for the experimental manipulation? If not, there is a confound in the experiment.
5. What type of design was used?

What other questions should be asked?

EXPERIMENT BRIEFS

1. An investigator attempted to ascertain the effects of hunger on aggression in cats. He took ten cats, kept them in individual cages, and put them on a food deprivation schedule such that at the end of two weeks the cats weighed 80 percent of their normal body weight. He then put the cats together in pairs for 15 minutes and watched to see if aggression or fighting would occur. In all cases, the cats showed the threat posture, and in most cases fighting occurred. The investigator concluded that hunger increases aggression in cats.
2. Psychologists working for food and beverage companies have always played a critical role in product development. In one experiment conducted by a consumer psychologist, subject preference for two types of cola was measured. The company had noticed that in one marketplace their brand of cola sold significantly fewer containers than their leading competitor. These data were particularly puzzling as on a nationwide basis their cola sold significantly better than their competitor's cola.

The researchers were concerned that some local condition may have contributed to the rejection of their cola and set out to test the hypothesis. The experimental design was a repeated measure design in which each subject tasted two colas. One cola was marked Q (the competitor's brand) and the other was marked M (the brand produced by the company doing the experiment). A random sample of citizens between the ages of 14 and 62 were asked to participate in the experiment. All subjects tasted brand Q and then brand M, giving their preference for each brand. Much to the surprise of the experiments, the subjects reported an overwhelming preference for the brand M company's brand of cola. The authors concluded that the sample preferred their company's brand and that the variable which must have contributed to the cola consumption of the other brand was the role of advertising. They therefore suggested a multimillion-dollar advertising campaign to rectify the situation.

3. An experimenter wished to examine the effects of massed versus distributed practice on the learning of nonsense syllables. She used three treatment groups of subjects, and subjects were randomly assigned to conditions. Group I practiced a 20-nonsense-syllable list for 30 minutes one day. Group II practiced the same list for 30 minutes per day for two successive days. Group III practiced the same list for 30 minutes per day for three successive days. The experimenter assessed each group's performance with a free-recall test after each group had completed its designated number of sessions. The mean recall of the 20 syllables for Group I was 5.2; for Group II, 10.0; and for Group III, 14.6. The means were significantly different from one another at the 0.01 level of significance, and the experimenter concluded that distributed practice is much superior to massed practice.

4. A certain psychologist was looking for the cause of college failure. He took a group of former students who had flunked out and a group of students who had received good grades. He gave both groups a self-esteem test and found that the group that failed scored lower on the test than the college success group. He concluded that low self-esteem is one of the causes of college failure and further suggested that the low self-esteem person probably expects to fail and exhibits defeatist behavior in college—which eventually leads to his or her failure.

5. A psychologist designed a study to determine if persons with high blood pressure could "learn" to control their blood pressure using biofeedback techniques. A device that records blood pressure is attached to the patient. The patient is given

"feedback" as to the level of blood pressure by a tone that decreases in loudness as the blood pressure decreases and increases in loudness as blood pressure increases. The patient is told to try to keep the tone as soft as possible. Five patients with high blood pressure each received 10 half-hour sessions using biofeedback. All five decreased their blood pressure level considerably over the 10 sessions, and the researcher claimed success for this method.

6. An experiment was designed to test a hypothesis that states that high-drive subjects would be able to learn a simple task much quicker than would low-drive subjects. The hypothesis further states that on a difficult task the opposite result would be found; that is, low-drive subjects would learn the task quicker. The experimenter's operational definition of drive was the subjects' scores on the Manifest Drive Scale. Twenty people who scored high on the scale (high-drive) and 20 people who scored low on the scale (low-drive) were given a difficult task to learn. The low-drive group learned the task quicker than the high-drive group, and the experimenter concluded that the hypothesis is correct.

7. An investigator set out to test the hypothesis that fear of punishment for poor performance has a detrimental rather than a facilitative effect on motor performance. As a measure of performance, the experimenter used a "steadiness" test in which the subject's task was to insert a stylus into a hole so that the stylus did not touch the sides of the hole. Each subject inserted the styles in 15 different holes. The experimenter manipulated fear by threatening the subjects with electric shock if they performed poorly on the task. He strapped an electric shock apparatus to the leg of each subject before he or she performed the task; however, he never shocked the subjects regardless of their performance. Subjects were randomly assigned to conditions. One group of subjects was threatened with 50 volts of electricity, and he called this the *mild-fear* condition. A second group of subjects was threatened with 100 volts of electricity, and he called this treatment the *high-fear* condition. Contrary to his hypothesis, the high-fear subjects did not perform any worse than the low-fear subjects—in fact, the means for both groups were approximately the same. Based on these results, the experimenter concluded that fear of punishment has little, if any, effect on motor performance.

8. An experimenter took 20 subjects who said they believed in astrology, gave them their horoscopes for the previous day, and asked them how accurate the horoscope had been in predicting

the previous day's occurrences. The subjects indicated their opinion on a six-position scale that ranged from subjects extremely accurate to extremely inaccurate. All 20 subjects reported their horoscopes as being accurate to some degree, and none reported his or her horoscope to be inaccurate. The experimenter concluded that horoscopes are accurate.

9. A 2 × 3 factorial design was used to evaluate the effect of dosage level of an experimental drug (Remoh) on the treatment of schizophrenia. Two patient classifications were used: (1) new admissions to a particular mental hospital and (2) patients who had been institutionalized for at least two years at that hospital. Both groups of patients had not been hospitalized previously. Patients received one of three levels of dosage—either 3 grams per day, 6 grams per day, or 9 grams per day. Within each patient classification group, patients were randomly assigned to one of the three dosage levels. There were 20 patients in each of the six groups. In addition to administering the drug, the experimenters also rated each patient each week as to the presence or absence of schizophrenic symptoms. After two months, it was found that very few (10 percent) of the long-term patients in each group had improved, regardless of dosage level. It was also found that approximately 50 percent of the newly arrived patients had improved in each of the three dosage level groups. The researchers concluded that (1) Remoh is effective only for new arrivals and not for chronic cases and (2) a dosage of 3 grams per day is sufficient to maximize the effectiveness of the drug.

10. An interesting controversy in clinical psychology is whether psychosis is inherited or is caused by environmental experiences. One "environmentalist" hypothesized that children who live with psychotic parents would be prone to having the same problem later in life. To test this hypothesis, she took 1000 psychotic adults and 1000 normal adults and traced back to see if either or both parents had been psychotic. Less than 1 percent of the normal adults, but over 30 percent of the psychotic adults, had had parents who were psychotic. Based on this result, she concluded that psychosis is not inherited but that childhood experiences with a psychotic parent make a person especially prone to the disorder.

11. A group of investigators suspected that rats trained to run on a wheel against a drag would run significantly faster if fed a 20 percent sucrose solution along with their daily rations than rats who received only the daily rations. One hundred Mayflower rats arrived from Plymouth Rock Animal Breeders (known in

the rat-running business as "designer rats") and were divided randomly into two groups. Fifty rats on a lab chow ran on the wheel against the drag followed by 50 rats on the sucrose and lab chow diet. The order in which the groups ran was determined by flipping a coin. The second group ran faster than the first group.[1]

[1]This example was suggested by Curt Mearns.

CHAPTER
5

Control of Subject Variables

EQUALITY OF SUBJECTS IN TREATMENT GROUPS

In psychological experiments the behavior of some species of animal is the focus of research. Since psychological research uses subjects (human or otherwise), it is not surprising that psychologists have devoted considerable attention to solving the problems of the control of those extraneous variables that are due to the characteristics of subjects. In fact, a set of specific techniques has evolved that are applicable to a wide variety of research situations. This chapter will discuss the essential features of these techniques.

In research, the performance of one treatment group is often compared with that of another group. These groups consist of subjects who differ on a variety of traits that could influence the results. It is important that all treatment groups of an experiment be approximately equal as to these various traits, so that whatever results are found are attributable to the independent variable and not attributable to the fact that subjects in one treatment were different in some trait (e.g., IQ) from subjects in another treatment.

Field studies provide the greatest possibility that the results are caused by subject differences rather than by treatment differences. In any study in which subjects are used in their natural groups or have volunteered specifically for one treatment or another, a major question one should ask is: How do the subjects in the various treatments differ? For example, a large manufacturing company held leadership training courses for lower-level employees. These courses were run on a volunteer basis and took place at night on the employee's own time. In evaluating the effectiveness of this course 10 years later, it was found that those persons who had taken the course had advanced further in the company than those who had not taken the course. While this result was interpreted as supporting the effectiveness of the course, an alternative explanation would be that the course may have attracted only those people who were highly motivated to advance in the company. Thus the "treatment" group may have consisted of highly motivated people, and the "control" group may have consisted of unmotivated people. If this explanation is valid, then the course may have had little effect on advancement in the company; the results may be due to differences in motivation of the two comparison groups.

The problem cited above can be avoided by having the researcher *assign* subjects to various treatment groups in a manner that insures that the subjects in the groups are approximately equal in all relevant characteristics. There are three general techniques for accomplishing this, which will be discussed in this chapter. The first technique is called a *random subjects design* because subjects are randomly assigned to the separate treatments. The random assignment of subjects allows the experimenter to be fairly certain that the subjects in all treatments are approximately equal as to subject variables. This design incorporates the features of Model 1 introduced in Chapter 3. The second technique is the *matched subjects design* (Model 2). Using this design, the experimenter first gets the scores for each subject on some task or test and then assigns subjects to the various treatment groups so that all of the treatment groups are equivalent with respect to these scores. The third technique is the *within-subject design* (Model 3). Using this design, all subjects participate in all experimental treatments, which insures that the treatment groups are equal as to subject variables. Finally, the fourth technique is called a *repeated measure design with sequence balanced* (Model 4). This design repeats a measure of the independent variable with the same subjects and also compensates for sequencing effects. In the first two techniques, random groups design and matched subjects design, *different* subjects are assigned to the different experimental treatments. These two designs are called *independent groups designs*. In the third technique, the within-subject design, the same subjects participate in all experimental treatments.

RANDOM GROUPS DESIGN: MODEL 1

The most common method of assigning subjects to treatments is *random* assignment. In its most rigid form this would mean that a selection procedure is used so that each subject has an equal opportunity (or probability) of being assigned to each treatment group. If the experiment consists of two treatment groups, one method of random assignment would be to use a table of random numbers to assign subjects to treatment conditions or to flip a coin for each subject when he or she reports to the laboratory for the experimental session. Using a randomization procedure such as this, the experimenter can be fairly certain that, as a group, subjects in treatment A and subjects in treatment B are approximately equal.

While the above procedure is simple to administer and is consistent with the definition of random assignment, it also presents a serious problem. With this procedure it is quite possible that the experimenter will end up with unequal numbers of subjects in the two treatments. For example, it would be possible to end up with 15 subjects in treatment A and only 5 subjects in treatment B. This unequal number of subjects would be undesirable, because the treatment mean based on the 5 would probably be less stable than a mean based on more subjects. A second consideration is that some of the statistical analyses of the results are simplified if an equal number (n) of subjects is used in each treatment group. Ideally, what is needed is a procedure that allows for some random assignment but also insures an equal n in each of the treatment groups. Such procedures are commonly used; and although they are not in accord with a rigid definition of random assignment, they are usually called *random assignment* or, perhaps more accurately, *unbiased assignment*.

In the Linder, Cooper, and Jones experiment, in which the subjects were given $.50 or $2.50 for writing an essay, each subject reported to the lab booth individually and was assigned to a treatment when he or she appeared (Chapter 2, under *Factorial Designs*). Experiments of this type usually take place over periods of weeks or months, and an unbiased selection procedure must take into account the possibility that subjects who report earlier for the experiment may not have similar characteristics to those who report later for the experiment. For example, subjects who report later may have heard something about the experiment from subjects who reported earlier, or perhaps the more motivated subjects appear earlier. One procedure to take this into account would be some form of *block randomization*. Using this procedure the experiment is run in a series of "blocks" over the time period so that all treatments are represented within each block, but the order in which they appear within the block is somewhat random. For example, in a two-treatment experiment the experimenter might flip a coin for the first subject who reports to the lab booth. If the coin lands "heads," the subject is assigned to treatment A, and if the coin lands "tails" the subject is assigned to treatment B. The next subject who reports to the lab booth is assigned to the treatment not assigned to the first subject. Thus if "heads" comes up for the first subject, he or she is assigned to treatment A, and the second subject is assigned to treatment B. The coin is flipped for the third subject, and the fourth subject is assigned to the other treatment. This process is repeated until all 20 subjects have participated in the experiment. With this procedure the two treatments are equally distributed over the time period in which the experiment is conducted, but within each block there is unbiased assignment.

If more than two treatments are used in an experiment, block randomization can be used by putting slips of paper representing each treatment into a container and drawing one out as each subject reports for the experiment. For example, if six treatments are used, six slips of paper, each with a different letter representing one of the six different treatments, are put into a container. As each subject reports for the experiment one slip of paper is drawn out of the container, and the subject is assigned to the treatment represented by the letter on that slip. The slip is not put back into the container. The second subject is assigned to the treatment by drawing from the five remaining slips. This procedure is followed until all six slips have been drawn from the container. At this point all the slips are put back into the container, and the procedure begins all over again.

The procedures covered in the preceding paragraphs are especially applicable to experiments in which subjects report to the experiment over an extended length of time. In another type of research, all subjects for the experiment are available to the experimenter at the same time. For example, in the Paul experiment comparing the effectiveness of insight and behavior therapy in treating speech phobia, the experimenter had the names of 67 students who exhibited speech phobia and had to assign them to the four experimental conditions (Chapter 2, under *Experimental and Control* Groups). In the experiment on employment opportunity training discussed later in this chapter, there were 60 applicants for the course offered by the center, and the director somehow had to assign half of these applicants to the treatment group and half to the control group. There are several ways of accomplishing an unbiased assignment for this type of problem, and only a few of these will be discussed.

Suppose an experimenter has the names of 60 subjects and wants to assign 15 subjects to each of four treatment groups. One procedure would be to use the slips of paper and container method, except in this case 15 slips of paper with A marked on them would be placed in the container as well as 15 slips of paper with B marked on them, and so on. The names of the subjects would be listed on a sheet in alphabetical order. As the experimenter went down the list, he would draw a slip of paper out of the container (but not replace the slip) for each name and assign that name to be in the treatment indicated by the letter on the slip. This same procedure might be simplified by arranging the names in random order (e.g., put each of the 60 names on a 3 × 5 index card and shuffle the cards in the same manner as you would playing cards) and then assigning the first name on the list to treatment A, the second name to treatment B, the fifth to treatment A, and so on.

In some experiments the experimenter might not have the subjects' names before the experiment begins; however, all subjects report to the experiment at the same time. Suppose there are 60 subjects in a large room, and they must be assigned to four treatments. Probably the simplest way would be to start at the front of the room and have the subjects count off by fours. All number ones go into one treatment group, all number twos into a second group, and so on. This would seem to be an unbiased selection, assuming that the subjects are not seated in the room in some systematic manner. Frequently, subjects meet in a large room, and the various treatments are represented in the materials in test booklets passed out to the subjects. For example, Johnson and Scileppi (1969) wanted to study change in attitude to plausible and implausible communications. These communications were in the form of written messages that appeared in different test booklets together with attitude scales to be filled out by the subjects. Groups of 10 to 20 subjects were tested at the same time in classrooms. The procedure used in this experiment was to shuffle the test booklets (as one would shuffle playing cards) before passing them out. In this way the experimenters were randomly assigning treatments to subjects rather than vice versa.

Even rather complex assignment problems can be handled fairly easily. For example, a group problem-solving experiment was conducted in which subjects met in groups of three to solve a particular human relations problem. The independent variable was two types of instructions to the group, which we shall call treatment A and treatment B. Thus the problem was not only to randomly assign three-person groups to treatments (A or B) but also to randomly assign subjects into three-person groups. To further complicate the assignment problem, one member of each three-person group was to be randomly assigned to be the leader of his or her group. Sixty subjects reported to a large classroom, and somehow the experimenter had to end up with 10 three-person groups in each of two treatments with a randomly assigned leader in each group. This assignment problem was easily solved by assigning each subject a number from 1 to 60 as he or she entered the room. The experimenter had a deck of sixty 3×5 index cards, each with a number from 1 to 60 written on it. The experimenter shuffled the cards and then drew three cards from the top of the deck. The subject represented by the number on the first card drawn from the deck was designated as the leader of the group, and the subjects represented by the other two numbers were designated as members of the group. The experimenter kept drawing blocks of three cards from the top of the deck until all 20 groups were formed. The groups formed by the first 10 blocks of three cards were placed in treatment A, and the groups formed by the second 10 blocks of three cards were put in treatment B. Thus a rather complicated assignment process was handled very quickly and in a reasonably unbiased manner.

MATCHED SUBJECTS DESIGN: MODEL 2

Another way of insuring that all the treatment groups are equal as to subject characteristics is called *matching*. With the matching procedure all subjects are measured on some test or task that is assumed to be highly related to the task used in the actual experiment. The subjects then are assigned to the various treatment groups of the experiment on the basis of this pretest measure, so that the treatment groups will be approximately equal with respect to pretest scores. With this assignment procedure the experimenter can be assured that all the treatment groups are equal with respect to one subject characteristic that is believed to be highly related to performance on the actual experimental task.

Before discussing matching techniques let us note the assumptions made when using this procedure. First, the researcher assumes that he or she knows what subject characteristic is highly related to performance on the experimental task, and second, he or she assumes that he or she can get scores for each subject on this characteristic. There is always a danger that the first assumption is invalid and that the second condition cannot be met either because there is no good measure for the characteristic or because the experimenter cannot get the subjects' scores on that measure.

Pretest tasks or tests can usually be divided into two major types. The first are tasks or tests that are quite different from the experimental task but are assumed to be highly related to the task. For example, it might be assumed that intelligence is highly related to a learning task, and the experimenter would want to match on intelligence. In this case the pretest is an IQ test, which is quite different from the actual experimental task. Another type of matching variable is a task that is quite similar (or identical) to the experimental task. Lambert and Solomon (1952) trained 20 rats to run down a runway at the end of which was a goal box containing food. After 30 acquisition trials the 20 rats were divided into two groups. One group of 10 rats was blocked (from going down the runway) at a point near the goal box (treatment A), and the other group of 10 rats was blocked near the beginning of the runway (treatment B). The dependent variable was the number of trials it took both groups of rats to extinguish their running response, and a response was considered extinguished if the rat did not leave the start box within a 3-minute time period. The researchers had recorded how long it took each rat to leave the start box on the last four acquisition trials. On the basis of these time scores they assigned (matched) the 20 rats to the two treatment groups, so that, as a group, the two treatments were approximately equal in time taken to leave the start box before the two treatments were initiated. In this example the matching was done on the same task as was used in the experimental session.

One technique for matching is through the use of *matched pairs*. With this technique the experimenter takes two subjects with identical scores on the matching variable and assigns one to treatment A and the second subject to treatment B. Suppose a two-treatment learning experiment is conducted, and it is assumed that IQ is highly related to performance on the learning task. First, IQ scores would be obtained for a large number of subjects. From this pool of subjects the experimenter would pick out two people with an IQ of 135 and put one in treatment A and one in treatment B. Then he or she would pick out two subjects with an IQ of 130 and put one in A and one in B. This process is repeated until the experimenter has the desired number of subjects in each treatment group. Here there is a perfect matching of subjects in the two treatments. The same procedure is used if there are more than two treatments; for example, the experimenter would pick three subjects with an IQ of 135 and assign one to each of the three treatment groups.

A difficulty with this technique is that it may be impossible to find subjects with matching scores. In our example it may be that only one subject with an IQ of 135 may be found. In such cases the researcher may choose to match on the basis of similar scores (e.g., a range of 130–140). Even this latter technique may prove to be difficult. Another example would be a drug therapy experiment on schizophrenics in which an experimenter might want to match on age. He or she would take two 50-year-old schizophrenics and put one in the drug therapy group and one in the placebo group. Next, he or she would take two 46-year-olds, and so on.

In the preceding paragraph *precise* matching of subjects was achieved. However, and perhaps more frequently, this may not be possible. Suppose that in a rat experiment similar to that of Lambert and Solomon described above the experimenter had eight rats and wanted to use a matched pair technique to assign them to two treatments (A and B) based on the length of time it took them to leave the start box on the last acquisition trial. The times of the eight rats were 20.5, 17.2, 10.7, 8.0, 7.2, 6.5, 4.3, and 3.2 seconds. It is obvious that precise matching cannot be done since no two scores are alike. The experimenter is forced to use an ad lib matching procedure in which he or she attempts to balance out the scores. The following grouping of scores seems to be the best matching possible for the eight rats:

Treatment A	Treatment B
20.5	17.2
8.0	10.7
7.2	6.5
3.2	4.3
$\bar{x} = 9.72$	$\bar{x} = 9.67$

The means of both groups are approximately the same, and both groups contain high-, moderate-, and low-scoring rats. This "solution" is not without its problems as data generated from such a technique are difficult to analyze with appropriate statistical techniques.

Another matching technique that may be used is a *random blocks technique*. Suppose that 80 schizophrenics were to be assigned to four treatment groups in a drug therapy experiment. The experimenter also wants to match them as to age, since he or she has some evidence that the older the patient, the less favorably he or she responds to therapy. Using a random block technique the experimenter would rank the ages of the 80 patients. He or she would then take the four oldest patients and *randomly* assign one to each of the four treatments; then he or she would take the four next-oldest patients and randomly assign each of them to one of the four treatments, and so on, until all 80 subjects have been assigned. Thus the experimenter is taking blocks of patients of approximately equal age and randomly assigning the patients within each block to a particular treatment.

Matching techniques can be very powerful techniques for eliminating any bias due to subject characteristics *if* the experimenter knows what subject variables are highly related to the experimental task and *if* he or she can get scores on these variables.

WITHIN-SUBJECT DESIGN: MODEL 3

Another way of insuring that each treatment group is equal with regard to subject variables is to have all subjects participate in all treatments. Since the same subjects are participating in all treatments, it can be assumed that the treatment groups are equal as to subject variables. This type of design has been called a repeated measure design or a within-subject design. The latter term seems preferable and derives its designation because comparisons between treatments are made within the same subject. This was used in the example experiment (Chapter 2, under *Bar Graphs*) exploring the relationship between the frequency of a tone and the absolute loudness threshold. In this experiment each subject was tested at ten levels of frequency (25, 50, etc.). The mean absolute threshold for the four subjects at each frequency level formed the points of the curve. It is in this type of an experiment that the within-subject design is most frequently used; that is, experiments in which subjects are required to make numerous judgments to different stimuli (on different occasions), and these different stimuli can be considered the independent variable.

REPEATED MEASURE DESIGN—SEQUENCE BALANCED: MODEL 4

The within-subject design insures that subject characteristics are equal in all treatments, but the use of this design brings up an additional problem with regard to the *order* in which treatments or stimuli will be presented to the subject. The results might change rather dramatically because the subject's judgment of a particular stimulus may be influenced by his or her previous judgments. For example, it can be easily demonstrated that a subject's judgment of the lightness or heaviness of an object is in part determined by the heaviness of previously judged objects. In discrimination experiments the subject may become more accurate or proficient at a task as he or she makes more judgments. This is usually called a *practice effect*. The opposite phenomenon, a *fatigue effect*, can occur if the subject becomes tired or bored and his or her proficiency decreases. To control for the effects of order of stimulus presentation, several designs are possible.

One method is the use of a *Latin-square design*. Suppose 15 subjects are to judge three different stimuli (A, B, and C). Using a Latin-square design, different orders of stimuli will be derived such that (1) each stimulus appears once within a single order and (2) each stimulus appears once for each position. This is illustrated in Table 5.1.

Five subjects receive the stimuli in the order A, B, C; five subjects receive a B, C, A order; and five subjects receive a C, A, B order. Note that stimulus A is judged or responded to once in position 1, once in position 2, and once in position 3. The same pattern is true for stimuli B and C. The means of the judgments of all 15 subjects to the three stimuli can then be compared. These means are independent of order effects, because each stimulus appeared equally in each order position. Should the experimenter be interested in order effects, these can be examined by looking at the subjects' judgments of each stimulus as a function of the order position.

Another procedure for controlling for order effects is through the use of a random blocks design similar to that described earlier in this chapter. Suppose that an experiment is conducted in which each subject is required to make ten judgments on each of seven different stimuli. The seven stimuli are treated as a block, and the order of presentation within the block is determined by some randomization procedure. The first block (of seven stimuli) is presented to the subject, and after the subject responds to them the seven stimuli are randomized again, and this new order constitutes the second block. This procedure is repeated eight more times to this subject until he or she has made ten judgments on each of the seven stimuli. The same procedure is also repeated for each subject used in the experiment.

Table 5.1   THE LATIN-SQUARE DESIGN

	ORDER OF STIMULUS PRESENTATION		
	POSITION 1	POSITION 2	POSITION 3
1. Five subjects	A	B	C
2. Five subjects	B	C	A
3. Five subjects	C	A	B

The within-subject design often appears very desirable, because the same subjects are used in all treatments and therefore the researcher is sure that treatments are equal as to subject variables. It also appears desirable because fewer subjects can be used. However, the within-subject design cannot be used in many experiments simply because the treatments themselves negate this possibility. If an experiment were performed to compare rats reared in isolation with rats reared in groups, it is obvious that the same subjects could not be in both treatments. If a study is performed comparing high-IQ subjects with low-IQ subjects, or comparing one method of teaching French with a second method, it is apparent that independent groups of subjects have to be used in each treatment. For other experimental problems it may be much easier to use independent groups to test the hypothesis without going into any complicated procedure that may be needed for a within-subject design. The within-subject design seems to be most applicable to research problems, such as the experiment on the relationship between frequency and absolute threshold, in which subjects are required to make judgments on numerous different stimuli (on different occasions), and these stimuli can be considered the independent variable. Another type of experiment that necessitates a within-subject design is one in which order effects are of major interest. Experimenters concerned with what happens when a subject switches from a high-reward situation to a low-reward situation (and vice versa) would need the same subjects to participate in both conditions.

SUBJECT LOSS (ATTRITION)

We have discussed steps that are taken to insure that the subject characteristics in each treatment group are approximately equal. A related problem deals with subject *attrition*, or loss of subjects, in the various treatment groups of an experiment. This problem occurs particularly in experiments in which subjects have to participate in more than one session, since subjects coming for the first session may not necessarily appear for the second session. Consider the following hypothetical example:

EXAMPLE

An "employed opportunities" center offered a four-week course designed to help unemployed young adults with the techniques and procedures for finding jobs. Classes were held each day and covered such topics as where to go for a job, what type of job to look for, and how to fill out applications, as well as practice in taking psychological tests used for employee selection. Of the 60 young adults who signed up for the course, the director of the center randomly selected 30 to be in the course and used the other 30 as a control group to test the effectiveness of the course. The control group had no contact with the center and were simply told that there was no room in the course for them and that they would have to find a job on their own. The dependent measure was the percentage of subjects in the two groups who had found jobs within a month after the course had ended. Of the 16 people (out of 30) who completed the course, 12 (75 percent) were employed within a month. Of the 30 control group subjects, 15 (50 percent) were employed within this period. The director of the center took these results as evidence in support of his program, noting that he had increased the number of employed by 25 percent.

Before criticizing this study, let us note that research of this type is difficult to do well; however, it is quite important research. Training programs such as these should be evaluated as to their effectiveness. There are many types of programs—therapy, counseling, remedial reading, leadership training, and so on—that are very popular, but so far there has been little research that examines their effectiveness.

The main concern in the above example is that the statistics are based only on the 16 subjects who completed the course out of 30 who started the course. The real question is: Who were the subjects who dropped out? They possibly were those who were less motivated about getting a job, or less intelligent and therefore unable to understand the materials in the course, or less emotionally stable and therefore unable to accept the routine of coming to the center every day. Thus the poor employment prospects may have been weeded out of the treatment group. On the other hand, it is difficult to drop out of a no-treatment control group, and all the poor employment prospects were still in that group. What the head of the center might have been doing is comparing a group of good employment prospects (the poor ones dropped out) with a control group that consisted of both good and poor employment prospects. If this explanation is valid, then the effectiveness of the course is highly questionable.

In reality it is not known if the above explanation is valid; however, it seems to be a plausible explanation for the results. Any time there is a loss of subjects in an experiment, particularly if the loss is greater in one treatment group than another, it is important to find out the characteristics of those subjects who dropped out. One possibility would be to look at the employment rate of those who dropped out. If it is only about 25 percent, then it would seem to support the explanation that the poor employment prospects were weeded out. It it is around 50 percent (or higher), then the course would seem to be fairly effective.

The primary danger with subject attrition is that a select subgroup—for example, low-IQ subjects—may drop out of one treatment group but remain in another treatment group. If this occurs, then the treatment groups cannot be considered equal as to subject characteristics, and the results obtained may be due to subject differences rather than to treatment differences. It is important to have as much relevant information as possible on the subjects who dropped out in order to ascertain if they are a select subgroup. If this relevant information indicates that the dropouts are not a select subgroup but, rather, a "random" sample of the original treatment group, then the researcher may feel fairly confident that his or her results are not due to any particular subject differences. On the other hand, if this information indicates that the dropouts are a select subgroup, it may be possible to eliminate that subgroup in the other treatment groups and thus make the various treatment groups somewhat comparable.

Researchers can sometimes avoid problems of subject attrition by having subjects participate in only a single session. Thus there is no loss since there is no second session. If subjects are needed for more than one session, a standard procedure is to inform them before they start the experiment that they will be needed for more than one session and ask that they participate only if they can attend a second session. Frequently a phone call will bring in those who have forgotten. In experiments in which subjects are needed for repeated testing for perhaps 10 days or so (as in some perception experiments), it is customary to pay the subjects, and the subjects are hired only on the condition that they complete the required number of sessions. In animal experiments attrition usually is not a problem, since the subjects reside in your laboratory colony and can be used at will. However, there is always a danger in animal experiments of loss of subjects due to sickness or death, particularly if deprivation or stress treatments are used. For example, Bayoff (1940) compared the performance of rats reared in isolation versus rats reared in groups in a highly competitive stress situation. Ten of the isolation-reared rats died before the experiment was over; only one of the group-reared rats died.

Subject attrition problems should be avoided if possible by using one of the techniques stated in the preceding paragraph. If the study occurs over long periods of time and attrition problems seem inevitable, it is important to get information on all subjects, including the dropouts. Ideally, this information would include (1) pretreatment information, which might include intelligence data, motivation data, adjustment data, and so on; (2) data on the progress of the subjects up to the point at which they dropped out, for example, learning or performance data; and (3) posttreatment data on all subjects, such as employment rates in the above example. With this information the results of the experiment may be more easily interpreted.

DEFINITIONS

The following is a list of experimental design- or procedure-related concepts that were used in this chapter. Define each of the following:

field study
random groups design
random assignment
block randomization
matched groups design
independent groups design
matched pairs design

repeated measure design
random blocks technique
ad lib matching
within-subject design
practice effect
fatigue effect
Latin-square design
subject attrition

CHAPTER
6

Design Critiques II

The following is another series of experiment "briefs." There is a design problem in each of the briefs. You should critique and redesign each experiment in a manner similar to that in Chapter 4. These briefs were chosen so that you need little expert technical knowledge of the research area being covered in the experiment.

EXPERIMENT BRIEFS

1. A teacher of statistics wanted to compare two methods of teaching introductory statistics. One method relied heavily on the teaching of the theory behind statistics (*theory method*). The other method was labeled the *cookbook method* because it consisted of teaching the students various statistical tests and informing them when to use each test. The researcher found that a leading engineering school was using the theory method in all its introductory statistics classes and that a state teachers college was using the cookbook method in all its classes. At the end of each semester he administered a standardized test on the applications of statistics to the statistics classes of both schools. The results of this testing indicated the classes that received the theory method were far superior to the classes that received the cookbook method. The researcher concluded that

the theory method was the superior method and should be adopted by teachers of statistics.

2. In an effort to determine the effects of the drug chlorpromazine on performance of schizophrenics, two clinical investigators randomly selected 20 acute schizophrenics from a mental hospital population. The task used was one in which several stimuli had to be put in order along a dimension; for example, eight stimuli had to be ordered as to their weight. There were several tasks of this sort. The investigators used a within-subject design in which all subjects first performed the tasks after being injected with a saline solution (placebo) and then performed the tasks again (several hours later) after having been injected with chlorpromazine. Results indicated that fewer errors were made in the chlorpromazine treatment, which suggested to the investigators that this drug facilitates more adequate cognitive functioning in this type of patient.

3. It was hypothesized that sensory deprivation inhibits the intellectual development of animals. To test this hypothesis an experimenter used two rats, each of which had just given birth to eight pups. One rat and her litter were placed in a large cage with ample space and with objects to explore. The pups of the second rat were separated from the mother, and each was placed in a separate cage. These cages were quite small, and the only objects they could see (or hear) were the four walls and the food dispenser. After five months, both treatment groups of rats were tested in a multiple-T maze using food as a reward. After 20 trials all of the nondeprived pups were running the maze without error. On the other hand, the deprived pups were still making several errors in the maze. This latter group of rats frequently "froze" in the start box and in the maze and had to be prodded to move. The experimenter concluded that sensory deprivation inhibits intellectual development such that deprived rats did not have the intellectual ability to learn even a simple maze.

4. During World War II an investigator attempted to examine the hypothesis that punishment is more effective for training people than reward. The problem he picked was the identification of enemy and of friendly airplanes. In his experimental situation, he had the subjects sit in front of what appeared to be a radar screen. Silhouettes of enemy and of friendly airplanes were flashed on the screen in very short exposures (one second). As each silhouette was flashed on the screen, the subject had to respond by pressing one of two buttons—one button was marked "enemy"; the other was marked "friendly." Each subject participated in the experiment

for two hours on five successive days. On the first day, as each silhouette was flashed on the screen, the subject pressed one of the buttons and then was told by the experimenter if he had been right or wrong in his identification. Starting on the second day, subjects were randomly assigned to one of two groups. The procedure was similar to that of the first day except that in Group A the subjects were given ten cents after every correct identification but were not punished for a wrong identification. In Group B, each subject received an electric shock after every wrong identification but did not receive anything for a correct identification. This same procedure was continued for days three and four. The fifth day was considered the "test" day, and the subjects followed the same procedure except that no reward, punishment, or information from the experimenter was given to the subject. The number of correct identifications per 100 silhouettes presented was considered a test of the effectiveness of each training method. As expected, there was some loss of subjects over the five-day period; about 5 percent of the Group A subjects and about 35 percent of the Group B subjects had dropped out of the experiment by the fifth day. Results indicated that on the 100 test trials given on the fifth day, the mean number of correct identifications for Group A was 80, and the mean for Group B was 92. The difference between the means was statistically significant. The experimenter concluded that his hypothesis had been confirmed and suggested that all training programs be based on punishment.

5. A YMCA official in a small town wanted some evidence to prove that his program was valuable in training future leaders. He went back to the membership records and got the names of those boys who were active members in his program 20 years before. He also took school records and got the names of boys who were not YMCA members. He compared the two groups as to present occupations, salaries, and so on, and found that the YMCA group was doing much better. He concluded that this result was due to the influence of his program.

6. A psychologist was interested in developing a test that would predict the success of prospective lawyers. She selected a random sample of lawyers listed in *Who's Who*, under the assumption that they would be "successful" lawyers. She then contacted them by means of a mail questionnaire that contained several hundred questions. The results were analyzed, and a profile of successful lawyers was compiled. The questionnaire was given to a group of prospective law students, and those students whose scores were significantly divergent from the

successful lawyer group were advised not to pursue a law career.

7. A certain psychologist had a theory that as members of a group get to know each other better, the productivity of the group will increase up to a point and then will start to decrease slightly. The decrease ("the honeymoon is over" effect) is a point at which group members stop acting in a highly cooperative manner and start jostling for power, and so on. To test this theory he formed groups of individuals who were strangers and had them work a series of tasks. There were five tasks, each taking 35 minutes to work, and he gave the groups a 5-minute break between tasks. His results indicated that group productivity increased with the number of tasks up to the fifth task, and for the fifth task there was a significant decrease in group performance. On the basis of the evidence he considered his theory supported.

8. A clinical researcher examined whether interviews with patients or objective tests are better in the diagnosis of the patient's problems and outcomes. The experiment took place at a large mental hospital. In one group, 10 clinical psychology students each interviewed 6 new patients (during their internship at the hospital). The length of each interview was one to two hours. Another group of 60 patients was given a battery of standardized psychological tests (e.g., the MMPI), and the test results were interpreted by three clinical psychologists who had several years of experience in interpreting tests for the hospital. Each psychologist interpreted the test results for 20 patients. Both groups were asked to list the patient's major problems and to assign the patient to a diagnostic category (e.g., process schizophrenic, reactive schizophrenic). They were also asked to predict how long the patient would be in the hospital before he or she would improve enough to be released. Results indicated that the interviews were 67 percent accurate in predicting diagnostic categories and 22 percent accurate in predicting length of stay. The "test" group was about 83 percent accurate in predicting diagnostic categories and 65 percent accurate in predicting length of stay. The experimenter concluded that interviews are of questionable value in either diagnosis or prediction of outcome and should be discontinued.

9. An experimenter wanted to test the hypothesis that males are more creative than females. He also hypothesized that the male superiority in creativity would be heightened under ego-involving conditions. The design used was a 2 × 2 factorial design in which one variable was sex and the other variable

was high and low ego involvement. He manipulated ego involvement by telling half the subjects that this task was a measure of how intelligent they were and that he would post their scores on a bulletin board (high ego involvement). He told the other half of the subjects that he was developing the task and wanted to check its reliability and further told them not to put their names on the answer sheets (low ego involvement). His test of creativity was an "unusual uses" test in which a person is given the name of an object (e.g., hammer) and has to write down as many different unusual uses for that object as he or she can in 5 minutes. Twenty-five males and 25 females participated in each of the two ego-involvement conditions. The males were members of a senior ROTC class and the females were obtained from sorority pledge classes. Two objects were used: (1) army compass and (2) monkey wrench. Subjects were given 5 minutes for each object. The results indicated that the mean number of unusual uses for the two objects for males was 4.1 under low ego involvement and 7.6 under high ego involvement. The means for the females were 3.2 under low ego involvement and 2.4 under high ego involvement. Since both the main effects and the interaction effect were statistically significant using analysis of variance, the experimenter concluded that his hypotheses were supported.

10. A researcher was asked to conduct a quick survey in three large cities to find out what political issues or problems were of primary importance to the voters. The results of the survey were to be used by a political candidate in developing her campaign. The researcher randomly selected names from the phone books and called as many of these people as could be reached between 9 A.M. and 5 P.M. on Monday and Tuesday. The results were collated and presented to the candidate on Wednesday with the statement that this was a valid representation of the attitudes of the voters in the three cities.

EXERCISE

In the exercise at the end of Chapter 4 you were asked to formulate a series of questions that can be used to critique experiments. We would like you now to add questions to that list based on the material discussed in Chapters 5 and 6.

C H A P T E R
7

Ethics of Experimental Research

Throughout the history of American psychology, the issue of research ethics has been a subject of concern and debate. On one hand, a psychologist is a scientist dedicated to the discovery of laws of behavior. The purpose of experimental research in psychology is to enhance our knowledge of the psychological characteristics of the human species. In order to do this, psychologists use human and animal subjects in their experiments. Some would argue that in certain instances the development of valid laws of psychology may require that a subject be deceived or in some way physically harmed. This argument is based on the principle that the pursuit of knowledge must continue unabated. In addition it is possible that the ultimate truths revealed in psychological research may be of great benefit to humankind. On the other hand, a psychologist is bound by ethical considerations and an experimenter is bound by a code of ethics in which the psychological and physical safety of the subjects is rigidly safeguarded.

In many psychological experiments it is possible to advance our understanding of the psychological characteristics of the species while clearly maintaining the psychological and physical well-being of the subject. While an experimenter should always be sensitive to the ethical problems that may arise in experiments, they do not pose as serious an issue as in many other experiments in which the subject may be in some way harmed.

In some previous research, experimenters have at times used ques-

tionable means to obtain results. Such abuses of human and animal subjects have caused great concern among psychologists as well as some
members of the general public. (See Larson, 1982, for a recent trial of
an animal researcher.) Because of these concerns within the community
of psychologists and the public, a Committee on Scientific and Professional Ethics has been formed by the American Psychological Association.
Through many years of work, this organization has developed a series
of ethical principles.

Each of the ethical principles of psychologists is presented below with
an additional comment by the authors of this text. Following each principle, we describe a case study in which the principle is illustrated. Some
of these cases (in our opinion) represent clear violation of the principle,
while others are considered not to contain unethical procedures. These
cases are presented for your analysis and possible class discussion. In
the original document, published by the American Psychological Association, further examples and comments are provided and should be con-
sulted when ethical questions are raised.

ETHICAL PRINCIPLES OF PSYCHOLOGISTS[1]

Preamble

Psychologists respect the dignity and worth of the individual and strive for
the preservation and protection of fundamental human rights. They are
committed to increasing knowledge of human behavior and of people's
understanding of themselves and others, and to the utilization of such
knowledge for the promotion of human welfare. While pursuing these objectives, they make every effort to protect the welfare of those who seek
their services and of the research participants that may be the object of
study. They use their skills only for purposes consistent with these values
and do not knowingly permit their misuse by others. While demanding
freedom of inquiry and communication for themselves, psychologists accept
the responsibility this freedom requires: competence, objectivity in the
application of skills, and concern for the best interests of clients, colleagues,
students, research participants, and society. In the pursuit of these ideals,
psychologists subscribe to principles in the following areas: (1) responsi-

[1]This version of the Ethical Principles of Psychologists (formerly entitled Ethical
Standards of Psychologists) was adopted by the American Psychological Association's
Council of Representatives on January 24, 1981. The revised Ethical Principles contain
both substantive and grammatical changes in each of the nine ethical principles
constituting the Ethical Standards of Psychologists previously adopted by the Council of
Representatives in 1979, and a new tenth principle entitled Care and Use of Animals.
Inquiries concerning the Ethical Principles of Psychologists should be addressed to the
Administrative Offices for Ethics, American Psychological Association, 1200
Seventeenth Street, NW, Washington, DC 20036. These revised Ethical Principles
apply to psychologists, to students of psychology, and to others who do work of a
psychological nature under the supervision of a psychologist. They are also intended for
the guidance of nonmembers of the Association who are engaged in psychological
research or practice.

bility, (2) competence, (3) moral and legal standards, (4) public statements, (5) confidentiality, (6) welfare of the consumer, (7) professional relationships, (8) assessment techniques, (9) research with human participants, and (10) care and use of animals.

Acceptance of membership in the American Psychological Association commits the member to adherence to these principles.

PRINCIPLE 1
RESPONSIBILITY

In providing services, psychologists maintain the highest standards of their profession. They accept responsibility for the consequences of their acts and make every effort to ensure that their services are used appropriately.

Comment: Psychologists as scientists must accept primary responsibility for the selection of research topics and the methods used in investigation, analysis, and reporting. This principle means that psychologists should not suppress data that may be contrary to the main results as well as acknowledge the existence of alternative hypotheses and explanations. Also, psychologists should take credit only for work they have done.

Case Study 7.1

An experimenter was concerned with personality traits of subjects who scored well in a test of ESP. The gist of the experiment was to have subjects (receivers) report their impressions of an ESP card that was being viewed by another person (sender). The cards were randomly ordered and the sender was completely concealed from the receivers. Several receivers scored well above chance in their guesses but did not differ on personality measures from the rest of the group. The experimenter nevertheless thought he had identified a group of subjects who were unusually sensitive to receiving ESP signals and conducted a series of four additional experiments with the subjects who had scored well on the first experiment. The first of the three experiments yielded results in which the subjects scored significantly more hits than would be expected by chance alone, but on the fourth experiment, the results were no greater than would be expected from chance. In reviewing the results, the experimenter decided to report only the first three experiments, attributing the experimental results on the fourth experiment to sender and/or receiver fatigue. This case raises both procedural problems and ethical problems.

PRINCIPLE 2
COMPETENCE

The maintenance of high standards of competence is a responsibility shared by all psychologists in the interest of the public and the profession as a

whole. Psychologists recognize the boundaries of their competence and the limitations of their techniques. They only provide services and only use techniques for which they are qualified by training and experience. In those areas in which recognized standards do not yet exist, psychologists take whatever precautions are necessary to protect the welfare of their clients. They maintain knowledge of current scientific and professional information related to the services they render.

Comment: With regard to psychological research, this principle suggests that researchers should accurately represent their competence, education, training, and experience. Part of this principle is directed toward clinical psychologists who might use psychological tests in their practice. Experimental psychologists also sometimes use these instruments and should be aware of a test's limitations and validity as well as its purpose.

Case Study 7.2

An experimental psychologist trained in research design and physiological psychology, while working at a major eastern university, is approached by a large company that produces a health food cereal. The company asks that the psychologist design an experiment that will demonstrate the effectiveness of a cereal in reducing common ailments (e.g., colds) and absenteeism. The company agrees to pay the researcher a large sum of money if he will design the experiment and lend his name to the conclusion in a subsequent advertising campaign. The psychologist agrees to do so, but stipulates that he will have final approval of the advertising copy. The psychologist will not be directly involved in the collection of data, but is assured that it will be done according to his exact specifications. Although the psychologist has little knowledge or training in nutrition, he feels that his background in experimental design and physiological psychology is sufficient to design a valid experiment.

Should the psychologist have accepted this offer? Why? Why not? (See also, Principle 4.)

PRINCIPLE 3
MORAL AND LEGAL STANDARDS

Psychologists' moral and ethical standards of behavior are a personal matter to the same degree as they are for any other citizen, except as these may compromise the fulfillment of their professional responsibilities or reduce the public trust in psychology and psychologists. Regarding their own behavior, psychologists are sensitive to prevailing community standards and to the possible impact that conformity to or deviation from these standards may have upon the quality of their performance as psychologists. Psychologists are also aware of the possible impact of their public behavior upon the ability of colleagues to perform their professional duties.

Comment: Researchers must observe the prevailing community standards, rules, and laws. Where a conflict exists between these standards and the American Psychological Association's standards, researchers are encouraged to resolve the conflict within the context of serving the public interest.

Case Study 7.3

In an investigation of "higher moral standards" an experimental psychologist is interested in the strength of subjects' moral convictions. An experiment is designed in which subjects are told that a child is in critical need of a drug that can be derived from a fungus growth found in a particular limestone cave. The fungus grows in abundance in this cave and only a small portion is needed for treatment. However, the owner of the cave refuses to allow anyone to obtain the fungus. The experimenter finds that a large percentage of subjects report that they would trespass on the property to obtain samples of the fungus. In a second part of the experiment, the researcher asks some of the subjects to obtain the fungus illegally. He argues that a "higher moral principle" is served and that the results will have a major impact on our knowledge of civil disobedience.

Did the researcher's plan conform to Principle 3? Can this research be modified to achieve the psychologist's aim and yet conform to the ethical principles?

PRINCIPLE 4
PUBLIC STATEMENTS

Public statements, announcements of services, advertising, and promotional activities of psychologists serve the purpose of helping the public make informed judgments and choices. Psychologists represent accurately and objectively their professional qualifications, affiliations, and functions, as well as those of the institutions or organizations with which they or the statements may be associated. In public statements providing psychological information or professional opinions or providing information about the availability of psychological products, publications, and services, psychologists base their statements on scientifically acceptable psychological findings and techniques with full recognition of the limits and uncertainties of such evidence.

Comment: This principle applies to psychologists involved in the offering of psychological services (e.g., a clinical psychologist on a radio talk show) as well as to research psychologists who may offer services, products or publications. These should not be misrepresented through sensationalism, exaggeration, or superficiality.

Case Study 7.4

A research psychologist, working on problems of human memory, developed a superior mnemonic technique. She decided to test the technique by becoming a contestant on a television quiz show. After several successful appearances, she decided to write a book on the memory method. Because she was well known, due to her appearance on the television quiz show, she agreed to have her picture on the cover and in the advertising for the book, along with the statement, "The extraordinary memory of Dr. Jane Brown can be learned by you!" What specific ethical issues are raised by this example?

PRINCIPLE 5
CONFIDENTIALITY

Psychologists have a primary obligation to respect the confidentiality of information obtained from the persons in the course of their work as psychologists. They reveal such information to others only with the consent of the person or the person's legal representative, except in those unusual circumstances in which not to do so would result in clear danger to the person or to others. Where appropriate, psychologists inform their clients of the legal limits of confidentiality.

Comment: In the course of experimental work in psychology, a researcher may collect personal information. If it is necessary to report this information, the researcher should either obtain prior consent or adequately disguise identifying information. When working with children or people unable to give informed consent, the psychologist should take special care to protect the best interests of the person.

Case Study 7.5

At a meeting of lawyers, a social psychologist was asked to present her recent results of research on the decision-making process of juries. In one of her studies, the psychologist interviewed each member of a jury involved in a celebrated murder trial. In reporting the study, the identity of each member of the jury was carefully concealed. However, the psychologist did discuss the deliberative processes of subgroups. For example, the jury had among its members seven women, two blacks, one foreign-born Italian-American, an architect, and a truck driver. The deliberative process of groups was specifically revealed. For example, the researcher referred to the voting and deliberative patterns of the two blacks, the architect, and so on. When questioned about the ethical propriety of the revealing of the findings, the psychologist answered that the names of the jurors had not been used and, in addition, the jurors were now public figures whose opinions were no longer private.

What are your views?

PRINCIPLE 6
WELFARE OF THE CONSUMER

Psychologists respect the integrity and protect the welfare of the people and groups with whom they work. When conflicts of interest arise between clients and psychologists' employing institutions, psychologists clarify the nature and direction of their loyalties and responsibilities and keep all parties informed of their commitments. Psychologists fully inform consumers as to the purpose and nature of an evaluative, treatment, educational, or training procedure, and they freely acknowledge that clients, students, or participants in research have freedom of choice with regard to participation.

Comment: In the conduct of an experiment, psychologists frequently are entrusted with the care of people and groups. It is essential that the integrity of the people be respected. In no way should a psychologist exploit students, clients, or subordinates in experimental work. Psychologists should attempt to avoid "dual relationship" in which the subject of an experiment is also a close friend or relative. The APA is also specific about sexual relations: "Sexual intimacies with clients are unethical."

Case Study 7.6

A Ph.D. candidate at a large midwestern university was completing his dissertation on the relationship between mothers' religious attitudes and bedwetting tendencies of their children. His sample consisted of 48 mothers between the ages of 20 and 28 who were also white and members of a specific religious group, and whose children were "healthy." He had nearly completed his study when four subjects failed to meet their appointments with him. Since the dissertation was due in a short time, he decided to recruit subjects from friends and wives of friends. He was very careful to make sure that all new subjects were identical on the designated attributes.

The dissertation was successful. The candidate was awarded a Ph.D. degree and is now a valued member of a department of psychology at a large midwestern university.

Did the candidate act unethically?

PRINCIPLE 7
PROFESSIONAL RELATIONSHIPS

Psychologists act with due regard for the needs, special competencies, and obligations of their colleagues in psychology and other professions. They respect the prerogatives and obligations of the institutions or organizations with which these other colleagues are associated.

Comment: Sometimes collaborative research is done by psychologists in which other professionals are involved in the research. In these circumstances, the psychologist is bound to honor the traditions and practices of other professional groups. At other times a psychologist may conduct research in an institution or organization other than his own. In such cases, the researcher must secure appropriate authorization to conduct such research. Also, proper recognition of the role of the institution should be given.

This principle also addresses the issue of collaborative effort in experimentation. If a collaborator makes a major professional contribution to a research project, the collaborator should share authorship on any subsequent paper. Minor contributions or clerical assistance may be recognized in a footnote or introductory statement. A psychologist who collects and edits material from others should be clearly identified as an editor.

Case Study 7.7

"One of the best graduate seminars I took was in industrial psychology from good ol' Professor B. J. Smith," a colleague told me. "Each member of the seminar was given a very specific hypothetical problem in the design of work spaces and anticipated productivity. We did an extensive review of the literature and designed an experiment. We even anticipated the results and analyzed and discussed them. The professor gave us the problems and we did all the work, but it was an excellent learning experience."

Several years later I saw the same colleague at a psychology meeting, and I asked him how things were going. "Do you remember the story I told you about good ol' Professor Smith?" he began. "Yes, I do . . . something about a good seminar you had with him." "Well, the old fraud," he said, seething. "He took our research ideas, put them into practice, and published the results in the latest issue of the *Journal of Important Industrial Research.* See, here is a copy."

The article was similar to the design submitted by my colleague and even cited the very sources in the introductory material that were contained in the student's paper.

Did the professor act unethically?

PRINCIPLE 8
ASSESSMENT TECHNIQUES

In the development, publication and utilization of psychological assessment techniques, psychologists make every effort to promote the welfare and best interests of the client. They guard against the misuse of assessment results. They respect the client's right to know the results, the interpretations made, and the bases for their conclusions and recommendations. Psychologists make every effort to maintain the security of tests and other assessment techniques within limits of legal mandates. They strive to ensure the appropriate use of assessment techniques by others.

Comment: Some experimental work in psychology calls for the use of psychological tests. When using these instruments, the psychologist should respect the right of the subject to have a full explanation of the nature of the test. In reporting the results of testing, psychologists should identify any reservations they may have regarding the validity or reliability of the instrument. Also, the use of psychological assessment techniques by unqualified persons should be avoided.

Case Study 7.8

Part of an experiment in social psychology was to use a test instrument called "Study of Basic Attitudes" that was to be related to scholastic achievement. The subjects were to report to an investigator in the department of psychology of a small liberal arts college to complete the attitude scale. Several times subjects would come to the department only to find the principal investigator absent. Because the test was simple to administer—in effect it was a self-administering test—the experimenter decided to leave the test with the secretary to administer. The secretary had no formal background in psychometrics but was briefed in proper procedure for the administration of the test. Following the collection of data, the psychologist debriefed the subjects, which included a discussion of the test and its results.

Is the procedure questionable from an ethical standpoint?

PRINCIPLE 9
RESEARCH WITH HUMAN PARTICIPANTS[2]

The decision to undertake research rests upon a considered judgment by the individual psychologist about how best to contribute to psychological science and human welfare. Having made the decision to conduct research, the psychologist considers alternative directions in which research energies and resources might be invested. On the basis of this consideration, the psychologist carries out the investigation with respect and concern for the dignity and welfare of the people who participate and with cognizance of federal and state regulations and professional standards governing the conduct of research with human participants.

[2]A 76-page booklet entitled *Ethical Principles In the Conduct of Research With Human Participants* was approved by the Council of Representatives of the APA in August 1982. This booklet is available from the American Psychological Association, 1200 Seventeenth St., NW, Washington, DC 20036. The principles contained in this document are printed at the end of this chapter. Researchers using human subjects should be familiar with the details of this document.

Comment: This principle bears directly on the conduct of research with human subjects. A researcher must always safeguard the rights of human participants. A primary ethical consideration is the potential risk that a subject might face. An experimenter must assume responsibility for the well-being of the subject as well as the conduct of assistants and employees, all of whom also incur obligations. In any experiment that might involve subject risk, the psychologist should clarify the role and responsibilities of each subject and inform the participant of potential risks. Research with children requires special safeguarding procedures.

In some instances an experiment may require some deception. In these instances the experimenters should determine if such techniques are necessary, and if not, they should provide another methodology in which deception is not used. However, in instances that require deception the participants should be given sufficient explanation as soon as possible. All subjects should have the opportunity to withdraw from an experiment at any time. Procedures that are likely to cause serious or lasting harm are not to be used except under extraordinary circumstances (e.g., the results have great potential benefit, alternative means are inappropriate, and the subject has voluntarily consented to participate). Following an experiment, the subject should be debriefed and the nature of the experiment should be revealed. Finally, when the results are presented, the confidentiality of the subject should be maintained, unless otherwise agreed upon in advance.

Case Study 7.9

The correct identification of criminal offenders by an eyewitness is considered an important social issue as well as an important psychological issue. After reviewing several alternative procedures, a researcher decides that the best procedure is to stage a crime in the presence of eyewitnesses and then ask them for a description of the perpetrator. The experiment is conducted in "the field" and the setting is a fast-food outlet. All employees are carefully rehearsed regarding the staged crime. The "crime" is committed by an actor who enters the store displaying an unloaded handgun and demands all the money from the cash register. He tells the employees not to call the police and, in making his getaway, shouts to the customers that "the first one out the door is going to get blown away." Directly after the "thief" leaves, the researcher and his associates enter the store with a questionnaire, which is distributed to the patrons. The questions deal with the physical appearance of the thief, whether or not he had a weapon, what he said, and a series of photographs in which identification was sought.

Each patron was thoroughly debriefed after the questionnaire was completed, and the important social and psychological issues were discussed. An opportunity was provided for further debriefing and counseling, but no subject indicated a need for further intervention.

Comment on the experiment from an ethical standpoint.

PRINCIPLE 10
CARE AND USE OF ANIMALS[3]

An investigator of animal behavior strives to advance understanding of basic behavioral principles and/or to contribute to the improvement of human health and welfare. In seeking these ends, the investigator ensures the welfare of animals and treats them humanely. Laws and regulations notwithstanding, an animal's immediate protection depends upon the scientist's own conscience.

Comment: When an experimenter uses animals as subjects, special ethical safeguards must be ensured. All laws regarding the use, care, treatment, and disposal of animals must be observed. The experimenter must supervise the care of laboratory animals, and appropriate consideration of their comfort and health should be made. When delegating responsibility, the psychologist is obligated to give explicit instructions regarding the treatment of animals. A special effort should be made to minimize discomfort, illness, and pain of animals. As with humans, a procedure for subjecting animals to harm should be done only when an alternative procedure is unavailable and the goal is fully justified. When an animal is to be killed, it should be done rapidly and painlessly.

Case Study 7.10

In a study of the effects of vitamin A on maze learning of rats in a semi-darkened environment, a researcher has reason to believe that the result would indicate an enhanced performance under minimum dosages, but at higher dosages performance would decrease. The experimenter selects four levels of vitamin A ingestion. The highest level of vitamin A ingestion has been shown by previous research to be toxic to rats. Nevertheless, the researcher argues that to demonstrate the hypothesized curvilinear performance curve on maze learning performance, such levels are necessary in this research and no other procedures are available.

---

[3]The American Psychological Association has prepared a pamphlet entitled *Guidelines for Ethical Conduct in the Care and Use of Animals,* which gives further details regarding personnel, facilities, acquisition of animals, care and housing of animals, justification of research, experimental design and procedures, field research, educational use of animals, and the disposition and killing of animals. The guidelines, approved by the APA Council of Representatives in August 1985, may be obtained (free) from the American Psychological Association, 1200 Seventeenth St., NW, Washington DC 20036. All psychology laboratories using nonhuman animals (vertebrates) should have a copy of the guidelines.

Furthermore, previous research has suggested that higher levels of vitamin A interfere with maze performance although such a hypothesis has not been tested empirically. Thus the results will tell us something new and are judged to have important scientific implications.

The animals are well cared for in matters other than the high level of vitamin A ingestion.

Upon collecting the minimum amount of data necessary for analysis, the experiment is terminated and the rats are rapidly and painlessly killed.

Does the procedure conform to principle 10?

As you can see from the above guidelines, it is impossible to cover *all* ethical questions that might arise in the course of experimental work in psychology, just as civil laws do not cover *all* contingencies of human conduct. These guidelines do provide a general structure of ethical principles that may be applied to a wide range of specific situations. These principles are subject to a degree of interpretation and, as such, may be interpreted differently by different investigators. When planning research that involves serious ethical questions, the researcher (especially the neophyte researcher) is advised to seek the opinion and interpretation of ethical principles from other scholars. Sometimes there is a fine line between ethical and unethical experimentation, and the advice of others may help determine the factors that determine the advisability of experimental work.

RESEARCH WITH HUMANS

The use of human subjects in a psychological experiment poses special problems. A psychologist is both a scientist and a member of society. Sometimes, in the zealous pursuit of scientific truths, experimenters may become so wrapped up in their research as to overlook some ethical considerations regarding human participants. This is a grievous mistake and will ultimately reflect poorly on experimenters and psychological research in general.

In the early days of psychological research, few ethical guidelines were available save the researcher's personal ethical code and the laws of society. Some research conducted during that period would be disallowed by current standards. Today, as a consequence, some prospective participants in psychological experiments are wary of volunteering for experiments. Some researchers have argued that the current standards are *too* restrictive and forbid the collection of important data. It is likely that new ethical standards, as "laws of the land," will evolve.

This section presents the principles that govern the conduct of research with human participants. We have also included an example of a consent form and one case study. It is suggested that each point be discussed in class and that students be encouraged to write "cases" of ethical and unethical research for each principle. Also, if contemplating research with human subjects, the complete Ethical Principles of Psychologists should be read thoroughly.

RESEARCH WITH HUMAN PARTICIPANTS*

The decision to undertake research rests upon a considered judgment by the individual psychologist about how best to contribute to psychological science and human welfare. Having made the decision to conduct research, the psychologist considers alternative directions in which research energies and resources might be invested. On the basis of this consideration, the psychologist carries out the investigation with respect and concern for the dignity and welfare of the people who participate and with cognizance of federal and state regulations and professional standards governing the conduct of research with human participants.

A. In planning a study, the investigator has the responsibility to make a careful evaluation of its ethical acceptability. To the extent that the weighing of scientific and human values suggests a compromise of any principle, the investigator incurs a correspondingly serious obligation to seek ethical advice and to observe stringent safeguards to protect the rights of human participants.

B. Considering whether a participant in a planned study will be a "subject at risk" or a "subject at minimal risk," according to recognized standards, is of primary ethical concern to the investigator.

C. The investigator always retains the responsibility for ensuring ethical practice in research. The investigator is also responsible for the ethical treatment of research participants by collaborators, assistants, students, and employees, all of whom, however, incur similar obligations.

D. Except in minimal-risk research, the investigator establishes a clear and fair agreement with research participants, prior to their participation, that clarifies the obligations and responsibilities of each. The investigator has the obligation to honor all promises and commitments included in that agreement. The investigator informs the participants of all aspects of the research that might reasonably be expected to influence willingness to participate and explains all other aspects of the research about which the participants inquire. Failure to make full disclosure prior to obtaining informed consent requires additional safeguards to protect the welfare and dignity of the research participants. Research with children or with participants who have impairments that would limit understanding and/or communication requires special safeguarding procedures.

*From *Ethical Principles in the Conduct of Research with Human Participants* (1982). Washington, DC: American Psychological Association.

CONSENT FORM

Principle D of the Ethical Principles has three basic components: (1) That the agreement is clear and explicit, (2) that the terms be fair and not exploitive, and (3) that the investigator honor the agreement. An example of an "informed consent" form is shown below:

Consent Form for "Farkle"

My name is Helen Zarkovic. I am a student working on an advanced degree in experimental psychology.

You have been asked to participate in a psychological experiment which we call "Farkle." The purpose of the experiment is to measure your reaction time to the matching of colors with color names, colors, or associates of colors. No deception is involved in this experiment. You will be asked to look at colors on a TV screen and press a key, which will measure your reaction time. The entire experiment will take 20 minutes or less. The task is similar to looking at TV and we foresee no risks or discomforts. You will receive class credit for your participation in the experiment.

The results of this experiment may be presented at professional meetings or published in the scientific literature. Your name will not be used in the reporting of results. Only group data will be used; however, your scores and name will be coded for a possible follow-up study or reanalysis of the data. All personal data will be held confidential.

If you wish to withdraw from the experiment you may do so at any time without penalty.

Following the experiment I will discuss the results of the experiment with you.

If you have any questions please feel free to ask me or the Supervisor of this research, Dr. Elizabeth Kane, Department of Psychology, University of Western California, phone 882-5968. Thank you.

I, _____ _____ understand that
   (First name)   (Last name) Please print.
my participation in this experiment is voluntary and that I may refuse to participate or withdraw from the experiment at any time without penalty.

_____
Signature Participant   Date

_____
Signature Experimenter   Date

E. Methodological requirements of a study may make the use of concealment or deception necessary. Before conducting such a study, the investigator has a special responsibility to (1) determine whether the use of such techniques is justified by the study's prospective scientific, educational, or applied value; (2) determine whether alternative procedures are available that do not use concealment or deception; and (3) ensure that the participants are provided with sufficient explanation as soon as possible.

F. The investigator respects the individual's freedom to decline to participate in or to withdraw from the research at any time. The obligation to protect this freedom requires careful thought and consideration when the investigator is in a position of authority or influence over the participant. Such positions of authority include, but are not limited to, situations in which research participation is required as part of employment or in which the participant is a student, client, or employee of the investigator.

G. The investigator protects the participant from physical and mental discomfort, harm, and danger that may arise from research procedures. If risks of such consequences exist, the investigator informs the participant of that fact. Research procedures likely to cause serious or lasting harm to a participant are not used unless the failure to use these procedures might expose the participant to risk of greater harm or unless the research has great potential benefit and fully informed and voluntary consent is obtained from each participant. The participant should be informed of procedures for contacting the investigator within a reasonable time period following participation should stress, potential harm, or related questions or concerns arise.

CASE STUDY:
THE EFFECT OF INDUCED ILLNESS
ON PERFORMANCE

In a recent experiment reported by Smith, Tyrrell, Coyle, and William (1987) in the *British Journal of Psychology*, the effects of experimentally induced colds and influenza on human performance was measured. The importance of this research, argue the researchers, is to determine whether minor illnesses "alter the efficiency of human performance."

The method involved recruiting volunteers who stayed at the Common Cold Unit for 10 days. They were housed in groups of two or three and isolated from outside contacts. Following a 3-day quarantine period the subjects were inoculated with nose drops containing the virus (or a placebo). An incubation period of 48–72 hours followed. Then each participant was assessed by a clinician who evaluated the severity of the illness. Objective measures included temperature, number of paper tissues used, and the quantity of nasal secretion.

Two performance tasks were done, in which subjects were to detect and respond quickly to target items that appeared at irregular intervals (a detection task), and a second task that is commonly referred to as a hand-eye coordination task.

The results indicated that influenza impaired the detection task but not the hand-eye coordination. The effects of colds were (generally) the reverse of the effects found for influenza.

The procedures were approved by the local ethical committee and the informed consent of the volunteers was obtained. All subjects were screened to exclude pregnant women, people taking sleeping pills, tranquilizers, and antidepressant medicines. The subjects had a medical examination and chest X-ray and any who "failed" the examinations at the outset of the trial were excluded. They were not paid but received food, accommodation, traveling expenses, and "pocket money." Other "clinical trials" were also conducted.

Discuss the ethics of this experiment. Did the experimenters follow acceptable (APA) standards? Were the subjects "coerced"? Was the risk/benefit ratio worthwhile? (Keep in mind that other tests were done.) Were there alternative means available to collect the data? Were the subjects treated in a way consistent with the Principles D and G? Would you volunteer for this experiment? Would you collect the data for this experiment? Why, Why not? We suggest that this case (or the original article) be read by class members and discussed in class.

H. After the data are collected, the investigator provides the participant with information about the nature of the study and attempts to remove any misconceptions that may have arisen. Where scientific or humane values justify delaying or withholding this information, the investigator incurs a special responsibility to monitor the research and to ensure that there are no damaging consequences for the participant.

I. Where research procedures result in undesirable consequences for the individual participant, the investigator has the responsibility to detect and remove or correct these consequences, including long-term effects.

J. Information obtained about a research participant during the course of an investigation is confidential unless otherwise agreed upon in advance. When the possibility exists that others may obtain access to such information, this possibility, together with the plans for protecting confidentiality, is explained to the participant as part of the procedure for obtaining informed consent.

CHAPTER
8

The Psychological Literature: Reading for Understanding and as a Source for Research Ideas

The scientific literature in psychology has expanded greatly during the last few years. In fact, the number of articles, books, and technical reports exceeds the capacity of any researcher to know *all* the sources, even within his or her own specialty, let alone the full range of papers in psychology. Because so much literature is available, and because the human mind is limited in its ability to collect and store information, the task of becoming an expert in a particular subject in psychology might seem to be impossible.[1] The task may be difficult but, given the guidelines in this chapter, not impossible.

As we shall see the literature in psychology is organized—and with the electronic storage of information in computer memories is becoming

---

[1] See R. L. Solso (1987a).

The expansion of scientific information and the dilemma created by such a wealth of information reminds us of a story about a KGB envoy and a CIA agent who met for lunch. The CIA agent complained that his job was like trying to fit together the pieces of an enormously complex jigsaw puzzle. "But you Soviets are *so* secretive it is as if I'm trying to create a picture with many pieces missing." The KGB operative retorted that he too was trying to put together a big jigsaw puzzle, but that his task was even more difficult. Referring to the great amount of information available in an open and free society, he said, "My problem is that I have too many pieces!"

Western scientific development is something like the CIA- KGB problem. In the early part of experimental psychology too few bits of information were available, so researchers set out to find the missing pieces. Today, with the explosion of information we face a problem of having too many pieces, at least for a researcher to read and understand. It is difficult to sift through the voluminous literature in psychology, reject the inappropriate, avoid the redundant, and focus on the relevant.

much more thoroughly organized. Given several principles of the way the literature is stored and organized, you will find it relatively easy to focus your reading of the literature, which is an important step toward the development of ideas in psychology.

IDEAS, HUNCHES, AND THE PSYCHOLOGICAL LITERATURE

In an earlier section of this book we wrote about the source of knowledge. Frankly, finding the source of the Nile or the Holy Grail or giving a complete description of the universe with two examples might seem to be easier than finding a truly original idea in science. Unfortunately no single, surefire pathway can be specified that will lead, inevitably, to the development of a new idea. We do have a suggestion that when combined with an inquiring mind will enhance the likelihood of an experimental hunch materializing. That suggestion is: *Read the psychological literature selectively and critically.*

First, consider the matter of selective reading. Suppose you are searching for knowledge and ideas about experimental psychology. You could read indiscriminately all the literature available, but the task would be impossible even for the most voracious reader. A far more efficient means is to select a topic of interest—be it perceptual motor processes, thumbsucking in children, language development in chimpanzees, the learning of mathematics, the use of personality tests in industry, or music and pain (see box on following page)—and then find relevant literature on that topic. Fortunately, the current literature in psychology (and other branches of knowledge) is organized.

ELECTRONIC LIBRARIES

Kriss Kline

Getting a comprehensive literature review requires paging through a library full of indexes, reference books, and thousands of journals. It may even require going to other libraries for references that are not in your library. This whole process can take several days or even several weeks to complete. With the advent of the personal computer and computerized databases many literature searches can now be completed in under an hour.

A computerized database can be defined as an organized collection of facts in computer-readable form. Imagine the complete *Psychological Abstracts* (1966–present) in a computer database. Using a computer or a terminal, and a modem (computer phone) call the electronic library (usually a local phone number). After "logging in," uncomplicated commands are used to access information (see sample search below). With these commands the database's contents can be searched at a rate of thousands of items per second. Most database services also provide on-line ordering procedures for references found during a computer search.

There are several electronic database vendors availale to the consumer (Dialog, The Source, Bibliographic Retrieval Service, CompuServe, etc.). These on-line services offer many different databases to choose from and valuable on-line help systems. When choosing a database service, several different aspects should be considered: what hours the database can be accessed (day or evening), what databases are needed (medical, engineering, psychological, etc.), and how charges are determined (per reference, computer time, evening vs. day rates, etc.).

The following is an example of a search using Dialog's night-time service. Knowledge Index:

```
? begin psycl
2/5/88 18:14:20 EST
Now in PSYCHOLOGY (PSYC) Section
  PsycINFO (PSYCI) Database
  (Copyright 1982 American Psychological Association)


? find music and pain


        3106  MUSIC
        5067  PAIN
  S1      21  MUSIC AND PAIN


? display s/1/1/1-21


  1/L/1
  74-16157
    Music therapy in pain management.
    Bailey, Lucanne M.
    Memorial Sloan-Kettering Cancer Ctr, New York, NY
    Journal of Pain & Symptom Management, 1986 Win Vol
    1 (1) 25-28 ISSN: 08853924
    Journal Announcement: 7406
    Language: ENGLISH Document Type: JOURNAL ARTICLE
    Suggests that music therapy can be used to treat
    pain and suffering by promoting relaxation, alteration
    in mood, a sense of control, and self-expression.
    Music therapy techniques are individually devised,
    with consideration given to the patient's physical,
    emotional, and psychological needs; coping abilities; and
    prior musical experiences. Active involvement is
    encouraged to facilitate cognitive and emotional
    expression, as well as involvement in pain management.
    (13 ref) (PsycINFO Database Copyright 1987 American
    Psychological Assn. all rights reserved) Descriptors: PAIN
    (36150); MUSIC THERAPY (32670); RELAXATION (43697);
    EMOTIONAL STATES (16900) Identifiers: music therapy,
    relaxation & mood alteration & control & self expression,
    pain patients Section Headings: 3300 (TREATMENT AND
    PREVENTION)


  1/L/2
```

The *Psychological Abstracts* and Other Indexes

The organization of knowledge in psychology is available through two sources. One is the collection of abstracted information that is accessible through computer-based information services (see box). These services are very useful and provide a wide range of information. Also, the search for selected topics is fast and convenient. They do require a computer, and most of the time a charge is made for on-line computer time. If you have a computer and modem, it is possible to "log on" to these systems. Sometimes departments of psychology or libraries will provide computer time or will do electronic searches for students. It is our impression that these computer-based indexes are the way information will be accessed in the future, and the student of experimental psychology is encouraged to explore the prodigious capacity of computer-based data stores.

The second way knowledge in psychology is organized is in library-based journals, books, and indexes. Books are excellent sources in which to find an overview of a selected topic. If you are interested in, say, thumbsucking in children, you might consult any one of a number of current books in developmental psychology and child clinical psychology. If you are interested in the basic research in that field, however, you will have to consult the current experimental literature in the field.

A primary index of the psychological literature—both domestic and foreign—is *Psychological Abstracts,* published by the APA each month, with an index volume published every six months and cumulative indexes every year. Using the abstracts is simple, and the more practice you have with them, the more efficient will be your search.

Suppose you are interested in perceptual motor processes, an example of which could be hand-eye coordination. In the subject index of the *Psychological Abstracts* you would find an entry called "perceptual motor processes" (see Figure 8.1). Following the entry, other related entries are suggested (e.g., "See also Perceptual Motor Coordination, Rotary Pursuit . . ."). Then a series of numbers are presented that correspond to the abstract of the article, which can be found in the main section of the abstracts. The abstract is only a bare summary of the article, but it should give you enough information for you to decide if the article is relevant to your interest or not. If the abstract is relevant, you should find the original article and read the details of the experiment. A complete description of the journal, year, volume, and page numbers is given.

Social Science Citation Index

Available in most college libraries, the *Social Science Citation Index* is a very useful library tool and yet is frequently unknown by many students. Every year it codes over 70,000 articles published in more than 2,000 of the world's journals. Such a comprehensive index would be impossible

BRIEF SUBJECT INDEX

Parkinsons Disease 9708, 10755, 10793, 10804, 11165, 11185
Parks (Recreational) [See Recreation Areas]
Parochial School Education [See Private School Education]
Parsons [See Ministers (Religion)]
Partial Hospitalization 11436, 11438, 11442, 11444, 11472, 11484
Partial Reinforcement [See Reinforcement Schedules]
Partially Hearing Impaired 10697, 10728, 10733, 10857, 10860, 10869, 10896, 11241, 11243, 11491, 11747, 11864, 11888, 11912, 11919, 11932
Participation [See Also Athletic Participation, Group Participation] 9913, 10124, 11070, 12145, 12230
Parturition [See Birth]
Passive Avoidance [See Avoidance Conditioning]
Passiveness 10179
Pastoral Counseling 10802, 10951, 11375, 11378, 11381, 11382, 11383, 11387, 11389, 11390, 11391, 11392, 11393, 11394, 11395, 11397, 11404, 11405, 11407, 11408, 11413, 11414, 11419, 11482
Pastors [See Ministers (Religion)]
Pathogenesis [See Etiology]
Pathology [See Psychopathology]
Perceptual Disturbances [See Also Agnosia, Auditory Hallucinations, Hallucinations] 10650
Perceptual Localization [See Auditory Localization]
Perceptual Measures [See Stroop Color Word Test]
Perceptual Motor Coordination 10782
Perceptual Motor Development [See Motor Development, Perceptual Development]
Perceptual Motor Learning [See Also Gross Motor Skill Learning] 9497, 11722
Perceptual Motor Processes [See Also Perceptual Motor Coordination, Rotary Pursuit, Tracking, Visual Tracking] 9271, 9275, 9278, 9643, 9725, 10649, 10781, 1148[illegible], 1221[illegible]
Perceptual Orientation [See Spatial Orientation (Perception)]
Perceptual Stimulation [See Auditory Feedback, Auditory Stimulation, Filtered Noise, Illumination, Olfactory Stimulation, Pitch (Frequency), Somesthetic Stimulation, Speech Pitch, Stereoscopic Presentation, Tachistoscopic Presentation, Visual Stimulation, White Noise]
Perceptual Style 9306, 9509, 11411
Performance [See Also Group Performance, Job Performance, Motor Performance] 9134, 9306, 9341, 9400, 9487, 9498, 9520, 9832, 10218, 10282, 10297, 10849, 11756, 11824, 12176, 12183, 12189, 12194
Personality Theory 10198, 10225, 10228, 10230, 10247, 10268, 11391
Personality Traits [See Also Aggressiveness, Androgyny, Assertiveness, Authoritarianism, Cognitive Style, Conformity (Personality), Conservatism, Courage, Creativity, Cynicism, Defensiveness, Dependency (Personality), Dogmatism, Egotism, Emotional Security, Emotionality (Personality), Empathy, Extraversion, Femininity, Honesty, Hypnotic Susceptibility, Impulsiveness, Independence (Personality), Initiative, Internal External Locus of Control, Introversion, Masculinity, Narcissism, Neuroticism, Openmindedness, Optimism, Passiveness, Pessimism, Psychoticism, Self Control, Sociability, Suggestibility, Timidity, Tolerance] 9263, 9385, 9894, 9933, 9943, 9963, 10206, 10223, 10227, 10232, 10250, 10268, 10282, 10304, 10306, 10312, 10412, 10499, 10605, 10606, 10975, 11136, 12045, 12115, 12120, 12139
Personality [See Also Related Terms] 9129, 10112, 10246, 10265, 10269, 10296, 10323, 10415
Personnel Development [See Personnel Training]
Personnel Evaluation [See Also Occupational Success Prediction, Teacher Effectiveness Evaluation] 12091, 12093, 12096, 12097, 12098, 12103
Personnel Management [See Also Career Development, Job Applicant Interviews, Job Applicant Screening, Labor Management Relations, Occupational Success Prediction, Personnel Evaluation, Personnel Placement, Personnel Recruitment, Personnel Selection, Personnel Termination, Teacher Recruitment] 12089, 12101, 12118
Personnel Placement 11845
Personnel Recruitment [See Also Teacher Recruitment] 12087
Personnel Selection [See Also Job Applicant Interviews, Job Applicant Screening] 11709, 12080, 12090
Personnel Termination 10652
Personnel Training [See Also Inservice Training, Management Training, Military Training] 11135, 12078, 12079, 12081, 12083, 12084, 12085, 12086, 12089
Personnel [See Also Related Terms] 11866, 12109, 12163
Perspective Taking [See Role Taking]
Persuasive Communication 10199, 12225
Pessimism 10277
Pets 11404
Phantom Limbs 11266
Pharmacology [See Also Psychopharmacology] 9705
Pharmacotherapy [See Drug Therapy]
Phencyclidine 9693
Phenelzine 10483
Phenomenology 9813, 10065, 11396
Phenothiazine Derivatives [See Chlorpromazine, Fluphenazine]
Phenotypes 9463, 10591, 10864
Phenoxybenzamine 9586
Phenylalanine [See Also Parachlorophenylalanine] 10366
Phenylketonuria 10722, 10764
Pheromones 9652
Philosophies [See Also Dualism, Epistemology, Logic (Philosophy), Mysticism] 9093, 9111, 10036, 10110, 10186, 10324, 11018, 11671
Phobias [See Also Agoraphobia] 9392, 10467, 11138, 11206
Phonemes [See Also Consonants] 10725
Phonetics 9269, 9779, 9868, 11241, 11753
Phonics 117[illegible]5

EXPERIMENTAL PSYCHOLOGY (HUMAN) 74: 9268–9277

Pearson approach to hypothesis testing is of no value, and hence a Fool's Type IIa error is irrelevant. If statistical testing errors are important and can be quantified, then adjustment for the Fool's Type IIa error region is equivalent to increasing the probability of making a Type I error. (8 ref)

EXPERIMENTAL PSYCHOLOGY (HUMAN)

9269. **Johnston, Rhona S. & McDermott, Erica A.** (MRC Cognitive Neuroscience Research Group, U St Andrews, Scotland) **Suppression effects in rhyme judgement tasks.** *Quarterly Journal of Experimental Psychology: Human Experimental Psychology,* 1986(Feb), Vol 38(1-A), 111–124. —Assessed the effects of paced vs rapid articulatory suppression on the ability to make rhyme judgments when pairs were presented either simultaneously or successively. Orthographic and phonological similarity were orthogonally manipulated in 4 experiments, each containing 12 university staff members and students. Findings show that there were consistent suppression effects on the accuracy of Ss' judgments to visually similar nonrhyming pairs (e.g., "pint–tint"), visually dissimilar rhyming pairs (e.g., "fare–wear"), and visually similar rhyming pairs (e.g., "fall–tall"), regardless of mode of presentation or speed of suppression. The size of the suppression effect was greatest for the visually similar nonrhyming word pairs. It is suggested that Ss need to carry out a recheck for phonological similarity when word pairs are visually but not phonologically similar and that encoding the words in articulatory form is particularly beneficial for making accurate rhyme judgments to such pairs. (14 rf) —*Journal abstract.*

9270. **Ruiz-Vargas, José M.; Zaccagnini, José L. & Delclaux, Isidoro.** (U Autónoma de Madrid, Spain) **El Procesamiento Humano de Información como Modelo de Conducta. [Human information processing as a behavioral model.]** (Span) *Boletín de Psicología (Spain),* 1985(Mar), No 6, 7–19. —Discusses epistemological differences between empirical and rational approaches to scientific psychology, and proposes human information processing as a model of human behavior. Importance of representational models of the world for scientific epistemologies is emphasized. (26 ref)

Perception & Motor Processes

9271. **Colley, Ann & Fossey, Jane.** (U Leicester, England) **Reproduction of complex movements: The effects of the presence of vision during encoding or at recall.** *British Journal of Psychology,* 1986(Feb), Vol 77(1), 75–84. —Investigated the role of vision in encoding and reproducing movement. Kinesthetic reproductions of a kinesthetically presented 2-dimensional movement were compared with reproductions for which vision was present either during the standard or at reproduction. Ss were 15 male and 15 female right-handed undergraduates. The presence of vision during the standard resulted in poorer accuracy and greater underestimation of movement size than when it was absent throughout or present during reproduction. The presence of vision during the standard resulted in less distortion of the linear components of movement shape, although no such effect was found for the angular components. The initial direction of movement was reproduced more accurately when visual experience of the movement was given during the standard or reproduction. (12 ref) —*Journal abstract.*

dence and that the increased error was the result of some psychological variable (e.g., inattentiveness, the underestimation of task difficulty, or both). (10 ref) —*Journal abstract.*

9273. **Heller, Morton A.** (Winston-Salem State U) **Effect of magnification on texture perception.** *Perceptual & Motor Skills,* 1985(Dec), Vol 61(3, Pt 2), 1242. —30 Ss viewed abrasive surfaces with the unaided eye, through low or high magnification, and matched them against an array of textures they touched. Results indicate that the texture of a surface can be accurately judged by both vision and touch and that it is difficult to disrupt perception with moderate magnification. (4 ref)

9274. **Holbrook, Morris B.; Greenleaf, Eric A. & Schindler, Robert M.** (Columbia U, Graduate School of Business) **A dynamic spatial analysis of changes in aesthetic responses.** *Empirical Studies of the Arts,* 1986, Vol 4(1), 47–61. —Suggests that models of evaluative judgment in applied aesthetics generally assume that movements of objects' positions in a multidimensional perceptual space will produce corresponding changes in preferences. However, such spatial representations have usually been tested using static designs or, at best, longitudinal studies that fail to tie perceptual movements to shifts in affective response. The present study conducted dynamic analysis of changes in aesthetic responses. 29 aesthetically naive undergraduates supplied perceptual and affective ratings of 20 art prints on 4 occasions spaced a week apart. Multiple discriminant analysis (MDA) of the perception data created an MDA space representing each S's perceptions of each print at the beginning and end of the period. Each S's preference function was constructed based on the MDA space, and beginning and ending perceived positions were used to compute changes in that preference function. These were regarded as predictions of changes in actual affect, and predicted and actual affective shifts were correlated to obtain a validity assessment of about $r^2 = .25$ for this dynamic spatial analysis of trends in tastes. It is concluded that static analyses may tend to lose about half their predictive power in dynamic applications. (40 ref) —*Journal abstract.*

9275. **Larish, Douglas D.** (Arizona State U, Tempe) **Influence of stimulus–response translations on response programming: Examining the relationship of arm, direction, and extent of movement.** *Acta Psychologica,* 1986(Jan), Vol 61(1), 53–70. —Examined whether the programming relationships of direction and extent of movement (Exp I) and arm, direction, and extent of movement (Exp II) are affected by a cognitive recoding process called a stimulus–response translation. The programming of these task-defined parameters was studied in 23 volunteers via the movement precue method. The effect of a spatial translation was studied by manipulating stimulus–response compatibility. Both experiments showed that the patterns of reaction time (RT) could be altered by indirect or noncompatible stimulus–response mappings. When stimulus–response compatibility was deficient, effects that might be attributed to programming processes were due instead to the translation process. Findings obtained from a spatially compatible stimulus–response ensemble demonstrated that the movement parameters of arm, direction, and extent could be selectively manipulated via the precue method. It is concluded that this method may be a useful tool in understanding how motoric decisions are made prior to movement if the spatial mapping among stimuli and responses is maximally compatible. (32 ref) —*Journal abstract.*

9276. **O'Connell, Bernard J.; Harper, Robert S. & McAndrew, Francis T.** (Knox Coll) **Grip strength as a function of exposure to red or green visual stimulation.** *Perceptual & Motor Skills,* 1985(Dec), Vol 61(3, Pt 2), 1157–1158. —Tested the color-strength hypothesis using colors (red and green) not associated with sex (e.g., blue for males) or differing in preferences between the sexes. 40 male

Figure 8.1 Montage of pages from *Psychological Abstracts*. Reprinted by permission.

without computers. The *SSCI* has four separate indexes: a subject index, an author index, a corporate index, and a citation index. Given the nature of your search, each of these indexes is useful. The subject index is an organized list of major words from the title of an article. For example, if an article is entitled "Reaction Time for Children in a Concept Formation Task," it might be indexed under the headings "Reaction Time," "Children" and "Concept Formation." The author index lists the name of the author (in our example K. Krushinski), the title of the article, the journal in which the article appears, the volume and page number of the article, the year the article appeared, the number of references, a brief address of the author, and a list of the sources cited in the article. The above-mentioned article might be listed as:

```
KRUSHINSKI, K.
     REACTION TIME FOR CHILDREN IN A CONCEPT FORMATION
     TASK
      J CHILD PSY      15  6  89    12R  N1
      PAX MEDICAL CENTER, LINDSEY AVE., HERSHEY, PA.
     LAIRD, R.    85    J EXP PSY     01  20   12
     (etc.)
```

The third index, the one we find to be most useful, is the *Citation Index*. This is a listing of the citation of articles and is based on the assumption that the theme of a published article is carried through in the work of subsequent researchers. As you become more familiar with specific areas of psychology, you will begin to recognize seminal papers in those fields. Contemporary work in the various areas of psychology frequently cites the original work. Thus, if you were interested in "iconic memory" and knew the original work was done by George Sperling, you could find out which contemporary researchers cited that article through the citation index, as shown in the page of the *Citation Index* reproduced in Figure 8.2. On this page you may be interested in the topic of mental rotation. Roger Shepard and Jacqueline Metzler published an important article on that topic in *Science* in 1971. The recent citation of that article is shown in the citation index. Following the date, journal, volume, and page of the cited article are the name, journal, volume, page, and year of the source in which the article was cited.

Several other indexes from areas related to psychology (e.g., biosciences, medicine, education, and periodic guide to general literature) can be found in most college libraries. One particularly useful index for biomedical and clinical psychology is *Index Medicus*. It does not contain abstracts, however: only titles and notation of articles. All major abstracts are now in computer databases and can be accessed on-line for a fee.

Reading and Understanding Psychological Articles

After you have found several articles pertinent to your interest, how do you go about digesting the material therein? Most psychological articles are technical reports littered with arcane jargon. You will need practice. On the brighter side, most articles are highly structured, and if you become familiar with the structure, comprehension is easier.

SHENFIELD S
85 B ATOM SCI 41 23
MARULLO S SOCIOL FOCU 20 135 87
87 NUCLEAR PREDICAMENT
(ANON) EUR ARCH R 42 451 87
SHENG
81 COLUM HUM RTS L REV 13 71
81 COLUM HUM RTS L REV 13 72
BENNETT WH CORNELL I L 20 1 87
SHENG YK
83 COMMUNITY PARTICIPAT
BENNINGE.CC HABITAT INT 11 145 87
SHENGELIA M
70 ISTORIIA GRUZINSKOI 21
GAGULASH.IS SOV ANTHR A 26 55 87
SHENGOLD L
72 FREUD HIS SELFANALYS 213
SCHWARTZ HJ J AM PSYCHO 35 657 87
79 J AM PSYCHOANAL ASS 27 533
85 PSYCHOANAL QUART 54 20
MEYER WS CLIN S WORK 15 136 87
SHENK A
79 CALCULUS ANAL GEOMET
NUGENT WR SOCIAL SE R 61 337 87
SHENK JC
74 ACA J CHIROPRACTIC 8 S 76
NYIENDO J MED CARE 25 516 87
SHENKAR O
84 83484 ISR I BUS RES
SHENKAR O PERS REV 16 30 87
SHENKEL WM
81 J PROP MANAGE 27 146
SHILLING JD J URBAN EC 22 90 87
SHENKEN
87 NZ MED J 100 23
CONLON GB NZ MED J L 100 466 87
SHENKER IR
77 CLIN PEDIATR 16 516
COMERCI GD J ADOLES H 8 356 87
SHENKER L
75 OBSTET GYNECOL 46 185
SHY KK ANN R PUB H R 8 165 87
SHENKER LJ
80 THESIS MCGILL U MONT
MCMILLAN MJ J ABN PSYCH 96 149 87
SHENKIN HA
51 J APPL PHYSIOL 3 465
WOLKIN A PSYCHOPHAR 92 241 87
64 NEUROLOGY 14 1112
ALMAY BGL PAIN 29 335 87
SHENKMAN L
73 J CLIN INVEST 52 205
LOOSEN PT PHARMACOPS 20 85 87
SHENNAN JH
68 PARLEMENT PARIS 318
FITZSIMM.MP AM HIST REV 92 269 87
86 LIBERTY ORDER EARLY
DUCHHARD.H HIST Z B 244 708 87
SHENNAN KIJ
83 J ENDOCRINOL 97 105
PEKARY AE J GERONTOL 42 447 87
SHENNAN S
82 SYMBOLIC STRUCTURAL 155
LEONE MP CURR ANTHR 28 283 87
SHENNAN SJ
83 RAIN 59 9
GAMBLE C PROG H GEOG R 11 227 87
SHENNUM WA
84 LONG TERM ADJUSTMENT
SHENNUM WA CHILD YOUT 16 81 87
SHENNUM WE
75 ED116176
GANSCHOW L PERC MOT SK 64 899 87

SHEPARD PT
85 SUSTAINABLE AGR INTE 244
SHEPARD PT SCI TECHNOL 12 19 87
SHEPARD R
70 THEORY COST PRODUCTI
POPE RD AM J AGR EC 69 378 87
SHEPARD RJ
77 ENDURANCE FITNESS
PARKES KR J APPL PSYC 72 275 87
82 RISK PASSIVE SMOKING
BECONA E REV LATAM P 19 19 87
SHEPARD RN
57 PSYCHOMETRIKA 22 325
ESTES WK J EXP PSY L 13 380 87
62 PSYCHOMETRIKA 27 125
BENSHAKH.G J EXP PSY G 116 91 87
CHOKOR BA ENVIR PL-A R 19 7 87
COURY BG ERGONOMICS 30 959 87
CROMLEY EK HEAL SERV R 21 499 86
HUBERT L PSYCHOL B 102 172 87
MULLEN K PSYCHOMETRI 52 235 87
62 PSYCHOMETRIKA 27 219
BENSHAKH.G J EXP PSY G 116 91 87
CHOKOR BA ENVIR PL-A R 19 7 87
HUBERT L PSYCHOL B 102 172 87
MULLEN K PSYCHOMETRI 52 235 87
63 HUMAN FACTORS 5 19
MULLEN K PSYCHOMETRI 52 235 87
63 J VERB LEARN VERB BE 2 93
SINGH SN J ADVER RES 27 23 87
64 HUMAN JUDGMENTS OPTI
YORK KM ORGAN BEHAV 39 303 87
64 J MATHEMATICAL PSYCH 1 54
SERGENT J J EXP PSY P 13 300 87
SHEPP BE J EXP C PSY 43 159 87
64 J ACOUST SOC AM 36 2345
DEUTSCH D PERC PSYCH 41 563 87
64 J ACOUST SOC AM 36 2346
BHARUCHA JJ PERC PSYCH 41 519 87
CUDDY LL " 41 609 87
PITT MA " 41 534 87
65 STIMULUS GENERALIZAT
DEUTSCH D PERC PSYCH 41 563 87
67 J VERB LEARN VERB BE 6 156
PEZDEK K CHILD DEV 58 807 87
SISSON N INT J MAN M 25 479 86
SPIRO RJ PSYCHOL R 177 87
70 COGNITIVE PSYCHOL 1 1
MARSCHAR.M PSYCHOL B 102 28 87
OZER DJ J PERS SOC 53 129 87
71 SCIENCE 171 701
BHARUCHA JJ BEHAV RE ME 19 131 87
BIEDERMA.I PSYCH LEARN 20 1 86
BOUMA A BRAIN COGN 6 295 87
BRUCE V APPL COGN P 1 109 87
COLLYER CE B PSYCHON S 25 189 87
EMMERTON J PERCEPTION 15 573 86
FOLK MD J EXP PSY P 13 395 87
JORDAN DS PERC PSYCH 41 489 87
LEVINE G PERS INDIV 8 599 87
MARSCHAR.M PSYCHOL B 102 28 87
NEIWORTH JJ J EXP PSY A 13 203 87
OZER DJ J PERS SOC 53 129 87
RHODES G PERC PSYCH 42 1 87
RICHARDS.JT BR J PSYCHO 78 189 87
ROBERTSO.LC J EXP PSY L 13 368 87
SHARPS MJ J GERONTOL 42 342 87
71 SCIENCE 191 952
MATTHYSS.S SCHIZO BULL 13 173 87
PARSONS LM J EXP PSY G 116 172 87
" J EXP PSY P 13 488 87
" PERC PSYCH 42 49 87
72 COGNITIVE PSYCHOL 3 228
PELLEGRI.JW BEHAV RE ME 19 231 87
72 MULTIDIMENSIONAL SCA 1
BENSHAKH.G J EXP PSY G 116 91 87
HAMMOND SM PERS INDIV 8 541 87
ROBINS CJ COGN THER R 11 197 87
72 MULTIDIMENSIONAL SCA 1 1

SHEPHARD RA
84 NEUROPHARMACOLOGY 23 677
84 PSYCHOPHARMACOLOGY 82 343
85 PSYCHOPHARMACOLOGY 86 313
86 NEUROSCI BIOBEHAV RE 10 449
SHEPHARD RA LIFE SCI 40 2429 87
SHEPHARD RJ
70 J SPORTS MED PHYS FI 10 206
FAULKNER RA CAN J PUBL 78 155 87
76 CAN MED ASSOC J 114 675
DESHARNA.R CAN J PUBL 78 161 87
GODIN G J BEHAV MED 10 241 87
77 ENDURANCE FITNESS
FAULKNER RA CAN J PUBL 78 155 87
78 HUMAN PHYSL WORK CAP
HARRISON GA CURR ANTHR E 28 209 87
79 BRIT J SPORTS MED 12 227
LEE C AUST PSYCHL R 21 427 86
79 S AFRICAN J SPORT SC 2 19
DESHARNA.R CAN J PUBL 78 161 87
GODIN G J BEHAV MED 10 241 87
80 CANADIAN J APPLIED S 5 69
80 J OCCUP MED 22 389
LEE C AUST PSYCHL R 21 427 86
81 J OCCUP MED 23 359
FALKENBE.LE ACAD MGMT R 12 511 87
82 CAN J PUBLIC HLTH 73 183
DESHARNA.R CAN J PUBL 78 161 87
GODIN G J BEHAV MED 10 241 87
82 CAN J PUBLIC HLTH 73 259
TSAI SP J OCCUP MED 29 572 87
83 PREVENTIVE MED 12 644
LEE C AUST PSYCHL R 21 427 86
83 PREV MED 12 664
PAVETT CM STRESS MED 3 87 87
85 J SPORTS MED 2 133
TSAI SP J OCCUP MED 29 572 87
86 EC BENEFITS ENHANCED
86 FITNESS HLTH IND
(ANON) ERGONOMICS B 30 876 87
86 SPORTS MED 3 26
ASTRAND PO ARCH PHYS M R 68 305 87
PEKKANEN J LANCET 1 1473 87
POWELL KE ANN R PUB H R 8 253 87
SHEPHARD RW
53 COST PRODUCTION FUNC
APOSTOLA.BE DEVELOP ECO 25 171 87
70 THEORY COST PRODUCTI
APOSTOLA.BE ENERGY 12 531 87
BREYER F J HEALTH EC 6 147 87
FARE R J I THEOR E N 143 343 87
FULLER DA RAND J ECON 18 124 87
KURODA Y AM J AGR EC 69 328 87
74 FRONTIERS QUANTITATI 2
MCELROY MB J POLIT EC 95 737 87
74 PRODUCTION THEORY 253
FARE R J I THEOR E N 143 343 87
SHEPHARD WO
75 J VOCATIONAL BEHAVIO 6 27
SANDBERG DE SEX ROLES 16 649 87
SHEPHER J
71 ARCH SEXUAL BEHAVIOR 1 293
BATESON P BEHAV BRAIN 10 71 87
KITCHER P " 10 61 87
83 INCEST BIOSOCIAL VIE
BLIZARD RA PERS INDIV B 8 597 87
KITCHER P BEHAV BRAIN 10 61 87
SHEPHERD A
86 LOCAL AUTHORITY APPR
BECKER S LOC GOVT ST 13 35 87
87 WATER PLANNING ARID
NORRIS M PUBL ADM D E 7 329 87
SHEPHERD B
85 ASSERTIVENESS
KILPATRI.R COMM DEV J 22 141 87
SHEPHERD CR
64 SMALL GROUPS SOME SO
KORZENNY F HISPAN J B 9 33 87

Figure 8.2 Unnumbered page of *Citation Index*. Reprinted by permission.

So, as you try to understand the peculiar language and unfamiliar structure of psychology we suggest two methods for better understanding:

1. Read with an inquiring mind.
   - The first stage in this process is to skim an article to gain an overview of the problem, the design, and the conclusions.
   - Read the article in detail.
   - Finally, review the article and identify the most important points.
   - Ask questions as you read. For example, you might ask in the first part of the title, what is the relationship between the past results and the current experiment? You may question the methods, the results, and the conclusions reached by the investigator.

2. Become familiar with the structure of articles in psychology. You will find an amazing consistency.
   - First is the title and affiliation of the authors.
   - Second, a brief summary or abstract is presented.
   - Third, there is the body of an experiment, which includes a review of previous studies.

- Fourth, there follows a description of the experiment and the results.
- Fifth, a discussion of the findings is presented with the references.

Look at an article and try to identify these components. Learn the structure.

A Form for Recording Articles

We have found that learning how to comprehend even complex experiments in psychology is a teachable skill. In addition to reading with an inquiring mind and learning the structure of articles, it is important to organize your notes and thoughts about an article. One way you can organize your notes about an article is on the form on the following page.

As you read an article fill out a copy of the form. Identify the topic and record the author(s), the title, and the publication from which the article was taken; the problem to be investigated and how the researcher(s) did the task; the results and interpretation; and, finally, your criticism of the experiment and what additional research you would do to answer some of the questions raised.

In addition to learning how to read and interpret an article in psychology, we have found that many students keep these brief reviews and refer to them later in their careers. Even professional psychologists have found this form useful. It is possible to copy this on a personal computer and record summaries of your readings, photocopy the form, or develop a form of your own.[2] If you habitually record your summaries of articles, in a short time you will have a sizable collection of reviews that may prove very useful in developing research ideas as well as offering a strong reminder of your previous readings.

[2]The author (RLS) would be interested in seeing some of the forms created by readers of this book.

REVIEW OF ARTICLE FORM

TOPIC _____ DATE _____

AUTHOR(S) _____

BRIEF TITLE _____

SOURCE _____

PROBLEM _____

_____

_____

HOW INVESTIGATED _____

_____

_____

_____

RESULTS _____

_____

_____

_____

_____

_____

INTERPRETATION _____

_____

_____

CRITICISM _____

_____

_____

ADDITIONAL RESEARCH TO DO _____

_____

_____

CHAPTER
9

Conducting Research and Writing a Research Paper

This chapter is about doing research and writing a report of your research. In one sense, these are the most complicated steps in the experimental process and, yet, if your experiment has been designed carefully, these steps follow logically. In Figure 9.1 the essential stages are shown. It should be noted, however, that these steps are only guidelines, and many times in the actual execution of a research program the researcher must backtrack, start over, reconsider new hypotheses, read new literature, go back to the "drawing board," redesign the experiment, take a long and thoughtful walk, seek help, abandon pet ideas, think new thoughts, consider the practical aspects of the proposal, and always apply logic, creative thought, and (it is hoped) wisdom throughout every part of the process. First, consider the matter of doing research.

DOING RESEARCH

Steps Involved in Doing Research

Before an actual experiment is undertaken, a researcher should be reasonably familiar with the previous findings in the field. It may be that someone else has done the experiment you are contemplating. Also, as mentioned in the previous chapter, from the vast amount of *literature* in psychology it is possible to know what has been done and, more important for our present discussion, through *creative, probing thought,*

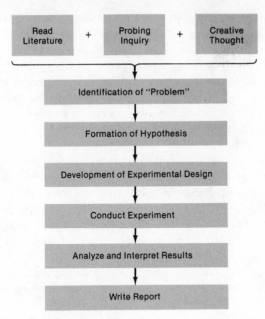

Figure 9.1   Steps involved in planning, doing, and reporting research.

what *needs* to be done. The information contained in previous studies may provide the identity of a *problem*—a term that means a question proposed for solution. Practically, a problem suggests that the current state of knowledge is either lacking in some critical dimension or that a gap in knowledge exists. An *hypothesis,* or testable proposition, can be constructed from a problem. Following the identification of a problem and the development of an hypothesis, a researcher should carefully *plan* an *experiment* that will test the hypothesis and solve the problem. Proper attention should be given to design and control (see previous chapters) in the planning of an experiment. The final step is *writing* a research paper.

As straightforward as the above scheme may sound—from literature, to needs, to problem, to hypothesis, to plan, to experiment, to report— we remind you of Murphy's original law: "If anything can go wrong (in research), it will," and its corollary, "Murphy was an optimist." Be prepared to have some disappointments: We know of no researcher who has not failed. You can, however, minimize the likelihood of a total fiasco through careful planning.

Plan Research Projects That Are Realistic

A key ingredient in planning an experiment, which you actually intend on completing, is the development of a manageable project. If you are interested in brain research involving human surgery and have neither skill nor equipment, or in the mating behavior of whales in their natural habitat and live in Nebraska, it is likely that your results will be publishable only in the *Journal of Frustrating and Quixotic Ventures*. It is not that bizarre thoughts are prohibited from the experimental laboratory or that one should not think expansively: The point is that if you plan a grandiose research program, then be prepared to endure the hardships involved in carrying out such a project, be it studying brain surgery or moving to California.

Numerous credible research projects involve fancy equipment that is unavailable in many laboratories. If the research is important and requires specialized equipment, it may be worthwhile to seek funds through grants or the institution's equipment moneys. Be forewarned that sometimes this process can take longer than doing the research. An alternative is to go to a laboratory (such as a medical school) or setting (such as the ocean) where your projects can be done.

Another practical consideration is the availability of subjects. Even though Nebraska has a "Navy"[1] there are no whales living in that state. Likewise there are few left-handed piccolo players in Minot, North Dakota, brown- and blue-eyed children in Sun City, Arizona, or Gaboon vipers in Amherst, Massachusetts. It may even be difficult for you to find usable rats, college-aged subjects, schizophrenics, bed wetters, or third graders in your area. Also, always be mindful of ethical considerations involved in working with subjects (see Chapter 7).

Can the proposed research be done within the time available? In our experience this component, more than any other, has caused students (and their professors) the most grief. Experimental projects usually take longer to complete than first anticipated. This may be due to equipment failure, the necessity to replan an experiment, the failure of subjects to show up for an experiment, natural disasters, unreliable coworkers, new discoveries reported in your field, power failures, spring break, and the like (see Murphy's laws). Good research requires time and dedication, and we know of no prominent researcher who has not devoted a major portion of his or her lifetime to science.

Do a Pilot Experiment

Even if your research project is methodically planned to the most minute detail, it is important to run test trials in order to become thoroughly acquainted with the technique and procedure. Also, it may be necessary to "debug" the methodology and/or equipment.

---

[1]One of the authors (RLS) is an honorary "Admiral" in the Nebraska "Navy," whose flagship is a prairie schooner.

Planning a Research Program

Throughout the years we have done research with students and collaborators we have found it useful to follow a prescribed form in planning and thinking about experimental research. One form used in one of the author's (RLS) laboratory is reproduced on the following page. It includes the name of the research, the experimenter or the experimenters in the case of a collaborative research project, the topic, the problem, and the significance. In order to avoid hassles with collaborators it is important to spell out in detail the responsibilities of each member of a research team. It is a good idea to discuss who will contribute as an author, the order of authorship[2], who will be mentioned in a footnote, or who will contribute in such a minor way that no acknowledgment need be given. A brief space is provided for previous results. A section is provided for a brief description of the experimental paradigm and the analysis of data. Critical to the experiment is a list of materials needed, including subjects. Finally, an anticipated timetable and the responsibilities of researchers are given.

This form is an outline that may be expanded. It may also be stored in a computer memory, which will increase the flexibility of its use.

In actual practice, we keep a file of these research proposals printed on two sheets of different-colored paper (we use blue and green). When contemplating a project, we begin by filling out the blue form. If the project develops to the point that it appears that we will actually do the experiment, it graduates to the green form (we go green). With the form we also keep a running file of reprints, diagrams, list of equipment, and data collected. We have found this form very useful and hope you do too.[3]

WRITING A RESEARCH PAPER

The final stage of experimentation is the preparation of a research paper. After a researcher has formulated a problem (usually after reading the literature in the field), developed and conducted an experiment, and analyzed the data, then he or she faces the problem of how best to communicate the findings to others. Results of scientific experiments that are not reported to the scientific community are of little use. Therefore,

---

[2] If two or more authors contribute equally to a project, order of authorship is sometimes determined by chance, as by flipping a coin. Sometimes members of an ongoing program will alternate the order of names on papers. The matter should not be one of ego, but of attributing credit (or blame) for the research in direct proportion to the actual contribution made by each member of the team.

[3] In the event you evolve a better form or one that you find more useful, we hope you will send it to author RLS so that we may share it with readers of a subsequent edition of this book.

RESEARCH PROPOSAL

NAME OF RESEARCH _____ EXPERIMENTER(S) _____
                                                    _____
                                                    _____

                                   DATE _____
TOPIC _____
_____
_____

PROBLEM _____
_____
_____

SIGNIFICANCE _____
_____
_____

PREVIOUS RESEARCH
    THEORY _____
_____
_____

    DESIGN _____
_____
_____

EXPERIMENTAL PARADIGM _____
_____
_____
_____

    ANALYSIS OF DATA _____
_____
_____

MATERIALS _____
_____
_____
_____

TIMETABLE AND RESPONSIBILITIES
PREPARATION OF MATERIALS
             PREPARATION OF MATERIALS _____ BY _____
PHASE 1 │ LITERATURE SEARCH _____ BY _____
             PILOT (DRY RUN) _____ BY _____
PHASE 2 │ RUN EXPERIMENT _____ BY _____
             ANALYSIS OF DATA _____ BY _____
PHASE 3 │ WRITE PAPER _____ BY _____
             SUBMIT FOR PUBLICATION _____ BY _____

© R. Solso

the investigator is encouraged to prepare his or her report in such a style that it is consistent with the current editorial policies of contemporary journals in psychology. Nearly all psychological journals subscribe to the style guide of the APA's *Publication Manual* (1983), and the serious researcher should obtain a copy of this guide. The style of journal articles discussed in this section is consistent with this manual.

Good scientific writing is a skill that can be learned through practice. It is unlikely that the first time you attempt to write a scientific paper the results will be acceptable, and you are encouraged to rewrite and revise your material frequently. If it is any solace to you, even experienced writers must constantly rewrite and revise their material. We will present some guidelines which will be of some assistance to you in preparing a scientific paper, but we also emphasize the importance of practice.

WRITING STYLE

Scientific writing must communicate ideas clearly. In order to achieve clarity, the APA *Publications Manual* suggests:

> To achieve clarity, good writing must be precise in its words, free of ambiguity, orderly in its presentation of ideas, economical in expression, smooth in flow, and considerate of its readers. A successful writer invites readers to read, encourages them to continue, and makes their task agreeable by leading them from thought to thought in a manner that evolves from clear thinking and logical development. (1974, p. 25)

A component of good scientific writing is the selection of words that convey a precise meaning. Ambiguous words should be avoided. In casual conversation it may be acceptable to use such expressions as "for the most part," "very few," "I would estimate," "an intelligence test was used," or "animals were deprived." However, in scientific writing the author should use words and phrases that can be operationally defined. For example, "very few" leaves the reader mystified as to how many; or "an intelligence test was used" does not specify *which* intelligence test was used.

Ideas should be developed logically. This principle is particularly important in the introduction to the research and in the discussion section, but it is also important in the description of scientific method. To illustrate this, consider the review of literature, which is normally the first major section of a research paper. The purpose of this section is to introduce the reader to the previous research that has been done on the subject so that the question you are addressing is placed in a meaningful context. It is impossible and distracting to recount all related research, so the writer must select the most pertinent studies and present them in such a way that the reader can understand the progressive development of previous experiments and/or theories. (In Part Two of this book, you will find many excellent examples of the way other authors have logically developed reviews of literature.)

Careful attention should be given to grammar and paragraph development. Avoid awkward word sequences. Sentences should be technically correct, but additionally sentences should state clearly the intended meaning.

A common error is to incorporate too many thoughts in a single paragraph. A paragraph should have a controlling idea which is supported by every sentence in the paragraph and finally, the organization of paragraphs should be unified around a major principle. If the topic is "the auditory feedback in bats in a dark room" or "children's dreams and bedwetting" or the "influence of green and blue packaging on the sales of laundry detergent," all components of the paragraph should revolve around that theme.

As the previous discussion suggests, the author has a great deal of freedom in writing *style;* however, the author of a scientific article in psychology has much less freedom in the overall *structure* of his report. The structure that has evolved in American psychology follows a tightly prescribed model, which is presented below (and further discussed in Part Two):

Title, author(s), and affiliation
Abstract
Review of literature, problem, and hypothesis to be tested
Method
    Subjects
    Apparatus or Materials
    Procedure
Results
Discussion (and sometimes Summary)
References

Consider each of these topics:

1. **Title, Authors, and Affiliation.** The title should be a clean brief statement of the subject under inquiry. Authors frequently suggest a statement of "the effects of _____ on _____ ," which gives the reader a clue as to the independent and dependent variables. The title is followed by the author's name and affiliation.

2. **Abstract.** In the abstract the author should summarize the essence of the article. Although the abstract usually contains less than 150 words, the researcher should try to identify the problem, method, results, and an abbreviated interpretation of the findings. The abstract is designed to interest and inform the

reader as well as present a structure that contributes to understanding the article.

3. **Review of Literature, Problem, and Hypothesis.** Articles typically begin with a brief historical review of major issues and previous results in the area. The results of other researchers are presented in a logical sequence so that the current experiment fits into a series of experimental developments. The statement of the problem and the statement of the hypothesis are frequently contained in a single paragraph, if not in a single sentence. The statement of the hypothesis may follow an "if _____ then _____ " structure. Frequently the independent and dependent variables are identified in this section.

4. **Method.** The method section describes the experimental design in sufficient detail to allow replication of the experiment. There are usually three parts: (1) subjects, (2) apparatus or materials, and (3) procedure. All independent variables and other variables that may affect the results are identified and, if necessary, controlled. If a procedure has been previously described by an earlier researcher, then it is permissible simply to cite the original source.

5. **Results.** In this section all relevant data generated by the experiment and the analysis of these data are reported. In addition to their presentation in the text, the data may be presented in tabular or graphic form. Because of space limitations, an attempt is made to present as much pertinent material as possible in the most succinct form. All graphs and tables must have some explanation within the text. It is usually impractical to present the data for every subject. Therefore, in summarizing the results an experimenter should employ those statistical techniques that most clearly convey the findings.

6. **Discussion.** The purpose of this section is to interpret the results of the experiment: to point out any reservations about the results, to note similarities or differences between these results and the findings of other investigators, to suggest future research, and especially to note the implications of the results to theory and/or practice in the specific research area.

7. **References.** All material cited in the article should be noted in this section. The general format is: author(s) name(s), title, journal, publication year, volume, and pages. (See the APA *Publication Manual* or examples at end of reviewed articles in Part Two for specific examples.)

Example of a Manuscript

In this section you will find an actual article that has been slightly changed from the published version. This article contains the essential ingredients of good experimental design. It is presented in the form of a manuscript prepared for submission to a professional journal and should provide you with a model of experimental design as well as APA style. As you read the manuscript, pay close attention to the (1) title, authors, and affiliation, (2) abstract, (3) review of literature, problem, and hypothesis, (4) method (subjects, apparatus or materials, and procedure), (5) results, (6) discussion, and (7) references. We have labeled these sections for you, but on the submitted manuscript they would not be labeled (see Figure 9.2).

Several features of the paper by Loftus and Palmer (1974) are particularly noteworthy. First, consider the abstract. The authors have done three essential things: They have told us something of the methodology, they have told us the basic results, and they have interpreted the results within a theoretical framework.

The review of literature is preceded by a series of questions designed to stimulate the reader's interest in the topic. This is not a required section, but in this experiment it serves its function well. The authors then present several findings of other researchers, followed by the problem. Then they briefly describe the method of the present article, followed by an informal hypothesis.

In the method section, Loftus and Palmer give us explicit information about what was done to whom. The results are presented verbally, statistically, graphically, and in the form of a table. Of particular interest in this manuscript is the discussion. The authors briefly review the findings and then give two interpretations of what these data mean. Then they discuss their findings and interpretation within the context of previous findings and theories. Their discussion is to the point and does not go beyond their results. The final section contains the references, which should serve as a model for your writings.

In the Loftus and Palmer experiment many of the essential features of design and style are clearly portrayed. In writing your own experiments for either class work or submission to a professional journal, you may wish to refer to this manuscript as a model.

Submission of Manuscript

After a manuscript has been written, the author submits it to a professional journal for review and possible publication. Selecting the appropriate journal for your manuscript can be difficult, especially for the beginning researcher. In general, one would prefer to publish his or her work in well-accepted, well-known, respected journals rather than "schlock" periodicals. Of course, knowing the difference between good and bad publications is a controversial matter. Journals devoted to empirical studies and published by the American Psychological Association (*Journal of Experimental Psychology*, which has four separate editions, *Child De-*

(Text continues on page 146.)

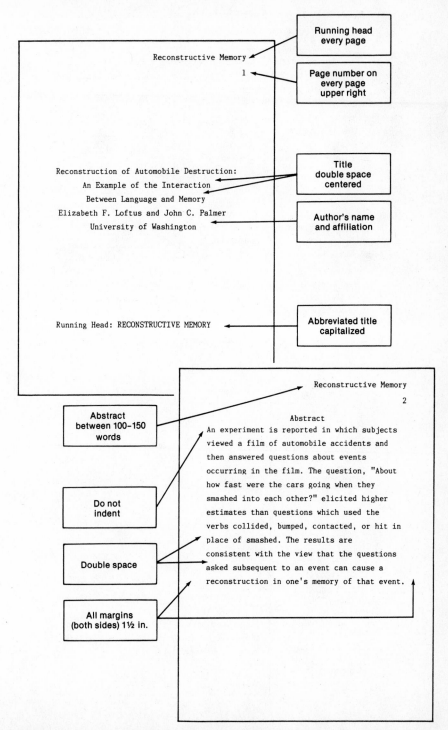

Figure 9.2   A model of experimental design and APA style (*continues on following 5 pages*).

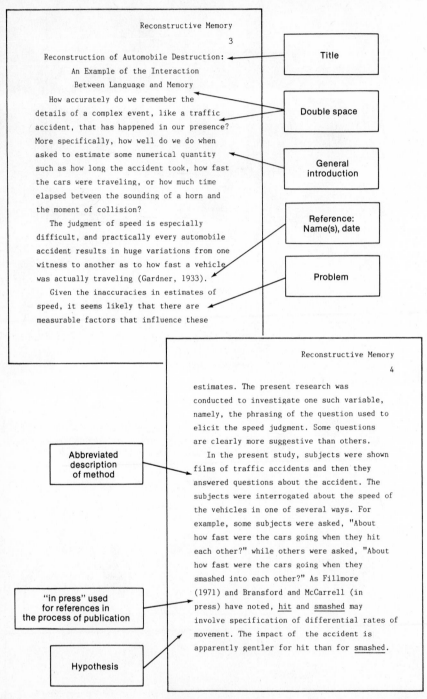

Reconstructive Memory

3

Reconstruction of Automobile Destruction:
An Example of the Interaction
Between Language and Memory

How accurately do we remember the
details of a complex event, like a traffic
accident, that has happened in our presence?
More specifically, how well do we do when
asked to estimate some numerical quantity
such as how long the accident took, how fast
the cars were traveling, or how much time
elapsed between the sounding of a horn and
the moment of collision?

The judgment of speed is especially
difficult, and practically every automobile
accident results in huge variations from one
witness to another as to how fast a vehicle
was actually traveling (Gardner, 1933).

Given the inaccuracies in estimates of
speed, it seems likely that there are
measurable factors that influence these

Title

Double space

General introduction

Reference: Name(s), date

Problem

Reconstructive Memory

4

estimates. The present research was
conducted to investigate one such variable,
namely, the phrasing of the question used to
elicit the speed judgment. Some questions
are clearly more suggestive than others.

In the present study, subjects were shown
films of traffic accidents and then they
answered questions about the accident. The
subjects were interrogated about the speed of
the vehicles in one of several ways. For
example, some subjects were asked, "About
how fast were the cars going when they hit
each other?" while others were asked, "About
how fast were the cars going when they
smashed into each other?" As Fillmore
(1971) and Bransford and McCarrell (in
press) have noted, hit and smashed may
involve specification of differential rates of
movement. The impact of the accident is
apparently gentler for hit than for smashed.

Abbreviated description of method

"in press" used for references in the process of publication

Hypothesis

Figure 9.2 (*continued*)

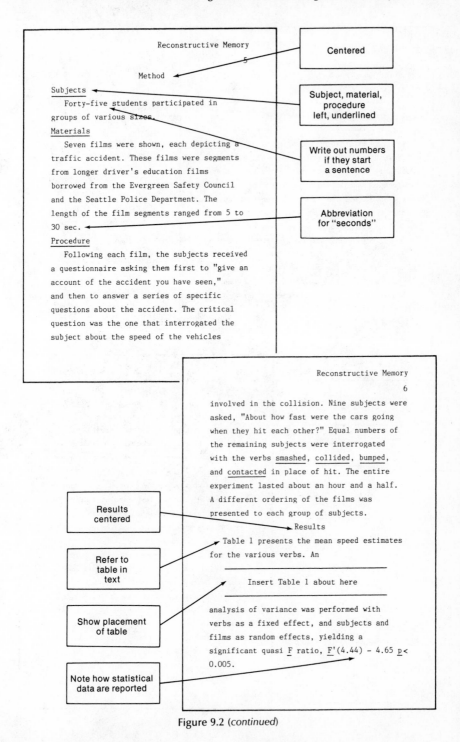

Reconstructive Memory

5

Method

Subjects

Forty-five students participated in groups of various sizes.

Materials

Seven films were shown, each depicting a traffic accident. These films were segments from longer driver's education films borrowed from the Evergreen Safety Council and the Seattle Police Department. The length of the film segments ranged from 5 to 30 sec.

Procedure

Following each film, the subjects received a questionnaire asking them first to "give an account of the accident you have seen," and then to answer a series of specific questions about the accident. The critical question was the one that interrogated the subject about the speed of the vehicles

Centered

Subject, material, procedure left, underlined

Write out numbers if they start a sentence

Abbreviation for "seconds"

Reconstructive Memory

6

involved in the collision. Nine subjects were asked, "About how fast were the cars going when they hit each other?" Equal numbers of the remaining subjects were interrogated with the verbs smashed, collided, bumped, and contacted in place of hit. The entire experiment lasted about an hour and a half. A different ordering of the films was presented to each group of subjects.

Results

Table 1 presents the mean speed estimates for the various verbs. An

_____

Insert Table 1 about here

_____

analysis of variance was performed with verbs as a fixed effect, and subjects and films as random effects, yielding a significant quasi F ratio, $F'(4.44) - 4.65$ $p< 0.005$.

Results centered

Refer to table in text

Show placement of table

Note how statistical data are reported

Figure 9.2 (*continued*)

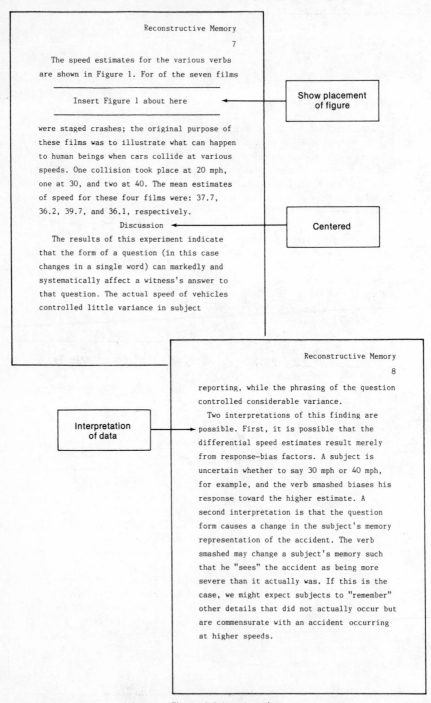

Reconstructive Memory

7

The speed estimates for the various verbs
are shown in Figure 1. For of the seven films

Insert Figure 1 about here

Show placement
of figure

were staged crashes; the original purpose of
these films was to illustrate what can happen
to human beings when cars collide at various
speeds. One collision took place at 20 mph,
one at 30, and two at 40. The mean estimates
of speed for these four films were: 37.7,
36.2, 39.7, and 36.1, respectively.

Discussion

Centered

The results of this experiment indicate
that the form of a question (in this case
changes in a single word) can markedly and
systematically affect a witness's answer to
that question. The actual speed of vehicles
controlled little variance in subject

Reconstructive Memory

8

reporting, while the phrasing of the question
controlled considerable variance.

Two interpretations of this finding are

Interpretation
of data

possible. First, it is possible that the
differential speed estimates result merely
from response-bias factors. A subject is
uncertain whether to say 30 mph or 40 mph,
for example, and the verb smashed biases his
response toward the higher estimate. A
second interpretation is that the question
form causes a change in the subject's memory
representation of the accident. The verb
smashed may change a subject's memory such
that he "sees" the accident as being more
severe than it actually was. If this is the
case, we might expect subjects to "remember"
other details that did not actually occur but
are commensurate with an accident occurring
at higher speeds.

Figure 9.2 (*continued*)

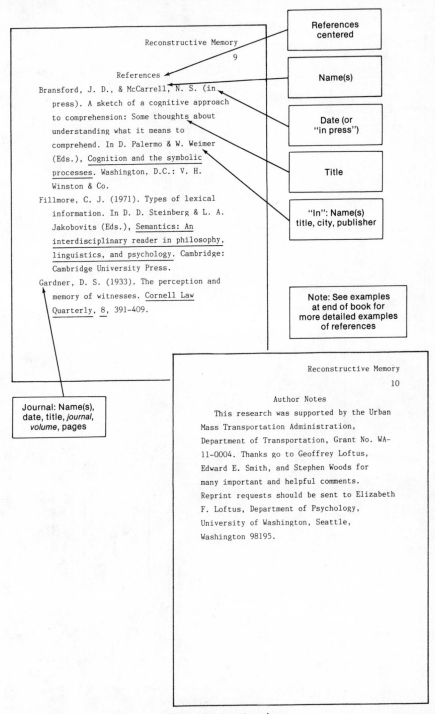

Figure 9.2 *(continued)*

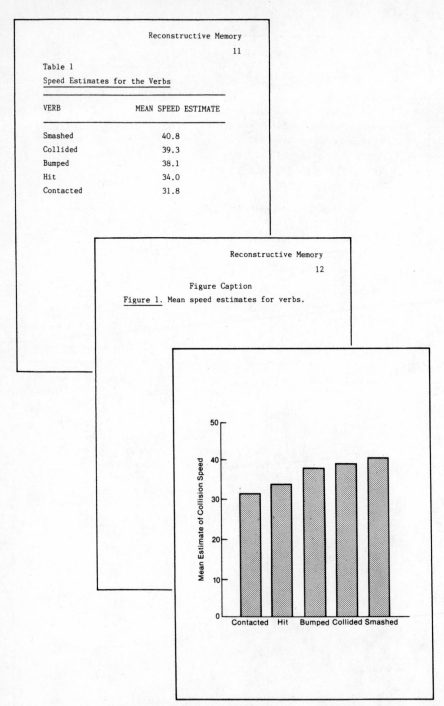

Reconstructive Memory

11

Table 1

Speed Estimates for the Verbs

VERB	MEAN SPEED ESTIMATE
Smashed	40.8
Collided	39.3
Bumped	38.1
Hit	34.0
Contacted	31.8

Reconstructive Memory

12

Figure Caption

Figure 1. Mean speed estimates for verbs.

Figure 9.2 (*continued*)

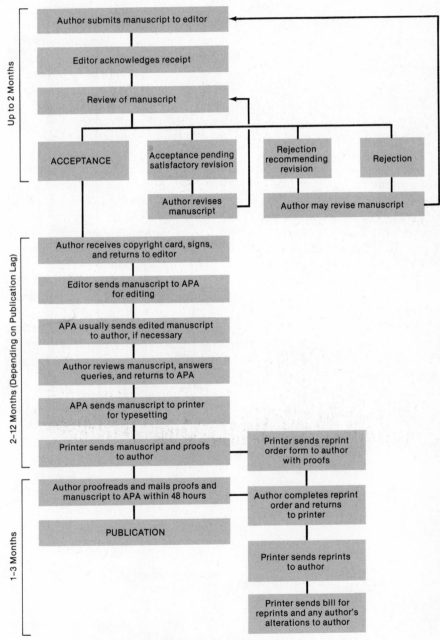

Figure 9.3   The APA publication process.

velopment, Journal of Personality & Social Psychology, Developmental Psychology, Journal of Comparative Psychology, and others) are unusually well regarded and widely read. However, numerous periodicals publish high-quality research articles. These include the publications of the Psychonomic Society (e.g., *Memory & Cognition, Perception & Psychophysics, Bulletin of the Psychonomic Society*); the British Psychological Society (e.g., *British Journal of Psychology, Quarterly Journal of Experimental Psychology*) and many other journals (e.g., *American Journal of Psychology, Cognitive Psychology, Neuropsychologia, Brain and Cognition, Cortex, Journal of Speech & Hearing Research, Science, Journal of Experimental Analysis of Behavior, Journal of Memory & Language* and so on).

This list is not inclusive. Many other high-level publications that deal with topics in experimental psychology are printed each year. It is the responsibility of researchers to find the most appropriate journals for their manuscripts. One clue to ascertaining the best journal for your manuscript is to find out where other researchers working on similar problems are publishing their work. You should be able to determine these sources through your readings and review of the literature. Many times you will be able to detect a common theme (or several themes) in a journal.

The review process may take up to six months, although many editors attempt to reach a decision on the manuscript within two months. An outline of the editorial process used by the APA is shown in Figure 9.3.

Part Two of this book deals with actual experiments carefully selected from the psychological literature, which demonstrate a variety of topics and experimental techniques.

PART TWO

ANALYSIS OF EXPERIMENTS

In Part Two several research articles are reviewed. These articles were selected to represent a variety of topics and design procedures. Having carried out an experiment, the psychologist's next obligation is to report its results to the scientific community. In practice, this means preparing a research paper for a journal read by other psychologists. The ability to write such papers and to read them with understanding is therefore one of the most important demands upon the professional psychologist.

Because journal space is limited and because there are so many other demands on the time of the scientist, a premium is placed on clarity and conciseness in the presentation of research results. At the same time, because the articles are aimed at other scientists, particularly those working in the same problem area, a knowledge of that area's research history and its methodologies is taken for granted, which makes journal articles difficult going, not only for students but sometimes for psychologists with different research specialties.

The articles are presented in a form slightly edited from the way they appeared in the psychological journals. Most articles are accompanied by an analysis that attempts to explain exactly why the researcher did what he or she did. By repeated exposure to, and careful analysis of, such papers, the student can come to feel comfortable in reading research reports and may develop an understanding of principles of experimental design in psychology. Three articles are not accompanied by an analysis; they are to be analyzed by the student.

Note that the article is always shaded and that the analysis corresponding to the appropriate section of the article appears below it. This format facilitates a continuous reading of the article and simultaneous analysis.

Many of the experiments in this section were selected because they provided us with an example of a *Special Issue*. These issues are set apart from the main text in our analysis of the article. We selected the special issues listed below and illustrated the way each one was treated in the corresponding article.

Special Issues	Articles
1. Control Problems	Cola Tasting—Chapter 10
2. Field-based Experiments ⎫	Distance and Rank—Chapter 12
3. Unobtrusive Measures ⎭	
4. Small *n* Experiments	Creative Porpoise—Chapter 14
5. Research with Animals	Maternal Behavior—Chapter 15
6. Attribution Theory	Humor—Chapter 16
7. Practical and Theoretical Problems	Russian Vocabulary—Chapter 18
8. Single-Subject Design in Clinical Studies	Therapy for Anger—Chapter 19
9. Multiple Experiments	Birdsong Learning—Chapter 22
10. Subject Variables	Note Taking—Chapter 23
11. Clinical Research	Weight Loss—Chapter 24

There is no "right" way to read an article, but listed below are some suggestions as to how you might try to understand research reports.

First, read the title and try to establish the general category within psychology that the study is investigating. For example, "Two-phase model for human classical conditioning" (Prokasy & Harsanyi, *Journal of Experimental Psychology*, 1968, *78*, 359–368). You probably have had some experience with "classical conditioning" and know something of the famous Pavlovian studies with dogs. These authors have apparently studied learning by using a conditioning method and have developed some model to describe their results. In addition to simply reading a title, it is suggested that you try to raise concrete questions about the problem and procedures. In the case of the example, you might ask: How did the authors condition a human—with a bell and meat powder? What are two possible phases to classical conditioning?

Most journal articles contain an abstract at the beginning of the paper. The major findings and method used are briefly described. A careful reading of this, coupled with an inquiring mind, will facilitate your understanding of the paper.

A quick scan of the entire paper, with greater emphasis on the review of literature, hypothesis, and discussion than on specific details of the methods and results sections, is suggested. Scanning the paper and the preceding two steps will allow you to get a general impression of the author's intent. You may want to return to the abstract to reestablish a point of view, but the critical aspect of scientific reading is a probing inquiry. As you review and read, *question!*

Finally, read the article in its entirety, sorting out data from discussion. During the reading, you should examine how the author controlled for variables and how he or she specifically treated the data.

In reading an article, you might ask yourself some of these questions, plus many more which would be determined, in part, by the responses to questions.

1. What is this research all about?
2. What is the general problem?
3. What are the results of others?
4. What is the hypothesis?
5. What type of materials did the author use?
6. How does he or she operationally define his or her variables?
7. What controls does he or she use?
8. How does he or she analyze his or her data?
9. How does he or she interpret his or her data?
10. Who does he or she cite as relevant investigators in the area?
11. Can I improve on this study?
12. What additional work needs to be done?
13. Can the data be interpreted in another way?
14. What have I learned from this article?
15. What new research can now be developed?

Some articles are so complex or use so much jargon that you may have difficulty in understanding them. This experience is not uncommon among new students (as well as some old students). Repeat the process of reading and questioning suggested above. Sometimes students find it helpful to review an article for a friend and allow him or her to ask questions.

For your convenience the table on the following pages summarizes some of the relevant characteristics of the articles in Part Two. We suggest you look at the table prior to reading each article in order to give yourself an overview of the article.

OVERVIEW OF CASE EXAMPLES

ABBRE-VIATED TITLE	PSYCHO-LOGICAL SUBJECT MATTER	SOURCE OF HYPOTHESIS	SETTING	TREATMENT OF SUBJECTS	STATIS-TICAL/ DESCRIPTIVE ANALYSIS	SUBJECTS	INDEPEN-DENT VARIABLE	DEPENDENT VARIABLE
Cola Tasting Chapter 10	Perception	Applied research	Laboratory	Paired comparisons within subjects	χ^2, table, percentages	Humans (college students)	Different colas	Identification of cola
Picture Memory Chapter 11	Cognitive memory	Hypothesis testing/theory	Laboratory	Random group (2 groups)	Table, percentages, t, z	Humans (college students)	Thematic clue	Memory for pictures
Distance and Rank Chapter 12	Personal space	Hypothesis	Field research	Designated groups (rank)	F, linear trend rank order correlation tables, χ^2	Humans (military personnel)	Rank	Distance when initiating a conversation
Seven Dwarfs Chapter 13 (to be analyzed by the student)								
Creative Porpoise Chapter 14	Creative learning	Hypothesis testing/theory (demonstration)	Laboratory/ seminatural setting	Single subject	Graphic, correlation	Porpoises	Reinforce-ment	Novel behavior
Maternal Behavior Chapter 15	Comparative motivation	Hypothesis testing/theory	Laboratory	Random group (4 groups)	F, Mann-Whitney, Duncan's Range Test, graphic	Rats	Injection of blood	Maternal behavior
Humor Chapter 16	Social	Hypothesis testing	Laboratory	Random group	F, table	Humans (college students)	Type of joke	Attitude attribution

Alcohol and Perception Chapter 17 (to be analyzed by the student)								
Russian Vocabulary Chapter 18	Learning	Theory/applied	Laboratory	Random group (contrast with a within-subject design)	Percentages table, F, figure, scatter plot	Humans (college students)	Keyword	Russian vocabulary memory
Therapy for Anger Chapter 19	Clinical	Applied clinical	Clinical	Single subject	Figures, correlation	Hospitalized Patients	Therapy	Anger control/reduction
Office Environments Chapter 20	Environmental psychology	Applied problem	Seminatural setting	Comparison between groups	Percentages, figures	Employees	Industrial settings	Types of activities
Prosocial Behavior Chapter 21 (to be analyzed by the student)								
Birdsong Learning Chapter 22	Intersensory processing	Hypothesis	Laboratory/ 2 experiments	Assigned groups	Table, F, t, percentages	Humans (college students)	Presentation means	Name birdsongs
Note Taking Chapter 23	Educational psychology	Hypothesis testing/applied	Natural setting	Random group (5 groups)	F, table, correlation, t	Humans (college students)	Note taking and note reviewing	Recall of lecture
Weight Loss Chapter 24	Clinical	Hypothesis testing/applied clinical	Natural setting/ minor laboratory intervention	Random group (5 groups)	F, Newman-Keuls, percentages, averages	Human adults	Type of "therapy"	Weight loss
Parents' Views Chapter 25 (to be analyzed by the student)								

10

Cola Tasting

INTRODUCTION

In a recent test of cola beverages conducted for a leading business newspaper 70 percent of the 100 tasters were mistaken as to which of four cola beverages they were tasting. Yet, if opinions about colas are asked, many people swear allegiance to one or another soft drink.

The first experiment in this section on analysis of actual experiments deals with the identification of cola beverages and presents some interesting design problems. As you read this experiment ask yourself what variables, both experimental and cultural, may influence a person's preference for one cola drink over another.

The purpose of including an experiment on the identification of soft drinks is to introduce you to some of the issues involved in controlling the potential variables in a psychological study and to heighten your awareness of how complex are the relationships between a "real world" stimulus (cola drinks) and the psychological evaluation of the stimulus.

SPECIAL ISSUES

Control Problems

It was previously suggested (Chapter 3) that there are two types of control. The first is when the experimenter makes something occur and the second is when he or she prevents something from occurring (extraneous variables). We have discussed the first type; let us now look at the extraneous variables that were controlled for:

Relations between pairs. One cola may be easy to identify when presented with a second cola but difficult to distinguish from a third. For example, Coke may be easy to identify correctly when it is paired with Royal Crown but difficult to identify when it is paired with Pepsi. To avoid problems such as this, each cola was paired with each other cola an equal number of times. Another problem is that within each pair presentation, the first cola may be easier to identify than the second cola, or vice versa. To eliminate this problem each cola could have been presented first and second in the pair an equal number of times. While Thumin did not systematically control for this, it could easily have been incorporated into the design to strengthen the procedure.

Order effects. It was previously pointed out (Chapter 5) that when subjects make a series of judgments, order effects can appear. The subject may become more sensitive with practice or may become fatigued and his or her judgments become poorer. In taste experiments it is very probable that the subject's judgments may become poorer since not all of the cola may be washed out of his or her mouth on each trial, and the residue may confuse later judgments. To control for this Thumin presented the stimulus pairs in random order. This procedure is described in Chapter 5 and is an effective method for controlling order effects.

Stimulus accumulation. Related to the preceding point, the subject washed out his or her mouth with water after each pair was judged. This control procedure attempts to eliminate the confusion of taste from one pair to the next. It is reasonable to assume that the subject could not have very sensitive judgment for a given pair if the taste of the previous pair were still in his or her mouth. Further, clean cups were used for each presentation to eliminate the possibility of an accumulation of cola in the cups.

Visual cues. Thumin was testing to see whether or not the subjects could identify the colas by taste. To do this he had to eliminate any other cues that might help the subject identify the cola. One such set of cues are visual cues; for example, Coke may look different from Pepsi. To eliminate the visual cues, the experimenters will blindfold subjects in experiments of this type to eliminate any possibility of visual cues.

Temperature. Most people drink colas that are cooled to a temperature of 5°C (46° F). Their experience with tasting cola is limited to this temperature; that is, the taste of Coke is really the taste of Coke at 5°C. To allow for this, Thumin presented all colas at a constant temperature of 5°C.

Guessing. In many judgment experiments, a problem arises with respect to the subjects' guessing at the identification. Some subjects will guess when they do not know the correct identification, and some subjects will simply say that they do not know. To avoid these problems Thumin had all subjects guess at the correct identification when they did not know. He is quite explicit about this in his instructions when he points out to the subjects that even if they are not sure of the brand they should tell him what brand they *think* it is.

Note also that Thumin has (in a sense) corrected for guessing in his statistical analysis (see footnote to Table 10.1). There are three choices: one choice is correct and two are incorrect. Thus the probability of being correct is one out of three. The statistic is testing to see if the subjects' identifications were more correct than would be expected by guessing alone (chance).

Subject variables. Chapter 5 presented a lengthy discussion of the problem of subject variables in research. Much of this discussion centered around the problem of having people in one experimental treatment who are different as to some subject characteristics from people in another treatment. Thumin has avoided this problem by having all subjects participate in all treatments—a *within-subject design*. It was also pointed out above that Thumin has controlled for some of the problems of the within-subject design by controlling for order effects. However, another subject variable problem could appear in this experiment. Suppose that only those subjects who are heavy cola drinkers are those who can distinguish between the different brands. Let us further suppose that the sample of subjects Thumin picked had a small proportion of heavy cola drinkers. Because the sample contains only a few heavy drinkers (who can make correct identifications), then the majority of light drinkers might make it appear as if the identification of cola beverages is little better than chance. Thumin protected himself from this subject variable problem by finding out how much cola each subject drank per week. It is apparent that this subject variable did not influence the data (see Table 10.2); however, Thumin took into account this possibility and did test to see what effect this subject variable had on the dependent variable.

The above discussion of design and control procedures should make it evident to the student that careful planning is needed before an experiment is executed. It was pointed out in Chapter 3 that control problems vary with each experimental area. The Thumin experiment illustrates some of the considerations that must be taken into account in experimenting with the identification of the taste of substances.

Identification of Cola Beverages

FREDERICK J. THUMIN
WASHINGTON UNIVERSITY

An attempt was made to overcome certain methodological inadequacies of earlier studies in determining whether cola beverages can be identified on the basis of taste. Some 79 Ss completed questionnaires on their cola drinking habits and brand preferences, then were tested individually on samples of cola beverages presented under methods of paired comparisons. Significant chi-square values were obtained for Coca-Cola and Pepsi Cola, due to the large number of correct identification for these brands. Correct identification of Royal Crown, however, did not differ from chance expectancy. No significant relationship was found between ability to identify cola beverages and degree of cola consumption; nor were Ss any better at identifying their "regular" brand than they were other brands.

Earlier studies attempting to determine whether cola beverages can be identified on the basis of taste[1] have, in the main, obtained negative results (Bowles & Pronko, 1948; Pronko & Bowles, 1948; Pronko & Bowles, 1949; Pronko & Herman, 1950; Prothro, 1953). These results may, in part, be attributed to certain methodological difficulties. For example, in the majority of these studies, **the subjects were not informed as to what brands they**

a **were attempting to identify. This lack of restriction encouraged guessing behavior, which resulted in the naming of irrelevant beverages (e.g., Dr. Pepper),** as well as relatively frequent mentions of the more heavily advertised brands such as Coca-Cola.

[1]In this report, the word "taste" is used in the broad sense—that is, to include gustation, olfaction, and possible tactual qualities as well.

SOURCE: Reprinted by permission from *Journal of Applied Psychology*, 1962, 46 (5), 358–360. Published by the American Psychological Association.

ANALYSIS

REVIEW OF LITERATURE, STATEMENT OF PROBLEM, AND HYPOTHESIS

The first paragraph attempts briefly to describe the previous research on cola identification and to note the scope of the present research. Although it is not common to define terms in research papers, the author uses an acceptable method to define "taste" in a footnote.

Three methodological factors (a, b, and c) were not well controlled in the previous literature, which may have an effect on cola identification. They are (a) not informing the subjects of the colas to be identified, (b) control over the subjects' experience with cola beverages, and (c) the

b
Moreover, the subjects were expected to identify the various colas on the basis of past experience, yet apparently **no attempt was made to determine whether the subjects had ever tasted these beverages, or to relate identification to degree of cola consumption.**

c
Each of these previous studies used essentially the same method of stimulus presentation; namely, **all beverages were presented simultaneously to the subject,** and only one such presentation was made. This technique, while satisfactory, would appear to be somewhat less sensitive than the method of paired comparisons, which requires the subject to identify each brand a number of times under various experimental conditions.

d
Thus, the purpose of the present study was to determine whether methodological inadequacies in the earlier studies may have contributed to the subjects' relative inability to identify brands. The primary modifications in experimental design were as follows: an indication of cola consumption habits was obtained, subjects were told in advance what beverages they were attempting to identify, and the method of paired comparisons was used for presentation of stimuli.

fact that the subjects received all the colas simultaneously. Following each of these possible contaminating factors, the author tentatively identifies the possible biasing (or confounding) result of these uncontrolled factors.

In (d) a succinct statement of the methodological differences between this and previous studies is made.

The method of paired comparison was used, which is a standard technique in which a subject is given two stimuli and asked to judge them. The judgment in the present study was a qualitative one; subjects were asked to identify specific colas by taste. (It should be noted that this method of comparison also permits quantitative judgments. Under these circumstances, the subjects would be asked to identify greater or lesser amounts of a certain quality.) A related procedure and one mentioned by Thumin in (c) as the predominant procedure used in cola tasting, is *multiple comparison*. In this method, subjects are asked to make judgments on a variety of stimuli.

This study does not clearly state a hypothesis to be tested, yet the reader can provide his or her own statement of a hypothesis with the material presented. How would you state the hypothesis?

METHOD

The method section of this report is fairly brief, but it contains the necessary information (including the exact instructions given to the subjects)

METHOD

Seventy-nine subjects were employed, all of whom were either college students or college graduates between the ages of 18 and 37 years. The subjects were first asked to fill out a questionnaire on their cola consumption habits and brand preferences. The cola beverages were presented to the subjects individually in an experimental room which was kept dimly lighted to eliminate possible visual cues. Instructions were as follows:

> I would like to have you taste and identify some cola drinks. I will place two cups at a time in front of you—one on your left and one on your right. Taste these two colas in any order you wish; then tell me what brand you think each one is. Be careful not to change the position of the cups while you are tasting them; that is, keep the left cup on the left, and the right cup on the right. Each time you finish with one pair of cups, rinse your mouth well by taking a few swallows of water from the water cup. When you have done this, I will give you the next pair.

> There are three colas involved in this study—Coca-Cola, Pepsi Cola, and Royal Crown. Even if you are not sure of the brand in some cases, I still want you to tell me what brand you think it is. The two members of a pair are always different brands; that is, a brand is never compared with itself. Are there any questions?

> Using the method of paired comparisons, six pairs of beverages were presented to the subject, one pair at a time. The subjects were exposed to each brand four times for a total of 12 judgments. The order of presentation of stimulus pairs was randomly determined. Stimulus cups contained 2 ounces of the beverages at an approximate temperature of 5° centigrade.

to allow the study to be replicated (e). However, there are some features of the design that the author does not make explicit, which we shall discuss here.

Type of Design

This design is quite similar to the design used for investigating the relationship between the frequency of a tone and the subject's sensitivity to that tone (Chapter 2) and is typical of the type of design found in these areas. In the present experiment the independent variable is three types of colas, whereas in the audition experiment the independent variable was tones of different frequencies. In the audition experiment the dependent variable was the absolute threshold of the tone, that is, some hypothetical point above which the subject can hear the tone and below which he or she cannot hear the tone. In the present experiment the dependent variable is the correctness of the identification of the cola; can he or she correctly identify it or not? In both experiments there were multiple presentations of the same stimulus; in the audition experiment there were six attempts at threshold determination for each tone, three

RESULTS

f

The chi-square was used to determine whether ability to identify brands differed significantly from chance expectancy. As Table 10.1 shows, the chi-square values for both Coca-Cola and Pepsi Cola were significant at the

g

0.01 level of confidence, while that for Royal Crown was not significant. Inspection of the data indicates that the significant divergencies obtained with Coca-Cola and Pepsi Cola are due to the large number of correct identifications of these brands; for example, more than twice the expected number of subjects were able to identify these brands correctly at least three times out of four.

---

using the descending method and three using the ascending method. In the cola experiment the subjects were exposed to each brand four times. By using several presentations of the same stimulus, the experimenter may get a more stable or reliable measure of each subject's judgments. The basic difference in the two experiments is that in the audition experiment the experimenter only presented one tone at a time. This was an appropriate method for that problem, because the experimenter was asking, "When can you hear the tone?" In the Thumin experiment, the problem is identifying the cola; "Which cola is it?" To get an answer to this question, Thumin argues that presenting two stimuli at the same time (paired comparison method) is a more sensitive method than the methods previously used.

RESULTS

The basic findings and summary of the statistical analysis are the principle part of the results section.

The χ^2 (chi-square) statistical procedure was used to treat the data (**f**). This is a relatively simple procedure in which the obtained results are compared with the results one would expect by chance alone. *Chance* is defined as the variation in the results that are due to uncontrolled factors such as guessing, experimental error, failure to perfectly mask stimuli, failure to achieve a perfect matching of subjects or randomization in experiments using different groups of subjects, and so on. In this experiment subjects who had no knowledge of the cola but simply guessed would be correct some of the time (i.e., correct by chance). The logic is that if the obtained results vary greatly from the results expected by chance, then probably some experimental conditions are responsible for this diversity.

The level of confidence or "level of significance" (**g**) is a reflection of the probability that the results would be obtained by chance alone. In the present study, the author establishes a level of confidence, or $p = 0.01$; that is, the probability that such results would be obtained by chance

Table 10.1   CHI-SQUARE FOR OBSERVED AND EXPECTED
FREQUENCIES OF BRAND IDENTIFICATION

BRAND OF COLA	OBSERVED AND EXPECTED FREQUENCIES	NUMBER OF CORRECT IDENTIFICATIONS				χ^2
		0	1	2	3 OR 4	
Coca-Cola	(f_o)	13	23	24	19	14.57**
Pepsi Cola	(f_o)	12	20	26	21	22.14**
Royal Crown	(f_o)	18	28	19	14	4.60*
All brands	$(f_c)^a$	15.6	31.2	23.4	8.8	

Note. For each comparison, $df = 3$.
[a]Expected values were obtained from the expression $N(\frac{1}{3}p + \frac{2}{3}q)^4$. Each sample had one chance in three of being identified correctly, and each brand was presented to the subject four times.
*$p < .05$.
**$p > .01$.

The results presented in Table 10.2 indicate that ability to identify cola beverages correctly was unrelated to degree of consumption; that is, correct identifications were essentially the same for heavy, medium, and light cola drinkers. Further analysis of the data showed that ability to identify a given brand was also unrelated to whether that brand was considered by the subject to be his or her "regular" brand.

By telling the subjects in advance what brands they were attempting to identify, irrelevant brand naming was eliminated as well as excessive naming of heavily advertised brands. Specifically, Coca-Cola was mentioned 317 times, Pepsi Cola 321 times, and Royal Crown 310 times.

DISCUSSION

h

The present study clearly demonstrated that certain brands of cola can be identified on the basis of taste. The significant chi-square values obtained with Coca-Cola and Pepsi Cola were due to the large number of correct identifications for these brands. **The subjects' inability to identify Royal**

i

Crown Cola can probably be attributed to a lack of _recent_ experience with this brand. Some 58 percent of the subjects said they had not had a Royal Crown for at least 6 months prior to the experiment.

No relationship was found between ability to identify cola beverages and degree of cola consumption (i.e., number of colas consumed in an average

---

alone is 1 in 100. Frequently, the level of confidence is set at 0.05 (or 5 in 100), and in some research it may be useful to demand even lower levels.

The author makes a clear summary statement of the identifiability of colas and also provides the reader with two tables which contain the data obtained. It is noted that two of the obtained χ^2 values exceed the 0.01 level.

Table 10.2   CHI-SQUARE FOR BRAND IDENTIFICATION
AS RELATED TO CONSUMPTION

NUMBER OF COLAS CONSUMED PER WEEK	NUMBER OF CORRECT IDENTIFICATIONS		
	0–3	4–6	7–12
Heavy	10	14	3
(7 or more)	(8.5)	(12.0)	(6.5)
Medium	7	9	10
(3–6)	(8.2)	(11.5)	(6.3)
Light	8	12	6
(0–2)	(8.2)	(11.5)	(6.3)

Note: Expected values appear in parentheses.
$\chi^2 = 5.44$; $df = 4$; $p < .05$.

week). Moreover, the subjects were no better at identifying their regular brand than they were at identifying other brands. Thus, it would appear that the subjects needed a certain minimal amount of recent experience with a brand in order to identify it, but beyond this minimal amount, additional experience (i.e., heavier consumption) did not help.

j   Within the framework of this study, **the method of paired comparisons proved to be sufficiently sensitive to detect small but significant abilities to identify cola beverages.** There appeared to be no problem with the development of sensory adaptation as successive pairs of stimuli were presented. Analysis of the data revealed that, as trials progressed, the subjects showed small (though nonsignificant) increases in ability to identify brands.

REFERENCES

Bowles, J. W., Jr., & Pronko, N. H. (1948). Identification of cola beverages: II. A further study. *Journal of Applied Psychology, 32,* 559–564.

Pronko, N. H., & Bowles, J. W., Jr. (1948). Identification of cola beverages: I. First study. *Journal of Applied Psychology, 32,* 304–312.

Pronko, N. H., & Bowles, J. W., Jr. (1949). Identification of cola beverages: III. A final study. *Journal of Applied Psychology, 33,* 605–608.

Pronko, N. H., & Herman, D. T. (1950). Identification of cola beverages: IV. Postscript. *Journal of Applied Psychology, 34,* 68–69.

Prothro, E. T. (1953). Identification of cola beverages overseas. *Journal of Applied Psychology, 37,* 494–495.

The author wishes to express his appreciation to A. Barclay who served as critical reader for earlier drafts of this paper.

DISCUSSION

In his discussion section the author reviews the major findings (**h**) and offers a plausible explanation for the lack of statistically significant findings regarding the Royal Crown condition (**i**).

Finally, the sensitivity of using a paired-comparison technique is mentioned (**j**).

QUESTIONS

1. In the design of this experiment (1) each cola was paired with each other cola an equal number of times and (2) the presentation of stimulus pairs was randomly determined. From the theory presented concerning within-subject designs (Chapter 5), lay out exactly how one sequence of 12 pairs might have been presented to a subject.

2. Why was a dimly lighted room used (e)? What effects would a red illuminated room have? A green illuminated room? A nonilluminated room? Would these conditions be worthy of research?

3. Why did the subjects rinse their mouths? Should they eat something neutral (e.g., a cracker) between tests? Do you think this is a critical variable?

4. What interpretation would you give to the fact that some subjects could not correctly identify the beverages? What factors may determine their preference for one cola over another? Would they be a more suggestible group?

5. No sex differences were noted in the selection of subjects. Why? Subjects were not asked to identify their "favorite" cola. What influence might this have had on the results?

6. Design an experiment to test whether or not subjects can correctly identify whole, skim, and powdered milk.

7. Design an experiment to test whether whole, skim, or powdered milk tastes better.

CHAPTER
11

Picture Memory

INTRODUCTION

Of all our perceptual attributes, identification of visual scenes must surely occupy a central position in our cognitive domain. In this clever experiment, Bower, Karlin, and Dueck demonstrate that contextual cues that aid in fitting the picture into a novel scheme greatly facilitate the way we learn and place in memory simple line pictures.

The technique of enhancing memory by "priming" a subject with mnemonic devices has been often used in the study of linguistic material. Consider the following syntactically correct but hard to understand and memorize sentence: The notes were sour because the seams split. Nonsense, you might think, but suppose you put this sentence into the context of "bagpipe." All of a sudden you easily understand why the notes were sour, and your ability to memorize that sentence is significantly enhanced. Do we memorize pictures, especially ambiguous pictures, in a similar way? Two well-designed and interesting experiments by Bower et al. indicate that we do.

Comprehension and Memory for Pictures

GORDON H. BOWER, MARTIN B. KARLIN,
and ALVIN DUECK
STANFORD UNIVERSITY

The thesis advanced is that people remember nonsensical pictures much better if they comprehend what they are about. Two experiments supported this thesis. In the first, nonsensical "droodles" were studied by subjects with or without an accompanying verbal interpretation of the pictures. Free recall was much better for subjects receiving the interpretation during study. Also, a later recognition test showed that subjects receiving the interpretation rate as more similar to the original picture, a distractor, which was close to the prototype of the interpreted category. In Experiment II, subjects studied pairs of nonsensical pictures, with or without a linking interpretation provided. Subjects who heard a phrase identifying and interrelating the pictures of a pair showed greater associative recall and matching than subjects who received no interpretation. The results suggest that memory is aided whenever contextual clues arouse appropriate schemata into which the material to be learned can be fitted.

The following experiments address the question of how people remember pictures. We may begin with the observation that pictures (drawings, diagrams, photographs) comprise a two-dimensional notational system which, like language, has both a "surface structure" (the medium) and a meaningful "deep structure" (the message). Like language, pictures have a terminal vocabulary (of strokes, shadings, etc.), sets of combination rules, often a referential field, and conventional rules for interpreting what a picture is about (see Gombrich, 1960; Goodman, 1968). Pictures, especially "realistic" ones, denote objects or scenes in a manner that parallels the symbolic way that words and sentences do. And just as language appears to be acquired as a perceptual motor skill, so also does it appear that children learn the conventional rules for interpreting the notational symbolism of pictures. These rules guide our construction of what a picture is about—what conceptual-

SOURCE: Reprinted by permission from *Memory and Cognition*, 1975, 3, 216–220.

ANALYSIS

REVIEW OF LITERATURE

In the present experiment, pictures were conceptualized as a "two-dimensional notational system" that shares many features of language structure and usage. As such, pictures, like language, are a communicative

izations it expresses or what objects it symbolizes. That we learn to interpret drawings is illustrated clearly by the difficulty novices have acquiring the symbolic system of their profession, such as ballet Labanotation, musical scoring, molecular structure, and the like.

a

We are interested in memory for pictures. **The hypothesis to be tested is that a major determinant of how well a person can remember a picture is whether or not he "understands" it at the time he studies it.** If he comprehends the picture—achieves a compact interpretation of it—then he should remember it much better than if he fails to comprehend it.

b

This hypothesis was suggested by the work of Bransford and Johnson (1972), Bransford and McCarrell (1974), and Doll and Lapinski (1974) on memory for *linguistic* material. They showed convincingly that a person's ability to recall a sentence depends on whether the sentence causes him to call to mind an appropriate referential situation. For example, consider causal sentences such as: (1) The notes were sour because the seams split; (2) The voyage wasn't delayed because the bottle shattered; (3) The haystack was important because the cloth ripped. Though simple in syntax and word meanings, such sentences prove difficult to understand and recall. The mind boggles because a causal connection is asserted to hold between two apparently unrelated events; the subject cannot call to mind an appropriate schemata (known scenario) into which the events can be substituted and thus related causally. But all difficulties dissolve if the subject is provided with a clue as to an appropriate causal schemata: The clue is a simple "thematic prompt" (for the three sentences above, *bagpipe, ship christening, parachutist*). The clue calls to mind a known scenario (see the "frames" theory of Minsky, 1974) into which the events mentioned in the sentence can then be fitted. The sentences then become comprehensible and memorable.

c

We wish to advance here a parallel argument for the role of comprehension in memory for pictures. **Our experiments will, therefore, expose subjects to pictures which are very difficult to "understand" unless one is given a**

medium; but the rules that govern pictorial memory and understanding may or may not be similar to the linguistic rules that govern its understanding. On an intuitive level our "understanding" of a picture at the time we see it ought to have a very strong influence on how well we remember that picture, and the researchers set out to see if that were true. The hypothesis of this study is clearly identified (and even labeled, in case you missed it) in (a).

The origin of that hypothesis was suggested by contemporary research on memory for linguistic material (b). If memory for pictorial material were similarly "primed," would the effect be similar? An abbreviated description of the overall experimental design is found in (c). Notice how clearly these psychologists set the stage for the experiments that

thematic clue; we then later test memory for pictures that had been shown with or without the clues. Of course, this means that we are investigating memory for "nonsensical" pictures, one for which subjects usually have no interpretation. But what makes a picture nonsensical or meaningless? There are doubtless several kinds of nonsense, but included would be pictures for which the viewer (1) does not know the conventions for interpretation (e.g., a musical score for a musician who only "plays by ear"), (2) does not know the conceptual denotations of the symbols (e.g., the step sequences corresponding to ballet Labanotation), or (3) knows both of the above but still can achieve no coherent understanding by applying the standard conventions because the picture does not supply enough interpretive clues. Examples of the latter kind, which we shall use in our research, occur with "impoverished" pictures: These are pictures which reduce or eliminate the salient clues or distinctive features of objects which typically guide our selection of their schemata from memory. Such pictures present fragments of hidden figures which may be seen only by suggestion. They appear uninterpretable until a clue retrieves from memory an appropriate conceptual frame which can then be fit onto the line fragments.

d

A curious side effect results from finally finding a conceptual schema which fits: The tension of "What is it?" dissolves with laughter into "Oh, now I get it!" Many of us became familiar with such visual jokes in the early 1960s with the "droodles" rage in America: A droodle was an uninterpretable drawing that turned out to have a funny interpretation (see Price, 1972). Figure 11.1 shows two of the examples used in our experiment.

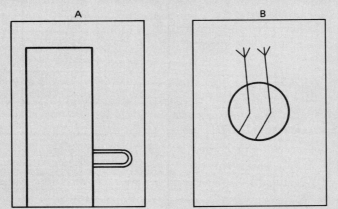

Figure 11.1 Droodles of Experiment I. Panel A: A midget playing a trombone in a telephone booth. Panel B: An early bird who caught a very strong worm.

are to follow. They must be convinced of the validity of the principle that memory is facilitated by presenting a cue that assists understanding. And if that is not enough to whet your appetite for what is to follow, Bower et al. tempt you further by illustrating the memory technique they use in terms of "droodles" verbally interpreted (**d**). One droodle we re-

EXPERIMENT I

The first experiment used free recall to assess the effects of comprehension on picture memory. Subjects studied a series of droodles pictures that they were to recall. Some subjects heard the interpretation as they saw each picture; other subjects, the controls, simply viewed the pictures without hearing any interpretive comment. The session ended with subjects drawing copies of those pictures they could recall.

e

A second, minor hypothesis tested was that subjects might distort their memory of the picture in a direction which provided a better fit to the prototype of the category used to interpret the picture upon original viewing. This "assimilation hypothesis" is an old one (see Carmichael, Hogan, & Walter, 1932; Riley, 1963), but the evidence regarding it has been equiv-

member, which is either a used lollipop or a filthy French postcard, side view, is shown below.

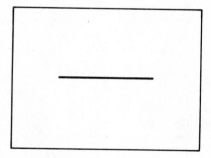

Once you "see" it, your memory for the object is significantly altered. Let's return to the experiment and see how Bower et al. use these simple line drawings to describe empirically the relationship between cueing of pictures and memory.

EXPERIMENT I

From (c) we already have a good idea what the design of the experiments will be, and understandably the first experiment used both a free-recall task immediately after presentation of all stimuli and a recognition memory test one week later as dependent measures. Subjects were assigned either to a group that heard an interpretation with each picture or to a group that saw only the ambiguous pictures.

In addition to the general hypotheses, the experimenters were trying to find support for the assimilation hypotheses (e), which states that subjects would distort their memory in the direction that would be more representative of the interpretation than the actual picture itself.

ocal. To test the hypothesis, we had subjects return after a week for a recognition memory test. Besides the correct picture, the multiple-choice set for each item contained two distractor pictures that were equally similar to the correct picture in terms of a line-overlap measure. One of these distractors exemplified a minor variation on the target which made it look even more like the interpreted prototype than did the original target (call this the "prototype" distractor). The other distractor involved a similarly minor line alteration but was done in such a manner as to violate the fit of the interpretive schema to the picture (call this the physically "similar" distractor). The expectation of the assimilation hypothesis is that, when recognition errors are made, subjects who learned the picture with a suggested interpretation will make relatively more errors on the prototype distractor than on the physically similar distractor. On the other hand, subjects who do not achieve the appropriate interpretation should tend to divide their errors evenly between the prototype and similar distractors.

Method

f   **The subjects were 18 undergraduates fulfilling a service requirement for their introductory psychology course. They were tested individually, assigned in random order to the "label" or "no-label" condition of the experiment.** All subjects studied a series of 28 simple droodles pictures shown on 3 × 5 in. cards at a rate of one every 10 seconds. As each picture was shown, its appropriate interpretation was given by the experimenter to the subjects in the label group but not to subjects in the no-label group. **Following presentation of the list, subjects had 10 minutes to draw all the pictures they could remember in any order they**

g   **wished.** The recall sheets were 8 × 11 in. papers marked off into a 3 by 3 matrix;

Method

In the experimental procedure all subjects are shown a series of 28 simple droodles pictures at the rate of one every 10 seconds. However, one group of subjects was given an appropriate interpretation after each picture (*label group*), whereas the other group of subjects was not (*no-label group*). In all other respects the two groups were treated exactly alike. Thus we have a simple two-group experiment, very similar to the type used in the Spallanzani experiment (Chapter 2).

The authors note that there were 18 subjects who were randomly assigned to either the label or the no-label group (**f**). The authors do not report the order in which the 28 pictures were presented to the subjects. However, it really doesn't matter in this experiment, because we are not concerned with any possible order effects of the cards (see within-subject design, in Chapter 5). The important point is that the order of presentation of the pictures to the subjects be exactly the same in both the label and the no-label group.

After the subjects had seen the 28 cards they were asked to recall as many of the pictures as they could in a *free-recall* situation (**g**); that is, the subject was instructed to recall by quickly sketching a recalled picture in one of the nine boxes on his recall sheet and to use as many sheets as necessary to complete his list recall. Before recall commenced, it was emphasized that the subject should aim for sketching the "gist" of the pictures recalled rather than for providing a lot of artistic detail of each picture. (The pictures could in fact be drawn very simply.) Following completion of the recall task, the subject was dismissed with an appointment to return the next week **"for other experiments."**

Upon returning the next week, subjects received the three-alternative multiple-choice test over 24 of the 28 pictures of the originally learned list (for four of the original pictures, we were unable to think up two similar distractors which met our criteria). The subject received a six-page booklet, with four multiple-choice triplets arranged in rows down each page. He was told that each triplet (row) contained one picture he had seen the week before as well as two closely similar pictures. **He was asked to rank order the three alternatives in each row, placing a 1 beside that test figure he considered most like the one he remembered seeing, a 2 beside the next most similar one, and a 3 beside that picture he considered least similar to the one he remembered.** The test was self-paced. Upon completing the test, the subject was debriefed and dismissed. One subject of the no-label group failed to return for the 1-week test, leaving eight subjects in that group at that point.

the subject had to sketch everything in any order from memory. Each subject was given a 10-minute time limit for this task. The experimenter has to set some sort of a time limit in tasks such as these, because the subjects could sit there for hours trying to recall another droodle. Usually the time limit is decided by having the researchers "pretest" the material, finding out how long it takes for most subjects to recall most of the pictures that they know.

Note that when the subjects were dismissed they were asked to return the next week "for other experiments" (h). If the subjects had been told that they were to come back to recall the droodles, many subjects would practice during the week. In order to eliminate these practice effects, the subjects were led to believe they were coming back "for other experiments."

Each subject returned a week after the original testing period and took a *recognition* test. Three versions of each droodle were shown to the subject—one was the original droodle, one looked more like the interpretation given the droodle to the label group than did the original (prototype distractor), and one was similar to the original but the interpretation didn't fit it. The subject was asked to rank order the figures (i). We assume the arrangement of the three pictures for each droodle was counterbalanced (or random) across each row. Note that subjects for both groups were treated exactly alike. The instructions and materials were the same for both groups. Enough of the experimental procedure is included to allow the replication of the experiment.

Results

Free Recall

A first noteworthy fact is that we had relatively few problems in scoring for "gist recall" of the sketches. We had anticipated severe problems produced by interfering or confused combinations of several pictures, or at least deletions causing the sketch to be unidentifiable. But subjects tended to recall (sketch) the pictures either relatively accurately or not at all. The primary result of interest is that an average of 19.6 pictures out of 28 (70%) were accurately recalled by the label group [standard error of the mean (SEM) = 1.25], whereas only 14.2 pictures (51%) were recalled by the no-label group (SEM = .92). **The means differ reliably in the predicted direction [$t(16)$ = 3.43, $p < .01$]. Thus, we have clear confirmation that "picture understanding" enhances picture recall.**

Recognition Memory

Despite the closeness of the distractors to the target, recognition of the correct target at the 1-week retention interval was very high. Subjects who received labels during study correctly recognized (gave a 1 rating to) a mean of 22.0 out of the 24 test triplets (92%); subjects receiving no labels during study

Results

In this experiment two groups were tested in the free-recall section— those who had the label and those who did not have the label when the droodles were presented. The hypothetical distribution of recall characters is portrayed in Figure 11.2. From these data the authors used the t-statistic to test for statistical "significance" and found (j) that the two means differ in the predicted direction at beyond the 0.01 level of probability, or that the observation would occur less than 1 time out of 100 by "chance" alone. Therefore, picture understanding through labeling enhances picture recall.

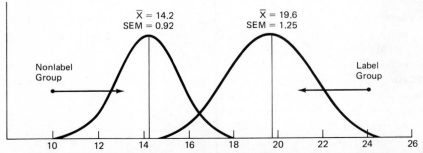

Figure 11.2   Number of pictures recalled (theoretical distribution based on data by Bower, Karlin, & Dueck, 1975).

correctly recognized a mean of 20.1 pictures (84%). With standard errors of .83 and 1.08, respectively, the means do not differ reliably.

Even noting the high levels of recognition accuracy, we may still ask whether the label and no-label subjects react differently to the prototype vs. the physically similar distractor. There are several indications that the label subjects considered the prototype distractor much closer subjectively to the target. First, considering only cases when the correct picture was not ranked first, the conditional probability that the prototype (rather than the similar distractor) was ranked first was .75 for the label subjects but only .38 for the no-label subjects. However, these conditional probabilities are based on very few observations. A more stable measure of differentiation is provided by the difference in rankings (on similarity to the remembered target) between the prototype distractor and the physically similar distractor. For the label group the mean rank assigned to the prototype was 2.17, whereas that for the similar distractor was 2.76, a difference of .59. In contrast, for the no-label group, the rankings of the two distractors was closer: 2.34 were for the prototype and 2.48 for the similar distractor, a difference of only .14. The difference in rankings is reliably larger for the label group than for the no-label group [$t(15) = 2.79$, $p < .02$]. **This result accords with the assimilation hypothesis: Subjects receiving the picture interpretation during study later reported that the distractor which moved in the direction of the interpreted prototype was closer to the target than was the distractor which involved a similarly minimal physical alteration but one which violated the interpretation given to the original target.** In contrast, the no-label subjects showed no comparable differentiation between the two kinds of distractors.

EXPERIMENT II

The initial experiment demonstrated the role of semantic comprehension in facilitating free recall of pictures. Having a meaningful name for a picture

k

The second part of the experiment involved recognition memory for the picture after a one-week interval. Both groups—those who originally had the labels and those who did not—recognized the original figures with a high degree of accuracy. The authors then speculate on the nature of the recognition task and the assimilation hypothesis (**k**).

EXPERIMENT II

The first experiment established that if a picture had a meaningful name, it helped in the recall of the picture. But what about more complex learning, as in the pairing together of two ambiguous pictures? Would a label that cleverly tied the two pictures together facilitate the association between two disparate pictures?

may facilitate recall because it provides a memorable summary or cue for later free recall. But an interpretation does more than provide a meaningful mnemonic label for a picture; it also causes unification or knitting together of the disparate parts of the picture into a coherent whole or schema.

l

The second experiment sets out to test more directly the influence of the unifying coherence of an interpretation upon picture memory. The subject was asked to study *pairs* **of nonsensical pictures and was later tested by cueing with one member of each pair for recall (in drawing) of the other**

m

member of the pair. Again, half the subjects received no interpretation of the pictures, whereas half heard a phrase that made both pictures and their pairing a meaningful sequence. Examples of picture pairs are shown in the three rows of Figure 11.3. Their interpretations are (from left to right panels in each pair): (1) rear end of a pig disappearing into a fog bank, and his nose coming out the other side of the fog; (2) piles of dirty clothes, then pouring detergent into the washing machine to wash the clothes; and (3) uncooked spaghetti, then cooked spaghetti and meatballs. **The hypothesis**

n

is that subjects hearing such interpretations during study will show much higher associative recall than will subjects who study the pictures without the interpretations.

Method

The subjects were 16 university students attending summer school. They were recruited by an advertisement and paid $1.50 for their participation. They were tested individually, assigned in random alteration to the label and no-label conditions ($n = 8$ per group). The subject was told to learn 30 picture pairs that were shown to him at a rate of one every 12 sec. The pairs were drawn and shown by means of 3×5 flashcards, one picture on a white card and its mate on a pink card. The subject had been told that in the later recall test he would be shown the picture on the white card and would have to recall (draw) its mate from the pink card. During presentation of each pair, the label subjects heard the experimenter supply an interactive interpretation to bind the picture-pair together. These were descriptions like those given for the three panels of Figure 11.3. Following presentation of the 30 pairs, the white deck was shuffled and presented as recall cues at a 20-second rate. "Gist" sketching of the recalled pictures was emphasized. If the subject had begun his drawing before 20 sec, he was allowed to complete it; otherwise, the next cue was presented after 20

The second experiment was designed to answer these questions (l). In (m) the experimenters give an abbreviated description of what they are going to do. Such a statement puts the reader "in the ball park" so that he or she can read the technical method section with a clear frame of reference. The hypothesis is stated in (n).

The method and procedure involved presenting pairs of nonsense pictures; one-half of the subjects received an interpretation that linked the

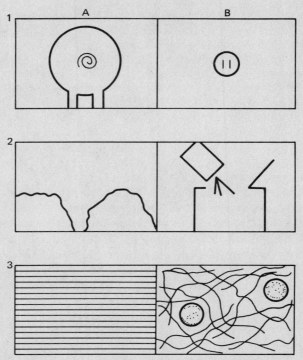

Figure 11.3   Pairs of nonsensical pictures used in Experiment II. See the text for an explanation of their contents.

o   seconds. **Subjects drew their recall sketches in numbered boxes, nine to a page;** they left blank any numbered box for which they could recall nothing to the corresponding cue.

After the cued recall test (conducted without feedback regarding the correctness

p   of subject's recall), **the subject received an associative matching test.** The 30 white and 30 pink cards were spread out in a random array over the table top. The subject was instructed to scan over the array, looking for the pairs of white and pink pictures he had studied. As pairs were recognized, the two cards were picked up by the subject and handed to the experimenter. The subject continued

pictures conceptually, and one-half of the subjects received no interpretation. It was hypothesized that subjects supplied with an interpretation would have better associative recall of the pairs than subjects who did not receive any interpretation.

The authors supply the details pertinent to this experiment: subjects, independent variables, description of stimulus materials, and procedures of the experiment. The dependent measures were an associative recall test (**o**) and an associative matching test (**p**).

this pairing until he had selected all pairs he could remember: because they were asked not to guess, many subjects stopped short of pairing off all members. The subject's associative matching score was simply the number of correct pairs he selected from the array before terminating. (The expected correct pairs obtainable by guessing in an associative matching test is about one, regardless of the number of pairs to be guessed at. See Feller, 1957, p. 97.) After completing the matching test, the subject was debriefed and dismissed.

Results

Associative Recall
Again no problem was encountered in scoring correct gist recall of the cued picture. **Cued recall averaged 21.75 pictures out of 30 (73%) for the label subjects (SEM = 1.78) and 13.13 (44%) for the no-label subjects (SEM = 2.41).** These percentages differ reliably [$t(14) = 2.87, p < .02$]. The effect is remarkably consistent over items, too: For 22 items the label subjects recalled more than the no-label subjects, for three items the no-label subjects recalled more, and there were five ties. The 22/25 predominance of items with more recalls by label subjects exceeds chance of 50% ($z = 6.55, p < .01$). Thus, the label group is uniformly superior in recall to the no-label group.

q

Associative Matching
The number of correct matches (of pairs) averaged 27.50 for the label subjects (SEM = .98) compared to 16.63 for the no-label subjects (SEM = 2.80). These differ reliably [$t(14) = 3.66, p < .01$]. Moreover, the relative gain in recognition performance above what could be recalled was much larger for the label subjects (70%) than for the no-label subjects (21%). The data show that the label subjects still exhibit superior associative coherence even when all the pictures are available and do not need to be recalled.

r

The influence on memory of labeling ambiguous pictures was statistically tested by means of a *t* test, which indicated that labeling strongly facilitates associative learning. The average number of correctly cued recall items was 21.75 for the label group and 13.13 for the no-label group (q). The hypothetical distribution of scores is shown in Figure 11.4.

The number of correct matches is reported in (r).

DISCUSSION

The results of these experiments are succinctly summarized (s), and the authors suggest that their results have empirically demonstrated what may have been suspected about human nature for some time.

The authors are very thorough in their discussion and very careful not

DISCUSSION

It has been argued that memory for a picture depends upon the subject achieving a conceptual interpretation of the picture as he views it. The hypothesis is the pictorial analog of that relating sentence recall to comprehension (e.g., Bransford & McCarrell, 1974). The point is intuitively obvious once it has been noticed (as are many other "facts" of psychology), and the experiments above are primarily demonstrational in nature. **Subjects provided with meaningful interpretations of single droodles show superior free recall. Subjects who hear an interpretation identifying and relating two pictures together show greater coherence of the pictures on later association tests.** Although control subjects were probably trying to come up with some sensible interpretation of the pictures, the difficulty of the task precluded much success. Presumably, if we had collected control subjects' attempts at interpretations (recording their "thinking aloud"), those pictures for which they achieved a meaningful interpretation would more likely have been recalled (see, e.g., Montague, 1972). The likelihood of this being the case remains to be checked.

One might question whether the associative coherence found in Experiment II is a result merely of identifying the objects in each picture or whether it depends in addition upon providing the meaningful linking relationship between the two pictures of a pair. For some pairs the linking relation was that the pictures in Figure 11.3 denoted different parts of the same object (the pig), different states of an object as it underwent changes (the spaghetti), or different objects associated with a common process (the clothes and washer). We feel these relations are very important for promoting associative coherence of the elements of the pair. To illustrate this point, four further subjects from the same source were tested under the same procedure as Experiment II, except that new pairs were constructed by re-pairing the old pictures in a random manner. As the pair was shown, each picture was separately interpreted (e.g., a pig's tail and detergent pouring into a washing machine). No linking relation other than contiguity was stated for connecting the two contents. These four subjects averaged only 7.75 correct in cued recall (SEM = 1.89) and 8.25 correct in associative matching (SEM = 1.11). If anything, the scores are lower than those for the controls who studied the original pairs without hearing the objects or relation identified. Quite possibly, this "mispairing" list was so difficult because semantically related objects appeared in different pairs, creating intrapair interference. Teasing out the several contributors to this poor learning would be a task for further experimentation. The significant fact we wish to glean from this poor recall

to come to wild conclusions, unsupported by the data in their experiments. In one instance they worry over the results of Experiment II in **(t)**. This may have been caused by labeling, which is independent of the "known" relationship between the pairs. The authors check this hunch by running a small group of subjects.

of mispaired items is that associative coherence depends heavily upon relating the two identified pictures and relatively little upon identifications per se that do not call to mind a known relationship between the two pictures.

How are our results to be related to previous work on picture memory? Previous work on learning of nonsense figures typically used recognition rather than reproduction measures and have been largely concerned with testing hypotheses of acquired distinctiveness or acquired equivalence of forms induced by learning different or the same arbitrary labels for the forms. Of more direct relevance to our results are those by Ellis (reviewed by Ellis, 1973), who found that the learning of "representative labels" to complex polygons enhanced their later recognition. Of course, the "representative labels" were simply a plausible name or interpretation of the figures. The fact that pairing with a representative label makes the pictures more memorable seems quite consistent with our hypothesis relating picture comprehension to memory. Since "association value" or "codability" has been a common variable in research on pictorial memory, one may ask whether our notion of a "semantic interpretation" of a picture is just a fancy name for an association to it. We think not. We intend "semantic interpretation" to be much more specific than the concept of "picture association" suggests. Associations may occur to many surface features of a picture or to fragments of it, all without improving memory for it. Presumably, picture memory would improve with greater "depth of processing," as does memory for words (Craik, 1973) or faces (Bower & Karlin, 1974). But this implies comprehending the picture, figuring out what conceptualization it expresses or what object it denotes: It means getting the "message" behind the "medium."

u

In the final paragraph (**u**) Bower et al. fit their newly found information into the larger structure of memory. The reader is urged to study this section for its clarity of writing and thought.

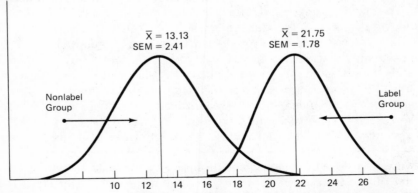

Figure 11.4   Number of pictures recalled (theoretical distribution based on data by Bower, Karlin, and Dueck, 1975).

REFERENCES

Bransford, J. D., & Johnson, M. K. (1972). Contextual prerequisites for understanding: Some investigations of comprehension and recall. *Journal of Verbal Learning and Verbal Behavior, 11*, 717–726.

Bransford, J. D., & McCarrell, N. S. (1974). In W. Weimer & D. Palermo (Eds.), *Cognition and the symbolic processes.* Hillside, N. J.: Lawrence Erlbaum Associates.

Bower, G. H., & Karlin, M. B. (1974). Depth of processing pictures of faces and recognition memory. *Journal of Experimental Psychology, 103*, 751–757.

Carmichael, L., Hogan, H. P., & Walter, A. A. (1932). An experimental study of the effect of language on the reproduction of visually perceived form. *Journal of Experimental Psychology, 15*, 73–86.

Craik, F. I. M. A. (1973). "Levels of analysis" view of memory. In P. Pliner, L. Krames, & T. Alloway (Eds.), *Communication and affect: Language and thought.* New York: Academic Press.

Doll, T. J., & Lapinski, R. H. (1974). Context effects in speeded comprehension and recall of sentences. *Bulletin of the Psychonomic Society, 3*, 342–345.

Ellis, H. C. (1973). Stimulus encoding processes in human learning and memory. In G. H. Bower (Ed.), *The psychology of learning and motivation* (Vol. 7). New York: Academic Press.

Feller, W. (1957). *An introduction to probability theory and its applications.* (Vol. 1. 2nd ed.) New York: Wiley.

Gombrich, E. (1960). *Art and illusion.* New York: Pantheon.

Goodman, N. (1968). *Languages of art* (2nd ed.). Indianapolis, Ind: Bobbs-Merrill.

Minsky, M. (1974). *Frame systems.* Unpublished manuscript. M.I.T. AI Project.

Montague, W. E. (1972). Elaborative strategies in verbal learning and memory. In G. H. Bower (Ed.), *The psychology of learning and motivation: In research and theory* (Vol. 6). New York: Academic Press. Pp. 225–302.

Price, R. (1972). *Droodles.* Los Angeles: Price/Stern/Sloan.

Riley, D. A. (1963). Memory for form. In L. Postman (Ed.), *Psychology in the making.* New York: Knopf.

NOTE: The authors thank Susan L. Karlin for creating most of the stimulus materials for Experiment II. The research was supported by Grant MH 13905-07 from the National Institute of Mental Health to Gordon H. Bower.

QUESTIONS

1. Devise another experiment to test the assimilation hypotheses mentioned in this experiment.
2. Why did the experimenters use two dependent measures in Experiment II: (1) a free-recall test and (2) a recognition test one week later?
3. The authors state that the finding that providing a conceptual interpretation for ambiguous pictures facilitates recall of the pictures is intuitively obvious. Why, then, must it be experimentally tested?

4. Design an experiment that tests whether groups of musical notes are easier to remember when accompanied by a verbal explanation.
5. Describe some practical and creative applications of this study.
6. The authors don't state specifically how they randomly assigned the 18 subjects into two groups (such that there are 9 subjects per group). How would you do it?

CHAPTER
12

Distance and Rank

INTRODUCTION

How close do you stand next to a person with whom you are talking? If you think about the answer, you will probably say, "It depends on several factors." Among the most apparent of these is familiarization. In general, we tend to stand closer to people we know (and like) than to those we do not know. But other factors also influence our behavior. An important consideration is the influence of our cultural background. Foreign diplomats report how misunderstandings develop from some people who stand too close to them during casual conversations, while others tend to stand too far away. An acquaintance of ours who was a member of the U.S. foreign service told of a diplomatic reception in a South American country in which the diplomat was "chased" around the room. While the diplomat was talking with his South American colleague, the colleague would edge a bit closer to the diplomat and the service officer would withdraw slightly. The South American would again move closer, the diplomat would withdraw, and so on. Throughout the reception the two could be seen advancing, retreating, advancing, retreating. If we could have interviewed the two, we would probably find the diplomat saying, "They are very 'pushy' people, those South Americans," while the South American would probably complain that the Americans are "stand-offish, distant, and aloof."

In the experiment by Larry Dean, Frank Willis, and Jay Hewitt, the influence of military rank on distance between members of different ranks and the same rank is studied empirically. We think this article may be interesting to you because the results may tell you something of your own behavior, but the purpose of including this article is to demonstrate two special issues: *field-based studies* and *unobtrusive measures*. Pay close attention to these issues before you read the article.

SPECIAL ISSUES

Field-Based Studies

Psychological research can be divided into two categories: laboratory experiments and field experiments. The majority of research reported in the scientific literature in psychology takes place in controlled laboratory settings. The reason psychologists tend to favor a laboratory setting in which to conduct their research is for methodological control. In the experimental laboratory the researcher can isolate the subject from the noisy world and precisely control the type of stimulation he or she receives. In effect, the experimental laboratory allows the researcher to create a microcosm in which only the factors that he or she wishes to influence the subject are allowed in; other cues can be either eliminated or brought under experimental control. Paradoxically, in this very strength of laboratory experimentation lies a weakness: the "artificial" nature of the laboratory setting. By removing the subject from his or her natural setting, as is done in laboratory experiments, the subject is deprived of the forces that are necessarily part of his or her normal life. These forces may constitute a sizable amount of stimulation that determines the subject's reaction. In technical language, some stimuli that are eliminated or controlled by the laboratory setting may be critical independent variables that significantly affect the dependent variable. A sophisticated researcher recognizes the relative virtues of laboratory experiments and field experiments and is able to select the appropriate method.

In making a selection, one critical question should be answered: Are the desired results of the experiment significantly related to the social situation? Many types of social and educational problems fall into this category. Consider mob behavior. The very presence of other members of the mob undoubtedly influences the behavior of any given individual, and his or her behavior may in turn influence the others. And how can an experimenter control for these pressures and other independent variables (e.g., the presence of an inflammatory stimulus, such as a lynched body)? It is simply impossible to recreate such a complex situation in a laboratory. Therefore, a researcher interested in mob behavior must either isolate some hypothesized factors of the larger issue for laboratory investigation, or turn to the field of action. A large number of worthy research projects must be studied in the realistic field in which they occur. In addition to the above example from social psychology, many educational problems are studied in the context of field experimentation. The present article and the research article by Fisher and Harris on note taking (Chapter 23) fall in the category of field studies.

SPECIAL ISSUES

Unobtrusive Measures

When subjects know they are being observed, they frequently behave differently than when they are not being observed. This generalization is true not only for human subjects in a psychological experiment but also in some cases of biological research in which animals are being observed. A partial explanation of this phenomenon can be attributed to the fact that the experimenter becomes part of the field whose influence he or she is attempting to measure. By way of analogy, consider the problem of making critical temperature measurements. A thermometer inserted into a substance not only reacts to the temperature of the substance but also reacts to its own temperature. In a similar way, an experimental psychologist is not only an observer of behavior but also becomes part of the environment to which the subject reacts.

Since precise control over stimulus variables in field-based experiments is sometimes unknown, the response variables are also sometimes questionable. As you may recall from our previous discussion of scientific theory and methodology, a cardinal principle of experimental research is to identify the precise cause of a special effect. If the causes of behavior (in experimental terms, the independent variables) are doubtful, then the behavior (the dependent variable) that is measured may or may not be related to the cause. The resolution of this profound dilemma is not easy; however, the issue has been reviewed in two influential books: Webb, Campbell, Schwartz, and Sechrest, *Unobtrusive Measures: Nonreactive Research in the Social Sciences* (1966), and Cook and Campbell, *Quasi-Experimentation: Design & Analysis Issues for Field Settings* (1979). In these sources (and other sources) experimental psychologists have grappled with the problem of the intrusion of an observer into the psychological field. To reduce the influence of an observer on a psychological experiment—especially a field-based experiment—the authors suggest a number of "unobtrusive" means experimenters can use to make measures on subjects. The interested student is referred to these sources for a more complete discussion of the issue.

Field-based experiments in which unobtrusive measures are made also raise questions of experimental ethics. Is it proper or ethical to skulk around with a clipboard, collecting data on unsuspecting people? (One experiment that we know of had an experimenter concealed in a public toilet, surreptitiously observing urinal preference of subjects. We are moved to identify the experimenter as a "stool" pigeon in this study.) Although questions may be raised about the propriety of field-based experiments, it would be erroneous to conclude that such experiments are unethical. We do, however, strongly recommend that in the planning of all experiments, including field experiments, the experimenter consult the APA ethical standards reprinted in Chapter 7.

Initial Interaction Distance Among Individuals Equal and Unequal in Military Rank

LARRY M. DEAN, FRANK N. WILLIS, and JAY HEWITT
UNIVERSITY OF MISSOURI AT KANSAS CITY

Interaction distances for 562 pairs of individuals in military settings were recorded at the moment a conversation was begun. It was found that interactions initiated with superiors were characterized by greater distance and, further, that this distance increased with the discrepancy in rank. When interactions were initiated with subordinates, interaction distance was unrelated to rank discrepancy. It was hypothesized that superiors are free to approach subordinates at distances that indicate intimacy, while subordinates are not at liberty to do so.

The increased interest in research in personal space is evidenced by three reviews of the literature in this area (Evans & Howard, 1973; Linder, 1974; Pederson & Shears, 1973). Research has been reported for both animal and human subjects. **Studies have been conducted using personal space as both an independent and a dependent variable.** As a dependent variable, personal

SOURCE: Reprinted by permission from *Journal of Personality and Social Psychology*, 1975, 32 (2), 294–299. Some portions of the original article have been omitted.

a

ANALYSIS

In this study, as in most articles, we suggest that you survey the entire article, not getting bogged down on the specific methodological issues but trying to gain an overview of the purpose, the technique, the results, and the conclusion. Having established this overview, you will see that the article is more than a simple experiment about how close people of unequal rank stand to each other, but that it deals with a broader issue of "personal space" and perceived status.

REVIEW OF LITERATURE

The development of this article follows a well-defined sequence of logically related studies of personal space. Of particular note in the introduction is the authors' statement that personal space has been treated as an independent *and* dependent variable (a). As you read this section can you think of other cases in which another variable could be treated as an independent or a dependent variable?

space has been related to the personal characteristics of initiators and recipients and to the relationships between initiators and recipients. The personal characteristics that have been related to their interaction distances have included the following: handicaps or stigma (Kleck, Buck, Goller, London, Pfeiffer, & Vukcevic, 1968), psychopathology (Horowitz, 1968), extroversion–introversion (Patterson & Sechrest, 1970), sex (Pellegrini & Empey, 1970), race (Jones & Aiello, 1973), and age (Willis, 1966).

Hall (1966) suggested four zones for social interaction: intimate, personal, social, and public. He asserted that the selection of a zone for a particular transaction was dependent upon the degree of intimacy of the relationship for the individuals involved.

In research involving personal space as a dependent variable, one characteristic of the relationship between subjects that has received considerable interest in animal studies and little interest in human studies is that of status or dominance. King (1965) reported that the distance a subordinate chicken remained from a stationary dominant bird was linearly related to the frequency with which the dominant had pecked the subordinate in the home coop. King (1966) also found a similar relationship with human children as subjects. Lott and Sommer (1967) reported that college upperclassmen sat closer to peers than to freshman "doing poorly in school" and sat nearer to peers than to professors. It is Sommer's (1969) opinion that studies relating spacing and dominance in humans have been far more infrequent than those with infrahumans "probably because this type of experimentation requires conditions that are uncommon in naturalistic settings" (p. 24). The lack of information in this area may be due to the difficulty in specifying observable dominance in humans. **The present study was designed to overcome this difficulty by observing interaction distances in military settings where dominance relations are indicated on the clothing of the individual.** If, as has been indicated in infrahuman studies, superiors are capable of inflicting aversive consequences, whereas peers and subordinates are not, then subjects might prefer greater interaction distances with superiors. If this is true, the relative status of the interaction initiator and recipient would be an important variable in determining their interaction distance. The following questions were of interest in the present study: (a) Would interaction distances be greater when people initiated an interaction with superiors than when they initiated them with peers? (b) Would interaction distances be

---

Dean et al. point out that studies of personal space have been done with animals and humans, and at least hint (**b**) that the critical factor in both human and animal studies might be related to dominance.

In section (**c**) the problem is stated, followed by the hypothesis in section (**d**). The "hypothesis" is presented in terms of an "if . . . then" proposition, which is supported by a series of specific questions (**e**) the research purports to answer.

less when people initiated an interaction with subordinates than when they initiated them with peers? (c) Would interaction distance increase monotonically with status discrepancy when people initiated interaction with superiors?

METHOD

Subjects

Subjects were 562 uniformed active duty United States naval personnel who were observed as they initiated an interaction with another uniformed member of the U.S. Navy.

Procedure

f_1

g

h

i

f_2

Observations were made at the U.S. Naval Station, Long Beach, California, by three uniformed naval personnel who were assigned to duties at that station. Subjects were observed during duty hours but in nonworking settings. These settings were chosen for the study in order to minimize the effects of regular working relationships upon the interaction distances. **The settings were as follows: the Navy exchange, the cafeteria, the lobby of the station hospital, and a station recreation center.** All subjects were standing when observed. Interactions were selected when one person was standing and was approached by another **who began a conversation.** At the moment the first words were spoken, observers **recorded the number of floor tiles between their nearest feet to the closest half-tile.** All settings had standard 22.86-cm. square floor tiles. Also recorded was the rank of the initiator and of the recipient of the interaction. The subjects were an incidental sample of all personnel in the settings meeting the above requirements during the hours of observation. The observers attempted to use each subject only once in the role of initiator. Interjudge reliability for the distances was so high in a pilot study that only one observer was used for each dyad. **Observers were not informed as to the hypotheses** of the study. **Notations were made unobtrusively,** and no subject was observed to react to the observers.

METHOD

There are several important features of the procedure section for our discussion of field research. First, the data were collected in a natural setting (f_1) with notations made "unobtrusively" (f_2). Second, the data (i.e., the distance between subjects) were collected at the moment conversation started (g). Third, the researchers used an inventive measure of distance: the number of floor tiles between the subjects (h). Finally, the observers were not told of the hypothesis (i), which is a form of a "double blind" design (see Chapter 2) in that both the experimenters

RESULTS

Peers Versus Subordinates and Superiors

For naval personnel of ranks of commander through third-class petty officer, information was available for interaction distance between these individuals (as initiators) and others of higher, equal, or lower status. For captains this information was available for others of equal or lower status. For seamen, this information was available for others of equal or greater status. The mean interaction distances are presented in Table 12.1. Combined across all ranks, mean interaction distance was greater when people initiated an interaction

j_1 with a superior than with a peer, $t(289) = 3.18, p < .01.$ When people initiated an interaction with a subordinate, however, mean initial interaction distance was not significantly different from that which existed for peer in-

j_2 teractions, $t(304) = .37, p < .05.$ Examining interaction distances for people

j_3 at each rank in Table 12.1, it can be seen that in **10 of the 11 ranks** for which information was available, the mean interaction distance was greater when the recipient was higher in status than when the recipient was equal in status. Finding such a pattern in 10 of 11 cases is significantly higher than

j_4 one would expect on the basis of chance $\chi^2(1) = 5.82, p < .05.$

Interaction Distance and Status Discrepancy Collapsed Across Ranks

[Another . . .] of investigating the relationship between status discrepancy and interaction distance is to compute the mean distance for people who initiate an interaction with someone one step above them in rank (e.g., a commander

and the subjects are unaware of the purpose of the experiment. (In the latter case, subjects in most field experiments are not aware they are even participating in an experiment.) This procedure section is particularly rich in methodological issues associated with field research and deserves your careful attention and reflection.

RESULTS

The overall results are initially presented in the form of tabular information (Table 12.1). We think the results could be effectively presented in graphic form. From the data in Table 12.1 we have composed Figure 12.1, where we can clearly see the essential results, which indicate that the distance between military personnel tended to be greater when a person of a lower rank approached another person of a higher rank than when peers approached each other or persons of higher rank approached a person of lower rank. This observation is confirmed by statistical analysis (j_1, j_2, j_3, and j_4). These data answer the first and second questions raised in section (e).

Table 12.1   MEAN INITIAL INTERACTION DISTANCE AND
NUMBER OF INTERACTIONS OBSERVED AS A FUNCTION OF
STATUS OF INITIATOR OF INTERACTION

STATUS OF INITIATOR	STATUS OF RECIPIENT					
	HIGHER		SAME		LOWER	
	$\bar{X}$	n	$\bar{X}$	n	$\bar{X}$	n
Captain	—	—	4.00	1	3.45	38
Commander	3.50	4	3.17	3	3.03	15
Lieutenant commander	3.60	5	3.00	2	3.54	27
Lieutenant	4.71	7	2.81	8	3.51	35
Lieutenant junior grade	4.50	8	3.40	5	3.12	17
Ensign	4.01	12	3.00	2	3.35	10
Warrant officer	3.79	21	3.33	3	3.15	20
Chief	3.98	27	3.55	11	3.57	36
1st-class petty officer	3.76	21	3.79	14	3.38	17
2nd-class petty officer	3.58	20	3.45	11	3.63	31
3rd-class petty officer	3.95	39	3.69	8	3.40	5
Seaman	4.06	45	3.46	14	—	0

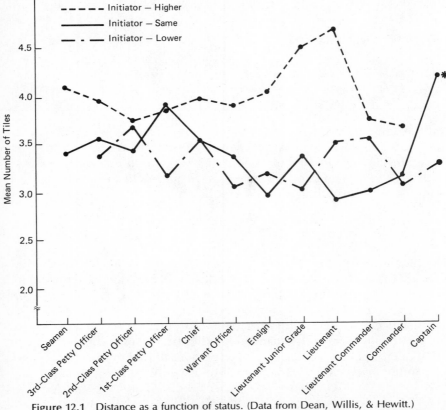

Figure 12.1   Distance as a function of status. (Data from Dean, Willis, & Hewitt.)
*$n = 1$

with a captain), the mean distance for people who initiate an interaction with someone two steps above them in rank, etc. As status discrepancy increases, interaction distance should also increase for interactions initiated upward.

The results of a one-way analysis of variance carried out on the means in the top half of Table 12.2 (interactions initiated upward) indicated that

Table 12.2   MEAN INITIAL INTERACTION DISTANCE BETWEEN INDIVIDUALS OF UNEQUAL RANK AS A FUNCTION OF NUMBER OF STEPS IN RANK SEPARATING INDIVIDUALS

| INITIATOR | STEPS IN RANK SEPARATING INDIVIDUALS | | | | | | |
	1	2	3	4	5	6–7	8+
Subordinate							
$\bar{X}$	3.54	3.88	4.38	3.82	3.97	3.96	4.59
n	55	36	21	28	16	26	27
Dominant							
$\bar{X}$	3.55	3.29	3.53	3.59	3.21	3.14	3.33
n	84	36	36	38	17	35	26

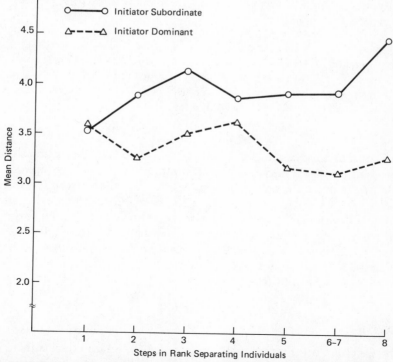

Figure 12.2   Distance as measured by mean number of tiles between subjects of varying military rank. (Data from Dean, Willis, & Hewitt.)

k_1
k_2
k_3

there was a significant effect of status discrepancy, $F(6,202) = 2.90, p < .05$. The linear trend was significant, $F(1,202) = y.84, p < .001$, as was the cubic, $F(1,202) = 5.76, p < .01$. It would thus appear that interaction distance tends to increase as a function of status discrepancy for interactions initiated upward. A rank-order correlation was carried out between the obtained rank order of the seven means and the rank order predicted by the model (increasing distance as a function of increasing status discrepancy). The rank-order correlation coefficient was .71, which is not significant with an n of 7.

DISCUSSION

l

The results of the present study support the proposition that interaction directed toward superiors is characterized by greater interaction distance than those directed toward peers and further that the distance is greater when the difference in rank is greater. The relationship does not obtain when interactions are directed toward individuals lower in rank. Previous research has shown the closer interaction distances are characteristic of intimacy (e.g., Willis, 1966). We might argue that a superior has the option of a formal (more distant) or more intimate (less distant) interaction, while a subordinate is usually required to initiate a formal interaction. Brown and Ford (1961) have stated this clearly:

m

If the person of lesser value were to initiate associative acts, he would run the risk of rebuff; if the person of higher value initiates such acts there is no risk. The superior then must be the pacesetter in progression to intimacy; the person of higher status is, we believe, the pacesetter not in linguistic address alone but in all acts that increase intimacy. (p. 383)

The results were analyzed in yet another way which compared the distance between ranks. A reader may ask, "As the difference in rank increases, is there also a measurable increase in the physical distance between individuals?" We have presented the results in the form of a graph (Figure 12.1). The results were statistically treated (k_1, k_2, and k_3). These tests suggest that distance tends to increase as steps between subjects increase. This tendency is shown in Figure 12.2 and answers the third question raised in (e).

DISCUSSION

The researchers briefly review their findings (l), relate the results to previous experiments (m), and suggest an alternate hypothesis (n).

Previous reports of studies of spatial relation have frequently failed to specify the initiated and recipient of the interaction. The results of the present study strongly suggest that these are important sources of information. If the interaction distance is great in a superior—subordinate interaction, our results indicate that it is the subordinate who is responsible for this distance.

Although our study indicated that interaction distances could not be predicted by status discrepancy when interactions were initiated by a superior, the average distance in one of the relationships in this setting may be of interest. The greatest interaction distance for junior officers initiating conversation with military subordinates was that for senior enlisted men. The average age and length of service is much greater for senior enlisted men than for junior officers. It is possible that dominance relationships even in a highly structured organization like the Navy are affected not only by the formal organization but also by norms regarding age and experience provided by society at large.

REFERENCES

Brown, R., & Ford, M. (1961). Address in American English. *Journal of Abnormal and Social Psychology, 62,* 375–385.

Evans, G. W., & Howard, R. B. (1973). Personal space. *Psychological Bulletin, 80,* 334–344.

Hall, E. T. (1966). *The hidden dimension.* New York: Doubleday.

QUESTIONS

1. In this article personal space was treated as a dependent variable. Briefly outline a design in which personal space would be an independent variable. Also, be sure to identify the dependent variable in your design.
2. The authors suggest one alternate hypothesis. What other possible causes of these data might there be?
3. Briefly define a procedure in which personal space could be empirically observed.
4. Review the basic problems associated with field-based experiments.
5. Because the experiment called for a judgment on the part of the experimenter (e.g., counting the number of tiles), wouldn't it have been better if two or more observers were used? If not, why not? What influence might multiple observers have on the results.
6. Does this experiment violate any ethical principle? Which principles apply?

Horowitz, M. J. (1968). Spatial behavior and psychopathology. *Journal of Nervous and Mental Disease, 146*, 24–35.

Jones, S. E., & Aiello, J. R. (1973). Proxemic behavior of black and white first-, third-, and fifth-grade children. *Journal of Personality and Social Psychology, 25*, 21–27.

King, M. G. (1965). Peck frequency and minimal approach distance in domestic fowl. *Journal of Genetic Psychology, 106*, 25–28.

King, M. G. (1966). Interpersonal relations in preschool children and average approach distance. *Journal of Genetic Psychology, 109*, 109–116.

Kleck, R., Buck, P. L., Goller, W. L., London, R. S., Pfeiffer, J. R., & Vukcevic, D. P. (1968). Effect of stigmatizing conditions on the use of personal space. *Psychological Reports, 23*, 111–118.

Linder, D. E. (1974). *Personal space.* Morristown, N.J.: General Learning Press.

Lott, D. F., & Sommer, R. (1967). Seating arrangements and status. *Journal of Personality and Social Psychology, 7*, 90–94.

Patterson, M. L., & Sechrest, L. B. (1970). Interpersonal distance and impression formation. *Journal of Personality, 38*, 161–166.

Pederson, D. M., & Shears, L. M. (1973). A review of personal space research in the framework of general systems theory. *Psychological Bulletin, 80*, 367–388.

Pellegrini, R. J., & Empey, J. (1970). Interpersonal spatial orientation in dyads. *Journal of Psychology, 76*, 67–70.

Sommer, R. (1969). *Personal space: The behavioral basis of design.* Englewood Cliffs, N.J.: Prentice-Hall.

Willis, F. N. (1966). Initial speaking distance as a function of the speaker's relationship. *Psychonomic Science, 5*, 221–222.

Requests for reprints should be sent to Frank N. Willis, Department of Psychology, University of Missouri, Kansas City, Missouri 64110.

Larry M. Dean is a lieutenant j.g. of the United States Navy and is presently head of the Fleet Problems Branch of the Operational Psychiatry Division of the Navy Medical Neuropsychiatric Research Unit, San Diego, California 92152.

CHAPTER
13

Long-Term Memory

INTRODUCTION

The next article, "Does It Pay to Be 'Bashful'?: The Seven Dwarfs and Long-Term Memory" by Meyer and Hilterbrand is a type of controlled "Trivial Pursuit" in which long-term memory for common events is examined.

This article is to be analyzed by the student. Some things you should look for include:

What is the theoretical motivation behind this experiment?
Who were the subjects?
What is the independent variable?
How would you classify this design?
What was the dependent variable?
What controls (e.g., subjects) were employed?
How could long-term memory be assessed?

The original article had three experiments. For the purpose of simplicity only the first experiment is presented in this section.

Does It Pay to Be "Bashful"? The Seven Dwarfs and Long-Term Memory

GLENN E. MEYER and KATHIE HILTERBRAND
LEWIS AND CLARK COLLEGE

Recall of the Seven Dwarfs of the Disney film was measured. The name Bashful was at a disadvantage in both percentage recall and position of recall. An analysis of errors from a "tip-of-the-tongue" viewpoint suggests this is due to retrieval difficulties.

In a classic review of literature, Miller (1956) pointed out that subjects were able to discriminate between seven plus-or-minus two different stimuli without making an error in various kinds of memory and discrimination tasks. He referred to the number seven as the "magical number seven" and questioned whether it is more than coincidence that there are seven objects in the span of attention, seven categories for absolute judgment, seven digits in the span of immediate memory, and the classical groups of seven, such as the seven wonders of the world, the seven levels of hell, the seven daughters of Atlas in the Pleiades, and the seven ages of man. Because most students (and probably faculty) are unaware of Atlas's daughters, we used as a class example a contemporary group of seven, the Seven Dwarfs, from the Walt Disney Studio film *Snow White and the Seven Dwarfs*. Upon asking people to name the Seven Dwarfs, we noted an interesting effect. The name Bashful appeared to be at a disadvantage and was forgotten most often. Similar experiments have examined recall for lists assumed to be common in memory such as the presidents (Roediger & Crowder, 1976) and popular songs (Bartlett & Snelus, 1980). In Experiment 1, we decided to verify and quantify the "Bashful effect" using a free-recall task.

METHOD

We gave 141 college students a form with seven spaces and told them to write the names of the Seven Dwarfs in the order that they were remembered.

RESULTS

The percentage recall and average position of recall are presented in Figure 13.1 for each dwarf. We first analyzed the percent correctly identified (Cochran's $Q = 138.03$, $df = 6,986$; $p < .0001$). Bashful was named by

SOURCE: Reprinted with permission of *American Journal of Psychology*, 1984, 97(1), 47–55.

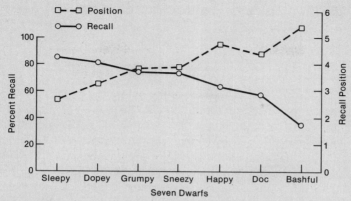

Figure 13.1   Recall percentage and average position of recall for the Seven Dwarfs.

only 35 percent of the subjects. Sleepy was remembered by 86 percent of the subjects.

The second analysis was performed on the average recall-position data because position might give another indication of long-term memory (LTM) accessibility. Because only 22 subjects correctly identified all seven of the dwarfs, we conducted several different analyses of position. In the first, all subjects were used, and we assigned average positions for the missing or nonrecalled dwarfs for each subject. When a dwarf was left out of a subject's list, he was assigned an average position on the blank spaces for that subject. If a subject correctly identified four dwarfs, the unnamed dwarfs would each be assigned an average position of six. Friedman's chi square was found to be significant ($\chi^2 = 36.62$; $df = 6$; $p < .001$). Again, Sleepy was strongest with an average recall position of 2.6, whereas Bashful was weakest with 5.39. Pearson's correlation between the average position of the dwarfs and the percentage correctly identified was $r = -.94$ ($df = 5$; $p < .01$). These data are presented in Figure 13.1. Also calculated was the average position for the 22 subjects who correctly recalled all dwarfs (Sleepy = 2.682, Grumpy = 3.318, Dopey = 3.409, Sneezy = 3.545, Doc = 4.791, Bashful = 4.909, Happy = 5.045). Again, Friedman's chi square was significant ($\chi^2 = 36.67$, $df = 6$, $p < .001$). We also calculated the average position of recall for all the subjects, excluding all those who had perfect recall (141 − 22 = 119). The mean positions were Sleepy = 2.150, Dopey = 2.768, Grumpy = 3.250, Sneezy = 3.253, Doc = 3.271, Happy = 3.687, Bashful = 4.643. The positions data scored by all three methods were highly correlated ($r > .81$, $p < .02$). Most interesting was that the data for percentage of correct recall was highly correlated with the position data from the 22 subjects who recalled all the dwarfs ($r = -.85$, $p < .01$). Thus, high recall is linked with early position of recall and vice versa. There were no differences in recall between those who had seen or not seen the movie ($p >$ .10).

DISCUSSION

The two analyses verify that Bashful is the poorest recalled of the Seven Dwarfs. Fewest remember him, and if they do he is usually recalled toward the end of the list. The major reason for this may lie in retrieval problems. Several factors support this view. First, we found that the dwarfs recalled best were cited first, because the position of recall by subjects having all correct correlated strongly with percentage correct. Thus, both measures indicate that these dwarfs are easily accessible. Second, our analysis of the incorrect guesses also suggested retrieval difficulties. We recorded 88 incorrect responses. Six were rejected as obviously facetious (i.e., the reporting of Fred and Ethel as one dwarf's name).

Our review of the incorrect guesses of the dwarfs' names suggests that many of the subjects were in a "tip-of-the-tongue" state. In an experiment designed to induce "tip-of-the-tongue" phenomena, Brown and McNeill (1966) found that recall for the ending and beginning letters of a word was better than that for middle letters. Our data parallel those findings. Of the Seven Dwarfs, the initial letters are B, D, H, S, and G, and 65.8 percent of the guesses started with these letters. We correlated the percentage recall of the dwarfs versus the number of guesses that had the starting letters of the dwarfs' names (percentages were averaged for names with the same starting letter) and found the relationship strong (Spearman's rho = .87; $df = 5$; $p = .027$). Thus, a letter with a strong recall of its dwarfs produced many guesses. Poor recall produced few guesses. Our subjects also kept the length of their guesses close to the length of the dwarfs' names. Mean length of the seven names is 5.2 letters and of the guesses it was 5.4, a nonsignificant difference ($t = 1.52$; $df = 81$; $p > .05$). Similarly, 5 of 7 (71 percent) of the dwarfs' names end in y and 74 percent of the guesses ended in y. That Doc and Bashful, the two least remembered dwarfs, do not end in y tends to support our position. Bousfield and Wicklund (1969) found that rhyming words tended to be recalled in a cluster in a free-recall task.

Semantics also seems to play a role. We divided the dwarfs' names into three categories: names and "neutral" adjectives, "good" adjectives, and "bad" adjectives (defined as a personal trait). Colleagues thought the categories reasonable. Of the dwarfs, the names fell into three categories: neutral, 1 (14.3 percent); bad, 4 (57.1 percent); good, 2 (28.6 percent). Of the guesses, 25 percent were name and neutral, 52 percent bad, and 23 percent good. Happy, Doc, and Bashful were the weakest dwarfs in recall and position. That the other four were negative adjectives might have aided their recall. Guesses therefore seemed concentrated around negative adjectives with similar phonetic structures. For example, the two most popular guesses were Dumpy (17 percent) and Grouchy (12 percent).

Apparently forces seen in the laboratory can be demonstrated in the internal structuring, acquisition, and recall of material in the real world (if the Seven Dwarfs represent the "real world"). The magical number seven seems prominent in contemporary as well as classical mythology, and retrieval factors operate in the magic number and the magic kingdom. Further, these data are readily replicable in a classroom and can be used for an easy demonstration of basic processes in memory. Asking a class to recall the dwarfs makes the differences between recall and recognition, TOT, organization, and so on readily apparent.

REFERENCES

Bartlett, J. C., & Snelus, P. (1980). Lifespan memory for popular songs. *American Journal of Psychology, 93*, 551–560.

Bousfield, W. A., & Wicklund, D. A. (1969). Rhyme as a determinant of clustering. *Psychonomic Science, 16*, 183–184.

Brown, R., & McNeill, D. (1966). The "tip-of-the-tongue" phenomenon. *Journal of Verbal Learning and Verbal Behavior, 5*, 325–337.

Miller, G. A. (1956). The magical number seven, plus or minus two: Some limits on our capacity for processing information. *Psychological Review, 63*(2), 81–96.

Roediger, H. L., III, & Crowder, R. G. (1976). A serial position effect in recall of United States presidents. *Bulletin of the Psychonomic Society, 8*, 275–278.

Sidey, P. (1980). Introduction. *Walt Disney's Snow White and the Seven Dwarfs*. New York: Harmony Books.

NOTE: Parts of these data were presented at the 1982 meeting of the Western Psychological Association. Requests for offprints should be sent to Glenn E. Meyer, Department of Psychology, Lewis and Clark College, Portland, OR 97219.

ADDITIONAL QUESTIONS

1. Offer an "alternate hypothesis" for the results.
2. Search the literature for other examples of long-term memory. How did other researchers investigate this phenomenon?
3. In what other ways could one measure the dependent variable in this experiment? (Hint: Look up the original published experiment with experiments 2 and 3.)
4. How general are these results? For memory research? For cross-cultural research?
5. What additional research is suggested by this experiment?
6. Replicate this experiment with different materials using your class as subjects. Analyze and report your results.
7. What individual personality factors might be operating?
8. Do a literature search for the relationship between knowledge for common events and intelligence.

14

Creative Porpoise

INTRODUCTION

Two models of learning dominated the research activities of learning psychologists during the early part of this century. One model was developed by Ivan Pavlov and is commonly called *classical conditioning;* the other model was suggested by E. L. Thorndike and refined by B. F. Skinner and is referred to as *operant conditioning.* The initial experiments of both groups attempted to identify the conditions for learning using infrahuman subjects. Pavlov used his famous salivating dogs, while Thorndike studied the effects of reward on the behavior of cats. Skinner employed rats in his early experiments, then pigeons, and finally, among other species, humans. The contemporary period has seen a great proliferation of species used in the learning laboratory, all with the basic purpose of establishing laws of behavior. The use of porpoises in the present study is a logical step to illustrate the effectiveness of operant conditioning in yet another species.

The model developed by B. F. Skinner has several basic components that the authors of the "The Creative Porpoise" have skillfully applied. As you read this experiment try to identify some of the following principles of operant behavior.

An important principle of operant conditioning is that responses that are followed by a reward or positive reinforcement tend to increase, while nonrewarded responses tend to decrease. Skinner demonstrated this principle in rats by measuring the increase in bar-pressing responses that were followed by reward. The apparatus developed by Skinner has been described previously, and it is suggested that you refer back to Chapter 2 for a better understanding of this article. The initial behavior of a rat placed in a "Skinner box" is normally exploratory in nature; it sniffs the corners, moves from one side to the other, examines the walls, and washes its face. Only a small number of these responses have anything to do with the response of bar-pressing. The skilled researcher can identify preparatory bar-pressing responses and reinforce them. The process of selectively reinforcing successive approximations of the principal response (bar-pressing) is called *shaping*. Gradually the animal moves closer to the bar, then places its paw close to the bar, touches it, and eventually depresses it.

Much information has been collected regarding the specific conditions that facilitate operant conditioning. For example, if the reward is presented immediately after the appropriate response, then conditioning is more rapid than if the reward is delayed. In reading the article by Pryor et al., notice how they try to present the reinforcement as soon after the response to be conditioned as is practical.

Other researchers have studied the role that secondary reinforcers (stimuli that have been associated with the primary reward) have on behavior. The general conclusions of these studies is that secondary reinforcers have strong rewarding properties. Consider the reward properties of the secondary reinforcement of money in many societies. In this study Pryor et al. use a distinguishable signal (a whistle) as a conditioned reinforcer.

The results of operating conditioning experiments lend themselves to graphic representation in the form of a cumulative frequency record. On a cumulative frequency graph, the subject's responses are accumulated and scaled on the ordinate, while time is recorded on the abscissa. Since the responses are accumulative, the response curve never goes down; nonresponding is depicted as a line parallel to the abscissa.

This article was selected not only to illustrate the principles of operant behavior but also to give the student a look at an experiment that is largely descriptive in nature. You note that only one subject was used, and the statistical portion of the paper is largely descriptive. In this article the researchers show how data from psychological experiments can be effectively portrayed in graphic form. From the graphs we can quickly grasp the results they obtained.

Another feature of this experiment is that it combines a laboratory setting with a natural setting. The experiment was conducted in a large artificial pool, but the habitat is designed to simulate, as close as is practical, the natural environment of the subject. To do this research in the natural field of the experimental animal (i.e., the ocean) is probably impractical (although not impossible, as the researchers cite research conducted in the open field), and to conduct the research in a confined and artificial environment may inhibit the responses of the porpoise.

Most research in psychology depends on large samples; however, to infer that psychological research must study samples in order to be valid is an unwarranted conclusion. Pryor, Haag, and O'Reilly describe the learning process with clarity by thoroughly examining the behavior of a single subject.

SPECIAL ISSUES

Small *n* Designs

Research that is based on a single subject or small sample is called small *n* or small number of subjects designs. Most research in psychology is based on a large sample of subjects that generate a large amount of data. These designs are sometimes called large *n* designs. Data gathered from these experiments are amenable to statistical analysis in which probabilistic conclusions about the source of the outcome are based. Because of the enormous popularity of large *n* experiments and their compatability with modern statistical techniques, it may seem that small *n* designs are methodologically inferior. However, three branches of psychology have resisted the trend toward large *n* experiments. The first of these groups is led by psychophysicists, another by clinical researchers (see Chapter 19, *Special Issues: Single-Subject Design in Clinical Studies*), and a third by operant learning psychologists.

Psychophysical experiments generally deal with the relationship between physical stimuli and a subject's perception of stimuli. In many psychophysical designs each subject may be exposed to dozens or even hundreds of different stimuli in a single experiment. Thus the number of observations in psychophysical experiments based on a few subjects, or even a single subject, may exceed the number of observations based on a large *n* experiment.

Studies of operant conditioning frequently use small *n* design and in some instances a single subject. In these experiments a single subject (or only a few subjects) is intensely studied. As in psychophysical experiments and clinical studies, the design is frequently a within-subject design, where subjects serve in each condition. A chief proponent of small *n* designs has been B. F. Skinner, who has observed the development of behavior in a single subject. The research presented here on the creative porpoise is a clear example of a single-subject experiment. As you study the experiment you will note that the authors make no use of inferential statistics but do use descriptive statistics. This is a general feature of small *n* experiments, but very sophisticated statistics are available to analyze the results obtained in single-subject experiments.

The Creative Porpoise: Training for Novel Behavior

KAREN W. PRYOR
OCEANIC INSTITUTE
RICHARD HAAG
MAKAPUU OCEANIC CENTER
JOSEPH O'REILLY
UNIVERSITY OF HAWAII

Two rough-toothed porpoises *(Steno bredanensis)* were individually trained to emit novel responses, which were not developed by shaping and which were not previously known to occur in the species, by reinforcing a different response to the same set of stimuli in each of a series of training sessions. A technique was developed for transcribing a complex series of behaviors onto a single cumulative record so that the training sessions of the second animal could be fully recorded. Cumulative records are presented for a session in which the criterion that only novel behaviors would be reinforced was abruptly met with four new types of responses, and for typical preceding and subsequent sessions. Some analogous techniques in the training of pigeons, horses, and humans are discussed.

a | The shaping of novel behavior, that is, behavior that does not occur, or perhaps cannot occur, in an animal's normal activity, has been a preoccupation of animal trainers for centuries. The fox terrier turning back somersaults, the elephant balancing on one front foot, or Ping-Pong playing pigeons (Skinner, 1962) are produced by techniques of successive approximation, or shaping. However, novel or original behavior that is not apparently produced by shaping or differential reinforcement is occasionally seen in animals. Originality is a fundamental aspect of behavior but one that is rather difficult to induce in the laboratory.

SOURCE: Reprinted by permission from *Journal of the Experimental Analysis of Behavior,* 1969, *12,* 653–661. Copyright 1969 by the Society for the Experimental Analysis of Behavior, Inc. Some portions of the original article have been omitted.

ANALYSIS

Other articles used in this book began with a review of other studies. Pryor, Haag, and O'Reilly introduce their paper with a general statement of the use of operant conditioning in a variety of species (a). In the following two paragraphs they describe some natural observations which

In the fall of 1965, at Sea Life Park at the Makapuu Oceanic Center in Hawaii, the senior author introduced into the five daily public performances at the Ocean Science Theater a demonstration of reinforcement of previously unconditioned behavior. The subject animal was a female rough-toothed porpoise, *Steno bredanensis,* named Malia.

Since behavior that had been reinforced previously could no longer be used to demonstrate this first step in conditioning, it was necessary to select a new behavior for reinforcement in each demonstration session. Within a few days, Malia began emitting an unprecedented range of behaviors, including aerial flips, gliding with the tail out of the water, and "skidding" on the tank floor, some of which were as complex as responses normally produced by shaping techniques, and many of which were quite unlike anything seen by Sea Life Park staff in Malia or any other porpoise.

b

To see if the training situation used with Malia could again produce a "creative" animal, the authors repeated Malia's training, as far as possible, with another animal, one that was not being used for public demonstrations or any other work at the time. **A technique of record keeping was developed**

c

to pinpoint if possible the events leading up to repeated emissions of novel behaviors.

METHOD

d

A porpoise named Hou, of the same species and sex as Malia, was chosen. Hou had been trained to wear harness and instruments and to participate in physiological experiments in the open sea (Norris, 1965). This individual had a large repertoire of shaped responses but its "spontaneous activity" had never been reinforced. Hou was considered by Sea Life Park trainers to be "a docile, timid individual with little initiative."

---

led to the hypothesis (b) and an abbreviated statement of their method (c). It is important to note that the development of a hypothesis may emanate from many sources. Although the most frequent source of a hypothesis is previous research, another significant source is the observation of behavior in a natural setting.

METHOD

Paragraphs (d), (e), and (f) contain the identification of the subject, a description of how conditioning was done, and an explanation of the technique used to record the data. In paragraph (e) the authors describe the use of a bell that was sounded at the beginning of the training session. The use of a signal at the outset of an experiment is called a *discriminative*

e | Training sessions were arranged to simulate as nearly as possible Malia's five brief daily sessions. Two to four sessions were held daily, lasting from 5 to 20 min each, with rest periods of about half an hour between sessions. Hou was given normal rations; it is not generally necessary to reduce food intake or body weight in cetaceans to make food effective as a reinforcer. Any food not earned in training sessions was given freely to the animal at the end of the day, and it was fed normal rations, without being required to work, on weekends. During the experimental period, no work was required of Hou other than that in the experiment itself. **A bell was rung at the beginning and end of sessions to serve as a context marker.** The appearance and positioning of the trainers served as an additional stimulus that the opportunity for reinforcement was now present.

f | To record the events of each session, the trainer and two observers, one above water and one watching the underwater area through the glass tank walls, wore microphones and made a verbal commentary; earphones allowed the experimenters to hear each other. The three commentaries and the sound of the conditioned reinforcer, the whistle, were recorded on a single tape. A typed transcript was made of each tape; then, by comparing transcript to tape, the transcript was marked at 15-sec intervals. Each response of the animal was then graphed on a cumulative record, with a separate curve to indicate each type of response in a given session (Figures 14.2 to 14.5).

stimulus, and it sets the occasion for the conditioned operant. The bell in this experiment signified to Hou that it was time for it to do its thing.

This paper may appear to be an observational study rather than an experiment. Upon closer observation it is apparent that there are independent and dependent variables, operational definitions of terms, hypothesis testing, and controlled conditions, to name but a few of the characteristics of experimentation.

This experiment differs from several other experiments reviewed in the following chapters because the dependent variable (novel behavior) is followed by the independent variable (reinforcement). Compare this sequence with Terkel and Rosenblatt's article, "Maternal Behavior Induced by Maternal Blood Plasma Injected into Virgin Rats" (Chapter 15), in which the dependent variable (maternal behavior) is *preceded* by the independent variable (maternal blood plasma). You may have observed in the Pryor et al. article that the porpoise's behavior and reinforcement was ongoing, and it could be argued that a chain of responses—reinforcement, response; reinforcement, response—would develop. The sequence of independent and dependent variable then becomes a matter of deciding which came first: the reinforcement or the behavior.

Several statements in paragraph (f) require comment. The researchers are careful to have two observers who tape-record their observations, which are then transcribed into a typed manuscript. They also have a precise copy of the time and frequency of the conditioned reinforcer (a whistle).

g

It was necessary to make a relatively arbitrary decision about what constituted a reinforceable or recordable act. In general, a reinforceable act consisted of **any movement that was not part of the normal swimming action of the animal, and which was sufficiently extended through space and time to be reported by two or more observers. Such behavior** as eye-rolling, inaudible whistling, and gradual changes in direction may have occurred, but they could not be distinguished by the trainers and therefore could not be reinforced, except coincidentally. This unavoidable contingency probably had the effect of increasing the incidence of gross motor responses. Position and sequence of responses were not considered. An additional criterion, which had been a contingency in much of Hou's previous training, was that only one type of response would be reinforced per session.

The experimental plan of reinforcing a new type of response in each session was not fully met. Sometimes a previously reinforced response was again chosen for reinforcement, to strengthen the response, to increase the general level of re-sponding, or to film a given behavior. Whether the "reviewing" of responses was helpful or detrimental to the animal's progress is open to speculation.

h

Interobserver reliability was judged from the transcripts of the taped sessions, in which a new behavior was generally recognized in concert by the observers. Furthermore, each new behavior chosen for reinforcement was later diagrammed in a series of position sketches. At no time did any of the three observers fail to agree that the drawings represented the behaviors witnessed. These behavior dia-grams were matched, at the end of the experiment, with film of each behavior, and were found to represent adequately the topography of those behaviors that had been reinforced (see Figure 14.1).

In (g) an operational definition of a reinforceable act is stated. In shaping behavior the act to be reinforced is sometimes ambiguous, and some researchers will "play it by ear"; that is, they will make a decision to reinforce or not reinforce on the basis of ongoing behavior. Since scientific procedure must be specified in sufficient detail to allow the experiment to be exactly replicated, Pryor et al. attempt to reduce the ambiguity of shaping by operationally defining reinforceable responses. In (h) they explain the measure of experimental reliability used.

A final procedural question is raised in (i), and some evidence is presented in answer to the question. Complexity of responses rather than novelty of responses may have been conditioned. The experimenters used a form of consentional validation in which observations made by one experimenter are compared with observations of another experimenter to resolve the issue.

RESULTS

The results reported in this article are generally self-explanatory, and it is suggested that you thoroughly read the original transcript. Several general remarks may guide your reading.

After 32 training sessions, the topography of Hou's aerial behaviors became so complex that, while undoubtedly novel, the behaviors exceeded the powers of the observers to discriminate and describe them. This breakdown in observer reliability was one factor in the termination of the experiment.

To corroborate the experimenters' observation that certain of Hou's responses were not in the normal repertoire of the species, and constituted genuine novelties, the diagrams of each reinforced behavior were shown or sent to the 12 past and present staff members who had had occasion to work with animals of this species. Each trainer was asked to rank the 16 behaviors in order of frequency of occurrence in a free-swimming untrained animal. The sketches were mounted on index cards and presented in random fashion to each rater separately. A coefficient of concordance (W) of 0.598 was found for agreement between trainers on the ranking of various behaviors; this value is significant at the 0.001 level, indicating a high degree of agreement (Siegel, 1956).

To test the possibility that the trainers were judging complexity rather than novelty in ranking, another questionnaire was prepared requesting ranking according to relative degree of complexity of action. Because some of the original group of 12 trainers were unavailable for retesting, the questionnaire was presented to a group of 49 naive students. The coefficient of concordance (W) for agreement between students was +0.295, significant at the 0.001 level. When the ranking for complexity and frequency were contrasted for each behavior, it was found that some agreement existed between the scores given by the two rating groups, Spearman Rank Correlation (RHO) +0.54, significant at the 0.05 level.

Thus, there seems to be some agreement between complexity and frequency, which should be expected, since complex behaviors require more muscle expenditure than simple ones. Furthermore, analysis was biased by the fact that the experienced group was asked to rate all behaviors serially, and had no way other than complexity to rate the several behaviors that many of them stated they had

First, you will notice the absence of F tests or t tests. Indeed, the paper is noticeably lacking in statistical analyses. In lieu of such analysis the writers present a well-documented protocol of the changing behavior of the porpoise. Finally, cumulative frequency graphs are skillfully utilized in this article. Pryor et al. provide a daily graph (Figures 14.2 to 14.6) in which the porpoise's behavior is charted using the coordinates of responses (vertical axis) and time (horizontal axis).

It is important for you to study these graphs, as they are used as the primary source of results. In Figure 14.2 there are three distinct components: (1) the number of actions, (2) time in minutes, and (3) behavior (inverted swim, porpoise, and corkscrew). Notice the relationship among the three behaviors and how the inverted swim rapidly increases. The same three components are in Figure 14.3, and numerous other behavior characteristics are observed. Trace the development of the flip in Figure 14.3.

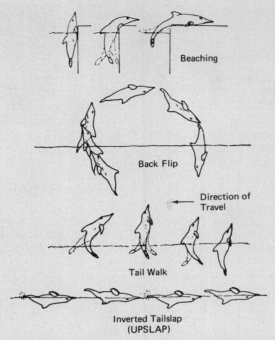

Beaching

Back Flip

Direction of
Travel

Tail Walk

Inverted Tailslap
(UPSLAP)

Figure 14.1   Four reinforced novel behaviors, including one shaped behavior—the tail walk.

never seen. However, the agreement between complexity and frequency was not as large between groups as it was within groups; allowing for the fact that the use of two rating groups makes it impossible to generalize the rating comparisons in a strict sense, the low frequency assigned to some noncomplex behaviors by the experienced group suggests that complexity and novelty are not necessarily positively correlated.

RESULTS

Sessions 1 to 14

In the first session, Hou was admitted into the experimental tank and, when given no commands, breached. Breaching, or jumping into the air and coming down sideways, is a normal action in a porpoise. This response was reinforced, and the animal began to repeat it on an average of four times a minute for 8 min. Toward the end of the 9-min session it porpoised, or leaped smoothly out of the water and in, once or twice.

Hou began the third session by porpoising; when this behavior was not reinforced, the animal rapidly developed a behavior pattern of porpoising in front of the trainer, entering the water in an inverted position, turning right side up, swimming in a large circle, and returning to porpoise in front of the trainer again. It did this 25 times without interruption over a period of 12.5 min. Finally, it stopped and laid its head against the pool edge at the trainer's feet. This behavior, nicknamed "beaching," was reinforced and repeated (Figure 14.2). Sessions 5, 6, and 7 followed the same pattern.

The trainers decided to shape specific responses in order to interrupt Hou's unvarying repetition of a limited repertoire. Session 8 was devoted to shaping a "tail walk," or the behavior of balancing vertically half out of the water. The tail walk was reinforced in Session 9, and Sessions 10 and 11 were devoted to shaping a "tail wave," the response of lifting the tail from the water. The tail wave was emitted and reinforced in Session 12.

At the end of Session 10, Hou slapped its tail twice, which was reinforced but not repeated. At the end of Session 12, Hou departed from the stereotyped pattern to the extent of inverting, turning right side up, and then inverting again while circling. The experimenters observed and reinforced this underwater revolution from a distance, while leaving the experimental area.

Although a weekend then intervened, Hou began Session 13 by swimming in the inverted position, then right side up, then inverted again. This behavior, dubbed a "corkscrew," was reinforced, and by means of an increasing variable ratio, was extended to five complete revolutions per reinforcement. In Session 14, the experimenters rotated their positions, and reinforced any

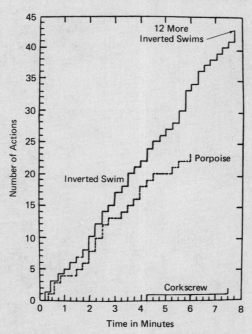

Figure 14.2 Cumulative record of Session 7, a typical early session, in which the porpoise began emitting the previously reinforced response. This response gradually extinguished when another response was formed.

descent by the animal toward the bottom of the tank, in a further effort not only to expand Hou's repertoire but also to interrupt the persistent circling behavior.

Sessions 15 and 16

The next morning, as the experimenters set up their equipment, Hou was unusually active in the holding tank. It slapped its tail twice, and this was so unusual that the trainer reinforced the response in the holding tank. When Session 15 began, Hou emitted the response reinforced in the previous session, of swimming near the bottom, and then the response previous to that of the corkscrew, and then fell into the habitual circling and porpoising, with, however, the addition of a tailslap on reentering the water. This slap was reinforced, and the animal then combined slapping with breaching, and then began slapping disassociated from jumping; for the first time it emitted responses in all parts of the tank, rather than right in front of the trainer. The 10-min session ended when 17 tailslaps had been reinforced, and other nonreinforced responses had dropped out.

Session 16 began after a 10-min break. Hou became extremely active when the trainer appeared and immediately offered twisting breaches, landing on its belly and its back. It also began somersaulting on its long axis in midair. The flip occurred 44 times, intermingled with some of the previously reinforced responses and with three other responses that had not been seen before: an upside-down tailslap, a sideswipe with the tail, and an aerial spin on the short axis of the body (Figure 14.3).

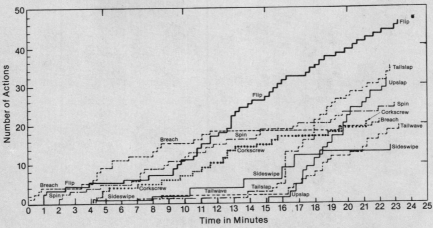

Figure 14.3   Cumulative record of Session 16, in which the porpoise emitted eight different types of responses, four of which were novel (flip, spin, sideswipe, and upslap).

This session also differed from previous ones in that once the flip had become established, the other behaviors did not tend to drop out. After 24 min, the varied activity—tailslaps, breaches, sidewipes with the tail, and the new behavior of spinning in the air—occurred more rather than less frequently, until the session was brought to a close by the trainer. The previous maximum number of responses in a given session was 110 (in Session 9, a 31-min session). In Session 16, Hou emitted 192 responses in a 23-min session, an average of 8.3 responses per min compared to a previous maximum average of 3.6 responses per min.

By Session 16, the experimenters had apparently been successful in establishing a class of responses characterized by the description "only new kinds of responses will be reinforced," and consequently the porpoise was emitting an extensive variety of new responses. The differences between Session 16 and previous sessions may be seen by comparing the cumulative record for Session 16 (Figure 14.3) with that of Session 7, a typical earlier session (Figure 14.2).

Sessions 17 to 27

In Sessions 17 to 27, the new types of responses emitted in Session 16 were selected, one by one, for reinforcement, and some old responses were reinforced again so that they could be photographed. Other new responses, such as unclassifiable twisting jumps, and sinking head downwards, occurred sporadically. The average rate of response and the numbers of types of responses per session remained more than twice as high as pre-Session 16 levels.

Hou's general activity changed in two other ways after Session 16. First, if no reinforcement occurred in a period of seven minutes, the rate and level of activity declined but the animal did not necessarily resume a stereotyped behavior pattern. Secondly, the animal's activity now included much behavior typically associated in cetaceans with situations producing frustration or aggressiveness, such as slapping the water with head, tail, pectoral fin, or whole body (Burgess, 1968).

Sessions 28 to 33

In all of the final sessions, the criterion that the behavior must be a new one was enforced. A new behavior that had been seen but not reinforced previously, the inverted tailslap, had been reinforced in Session 27. Session 28 began with a variety of responses, including another that had been seen but not reinforced before, a sideswipe at water surface with the tail, which was reinforced. In Session 29, Hou's activity included an inverted leap that fulfilled the criterion (Figure 14.4). In Session 30, Hou offered 60 responses

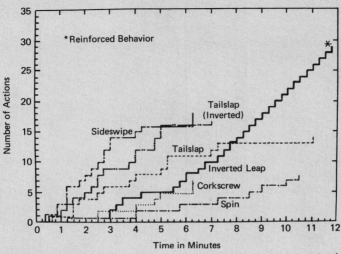

Figure 14.4 Cumulative record of Session 29, in which the porpoise emitted the three most recently reinforced responses initially, but soon emitted a novel response. When this response was reinforced, the others extinguished.

over a period of 15 min, none of which were considered new and were not therefore reinforced.

In Sessions 31, 32, and 33, held the next day, Hou's behavior was more completely controlled by the criteria that only new types of responses were reinforced and that only one type of response was reinforced per session. In Session 31, Hou entered the tank and, after a preliminary jump, stood on its tail and clapped its jaws at the trainer, who, taken by surprise, failed to reinforce the maneuver. Hou then emitted a brief series of leaps and then executed a backwards aerial flip that was reinforced and immediately repeated 14 times without intervening responses of other types. In Session 32, after one porpoise and one flip, Hou executed an upside-down porpoise, and, after it was reinforced, repeated this new response 10 times, again without other responses (Figure 14.5).

In the third session of the day, Hou did not initially emit a response judged new by the observers. After 10 min and 72 responses of variable types, the rate of response declined to 1 per min and then gradually rose again to seven responses per minute after 19 min. No reinforcements occurred during this period. At the end of 19 min, Hou stood on its tail and clapped its jaws, spitting water towards the trainer; this time the action was reinforced, and was repeated five times.

Hou had now produced a new behavior in six out of seven consecutive sessions. In Sessions 31 and 32, Hou furthermore began each session with

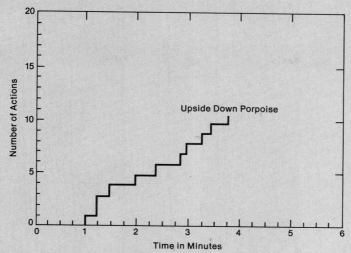

Figure 14.5 Cumulative record of Session 32. The porpoise emitted only a novel response in this session.

a new response and emitted no unreinforceable responses once reinforcement was presented. This establishment of a series of new types of responses was considered to be the conclusion of the experiment.

DISCUSSION

Over a period of 4 yr since Sea Life Park and the neighboring Oceanic Institute were opened, the training staff has observed and trained over 50 cetaceans of seven different species. Of the 16 behaviors reinforced in this experiment, five (breaching, porpoising, inverted swimming, tail slap, sideswipe) have been observed to occur spontaneously in every species; four (breaching, tail walk, inverted tailslap, spitting) have been developed by shaping in various animals but very rarely occur spontaneously in any; three (spinning, back porpoise, forward flip) occur spontaneously only in one species of *stenella* and have never been observed at Sea Life Park in other species; and four (corkscrew, back flip, tailwave, inverted leap) have never been observed to occur spontaneously. While this does not imply that these behaviors do not sometimes occur spontaneously, whatever the species, it does serve to indicate that a single animal, in emitting these 16 types of responses, would be engaging in behavior well outside the species norm.

A technique of reinforcing a series of different, normally occurring actions, in a series of training sessions, did therefore serve, in the case of Hou, as with Malia, to establish in the animal a highly increased probability that new types of behavior would be emitted.

j This ability to emit an unusual response need not be regarded as an example of cleverness peculiar to the porpoise. It is possible that the same technique could be used to achieve a similar result with pigeons. If a different, normally occurring action in a pigeon is reinforced each day for a series of days, until the normal repertoire (turning, pecking, flapping wings, *etc.*) is exhausted, the pigeon may come to emit novel responses difficult to produce even by shaping.

k A similar process may be involved in one traditional system of the training of five-gaited show horses, which perform at three natural gaits, the walk, trot, and canter, and two artificial gaits, the slow gait and the rack. The trainer first reinforces the performance of the natural gaits and brings this performance under stimulus control. The discriminative stimuli, which control not only the gait, but also speed, direction, and position of the horse while executing the gait, consist of pressure and release from the rider's legs, pressures on the reins and consequently the bit, shifting of weight in the saddle, and sometimes signals with whip and voice. To elicit the artificial gait, the trainer next presents the animal with a new group of stimuli, shaking the bit back and forth in the horse's mouth and vibrating the legs against the horse's sides, while preventing the animal from terminating the stimuli (negative reinforcement) by means of the previously reinforced responses of walking, trotting, or cantering. The animal will emit a variety of responses that eventually may include the pattern of stepping, novel to the horse though familiar to the trainer, called the rack (Hildebrand, 1965). The pattern, however brief, is reinforced, and once established is extended in duration and brought under stimulus control. (The slow gait is derived from the rack by shaping.)

DISCUSSION

The implication of training creative behavior in the porpoise is discussed in relationship to achieving similar results with the pigeon (j) and the horse (k), and possible common features of the present study and previous work by Maltzman are suggested (l).

QUESTIONS

1. In this study identify:

   conditioned reinforcer   cumulative frequency graph
   discriminative stimulus   shaping
   operant   creativity (operational definition)
   drive   response-reinforcement latency

1   Comparison may be made here between this work and that of Maltzman (1960). Working in the formidably rich matrix of human subjects and verbal behavior, Maltzman described a successful procedure for eliciting original responses, consisting of reinforcing different responses to the same stimuli, essentially the same procedure followed with Hou and Malia. It is interesting to note that behavior considered by the authors to indicate anger in the porpoise was observed under similar circumstances in human subjects by Maltzman: "An impression gained from observing Ss in the experimental situation is that repeated evocation of different responses to the same stimuli becomes quite frustrating; Ss are disturbed by what quickly becomes a surprisingly difficult task. This disturbed behavior indicates that the procedure may not be trivial and does approximate a nonlaboratory situation involving originality or inventiveness, with its frequent concomitant frustration."

Maltzman also found that eliciting and reinforcing original behavior in one set of circumstances increased the tendency for original responses in other kinds of situations, which seems likewise to be true for Hou and Malia. Hou continues to exhibit a marked increase in general level of activity. Hou has learned to leap tank partitions to gain access to other porpoises, a skill very seldom developed by a captive porpoise. When a trainer was occupied at an adjoining porpoise tank Malia jumped from the water, skidded across 6 ft of wet pavement, and tapped the trainer on the ankle with its rostrum, or snout, a truly bizarre act for an entirely aquatic animal.

Individual differences in the ability to create unorthodox responses no doubt exist; Malia's novel responses, judged *in toto*, are more spectacular and "imaginative" than Hou's. However, by using the technique of training for novelty described herein, it should be possible to induce a tendency toward spontaneity and creative or unorthodox response in most individuals of a broad range of species.

2. Can you suggest another technique for measuring the responses?
3. Take another species—for example, dog, goldfish, squirrel—and write a reinforcement schedule for the training of novel behavior. Operationally define your terms.
4. Why did the researchers avoid extensive statistical analysis in this paper?
5. Write a brief essay describing the worth of naturalistic observations. What are their strengths? Their weaknesses?
6. Write a lesson plan for 6-year-old children in which creative reactions are shaped and reinforced. What limitations to your lessons do you see?
7. From this research, have you changed your definition of "creativity"? If so, how?
8. List the control measures used in this experiment.

REFERENCES

Burgess, K. (1968). The behavior and training of a Killer Whale at San Diego Sea World. *International Zoo Yearbook, 8,* 202–205.

Hildebrand, M. (1965). Symmetrical gaits of horses. *Science, 150,* 701–708.

Maltzman, I. (1960). On the training of originality. *Psychological Review, 67,* 229–242.

Norris, K. S. (1965). Open ocean diving test with a trained porpoise *(Steno bredanensis). Deep Sea Research, 12,* 505–509.

Siegel, S. (1956). *Nonparametric statistics for the behavioral sciences.* New York: McGraw-Hill.

Skinner, B. F. (1962). Two synthetic social relations. *Journal of the Experimental Analysis of Behavior, 5,* 531–533.

NOTE: Contribution No. 35, the Oceanic Institute, Makapuu Oceanic Center, Waimanalo, Hawaii. Carried out under Naval Ordinance Testing Station Contract # N60530-12292, NOTS, China Lake, California. A detailed account of this experiment, including the cumulative records for each session, has been published as NOTS Technical Publication # 4270 and may be obtained from the Clearing House for Federal Scientific and Technical Information, U.S. Department of Commerce, Washington, D.C. A 16-mm film, "Dolphin Learning Studies," based on this experiment, has been prepared by the U.S. Navy. Persons wishing to view this film may inquire of the Motion Picture Production Branch, Naval Undersea Warfare Center, 201 Rosecrans Street, San Diego, California 92132. The authors wish to thank Gregory Bateson of the Oceanic Institute, Dr. William Wesit of Reed College, Portland, Oregon, and Dr. Leonard Diamond of the University of Hawaii for their extensive and valuable assistance; also Dr. William McLean, Technical Director, Naval Undersea Research and Development Center, San Diego, California, for his interest and support.

CHAPTER
15

Maternal Behavior

INTRODUCTION

Maternal behavior (i.e., nest building, care of the young, retrieving, nursing, etc.) has long been considered an "instinct." However, in recent years it has become apparent that calling a behavior an "instinct" tells us little about its causes. Psychologists have therefore begun to carry out studies of the factors underlying such behavior. These factors include stimuli coming from the nest or the young, brain mechanisms, and past experience. The suggestion that hormones, too, may play a role in controlling such behavior comes from the fact that maternal behavior develops very gradually and fades away in the same manner; that is, nest building and the bodily changes accompanying maternity take place prior to birth and disappear gradually as the young grow older. This characteristic of the behavior suggests that it may be related to the gradual buildup and decline of some chemical substance in the blood. One way to demonstrate the involvement of a hormone in the control of a behavior is to inject the hormone and see whether it can induce the behavior. Previous studies of this type have failed to produce clear-cut results, and the present study is an attempt to approach the problem in a somewhat different fashion.

The inclusion of this article on maternal behavior illustrates several important issues in experimental design. Among these issues is the problem of control. Several types of control were considered. The experimental animals were divided into equal-sized groups (presumably in a random fashion). Some groups were defined as control groups, and other groups were designated the experimental groups. All animals were treated identically, with the exception of the type of chemical injection they received. And another type of control was in the preparation of substances used in the injection.

This experiment has a distinctive independent variable and a control group. The experimenters measured the reaction of an experimental group against the reaction of a control group. You should note that something was done *to* the subjects. They were injected with various chemicals.

SPECIAL ISSUES

The Use of Animals in Psychological Research

Some of the earliest experiments in psychology were done with nonhuman animals. Ivan Pavlov used dogs, E. L. Thorndike used cats, William James used chickens, Harry Harlow used monkeys, the Gardners used chimpanzees, and B. F. Skinner used rats. Other psychologists have used guinea pigs, seals, porpoises, monkeys, elephants, whales, planera, bees, pigeons, sheep, pigs, horses, rabbits, etc. One might think psychologists were more interested in the behavior of nonhuman creatures than in the behavior of *homo sapiens*. However, in most instances experimental psychologists—even those who restrict their studies to nonhuman animals—strongly assert that their experiments are designed to lead to a better understanding of the human animal.

The basic premise on which psychological experimentation is predicated is that conclusions reached through the study of nonhuman subjects is in some way applicable to an understanding of behavior. There are many reasons for using animals in psychological research. In general, animals are more available than human subjects. It is possible to keep a nonhuman animal under observation 24 hr a day, seven days a week for months, if necessary. In addition, it is possible to perform procedures with animals which are impossible to do with humans. These types of research may involve the use of noxious stimuli, prolonged periods of deprivation, psychosurgery, the use of experimental drugs, etc. It is almost impossible to get a college sophomore to volunteer for some of these experiments, and even if subjects were available, many of the experiments would constitute a serious breach of ethics. No less important is the maintenance of ethical restraints on nonhuman subjects. The experimental psychologist who uses animals is honor-bound to conduct his or her research within the guidelines of the ethical canons described in Chapter 7 (see especially Principle 10 and the APA "Guidelines").

The experiment by Joseph Terkel and Jay Rosenblatt on maternal behavior induced by injecting maternal blood plasma into virgin rats was selected for inclusion in this section (along with the article on the creative porpoise) as it illustrates a type of experiment in which nonhuman animals serve as the experimental subjects. A similar experiment with humans would not be possible, and yet the conclusions reached in this experiment may give us a clue as to the source of maternal behavior in humans.

Maternal Behavior Induced by Maternal Blood Plasma Injected into Virgin Rats

JOSEPH TERKEL
INSTITUTE OF ANIMAL BEHAVIOR
JAY S. ROSENBLATT
RUTGERS UNIVERSITY

Induction of maternal behavior (i.e., retrieving) in virgins by exposure to young pups was studied to investigate effects of blood plasma from a postparturient female. Control groups were injected with blood plasma from nonmaternal females in proestrus and diestrus phases of the vaginal estrous cycle and with saline solution. Virgins injected with maternal blood plasma had significantly shorter latencies of maternal behavior than other groups. Injections of saline and proestrus blood plasma had no effect on maternal behavior. Virgins injected with diestrus blood plasma were significantly delayed in displaying maternal behavior. The findings indicate that there is a humoral basis for the appearance of maternal behavior after parturition.

Several recent attempts to induce maternal behavior in the rat by means of various hormones (i.e., estrogen, progesterone, and prolactin) injected directly into virgin or experienced females have not yielded results that would increase our undertanding of the hormonal basis of this behavior (Beach & Wilson, 1963; Lott, 1962; Lott & Fuchs, 1962). In this laboratory injected hormones (prolactin and oxytocin) have failed also to maintain maternal behavior in females that have become maternal after parturition or have been made maternal by Caesarean-section delivery of their fetuses several days before parturition. However, the conviction that maternal behavior in the rat is based upon hormones is supported by the success in inducing nest building in the mouse with progesterone (Koller, 1952, 1955) and in the hamster with estrogen and progesterone (Richards, 1965). Some success has

SOURCE: Reprinted by permission from *Journal of Comparative and Physiological Psychology*, 1968, 65 (3), 479–482. Published by the American Psychological Association.

ANALYSIS

REVIEW OF LITERATURE

In the review of literature, Terkel and Rosenblatt carefully present two sides of a controversy surrounding the hormonal basis of maternal behavior: While some researchers have found strong evidence for the hor-

been reported in inducing maternal nest building in rabbits using a combination of hormones (e.g., stilbestrol, progesterone, and prolactin were used by Zarrow, Sawin, Ross, & Denenberg, 1962).

a

In view of the difficulty of inducing maternal behavior in female rats with injected hormones, a difficulty no doubt based upon failure to introduce either the proper hormone or hormones in the proper order and dose level, we have attempted a different approach to the problem. Remaining close to the natural conditions under which maternal behavior normally appears at and shortly after parturition, we have attempted to transfer blood plasma from postparturient females that have become maternal within the past 48 hr to virgins, hoping thereby to induce maternal behavior in the latter. **Establishing that maternal blood plasma carries a substance or substances capable of inducing maternal behavior in virgins would be a first step in identifying the humoral basis of maternal behavior.**

b

We have shown recently that virgin females can be induced to show maternal behavior when they are exposed to young pups continuously for about 5 days (Rosenblatt, 1967). Cosnier (1963) reported a similar finding using shorter daily exposures. Both studies confirm an early suggestion by Wiesner and Sheard (1933) which was only partially verified in their own studies. Maternal behavior induced under these conditions appears to be of nonhormonal origin since ovariectomizing or hypohysectomizing virgins before exposing them to pups did not prevent the appearance of maternal behavior or alter, significantly, latencies for the onset of a major item of maternal behavior, namely retrieving (Cosnier & Couterier, 1966; Rosenblatt, 1967). In this study, therefore, we observed whether the latency for the appearance of retrieving (and other items of maternal behavior) by virgins exposed to young pups was significantly reduced by prior injection of maternal blood plasma as compared to prior injection of blood plasma taken from virgins in the proestrus or diestrus phases of the vaginal estrous cycle, or prior injection of saline solution.

c

---

monal basis of maternal behavior, others have not. In (a) the authors speculate about the ambiguity of previous experimental results by suggesting that such studies have not employed the proper hormones in the proper order and at proper dose levels. However, if the blood conditions in virgin rats (presumably behaviorally nonmaternal) were nearly identical to the blood conditions of rats who had recently delivered a litter of pups (presumably behaviorally maternal), then a reliable measure of the serological basis of maternal behavior would be possible. The authors suggest in (b) that we should identify the gross phenomenon of the hormonal basis of maternal behavior as a first step that would presumably lead to a more discrete analysis of specific chemicals responsible for this behavior.

METHOD

Subjects

Thirty-two virgin females, 60 days of age at the start of the experiment, were obtained from Charles River Breeding Farm, Dover, Mass. Twenty-four additional Ss of the same age provided blood plasma for the injections. Other rats provided the pups used in the maternal behavior tests. The Ss were housed individually in 45 × 50 × 40 cm. rectangular cages, each with transparent Plexiglas walls, grid floor, wall feeder, water bottle, and two bins containing hay and coarse wood shavings for nesting material. They were fed Purina chow and water ad lib supplemented twice weekly with vitamin-enriched bread, carrots, and lettuce.

The Ss were divided into four equal-sized groups. One group received plasma taken from maternal Ss; injections were given when Ss were in various unspecified phases of the vaginal estrous cycle. One control group consisted of Ss in proestrus that received plasma taken from females that were also in proestrus, and a second control group consisted of Ss in diestrus that received plasma taken from females that were also in diestrus. The fourth group of Ss received an injection of 0.9 percent saline solution at various unspecified phases of the vaginal estrous cycle.

The last sentence of the review (c) contains the hypothesis to be tested and a brief version of the research plan. In effect, the researchers state the dependent variable, latency of retrieving pups, and the independent variable, injection of different groups with (1) maternal blood plasma, (2) blood from animals in the proestrus phase, (3) blood from animals in the diestrus phase, and (4) saline solution. A fifth group was not injected and served as a control. The proestrus phase in rats is the period immediately prior to estrus ("heat"), and the diestrus period is the period that follows estrus. Presumably, these phases were selected to reduce the possibility of hormones present during the estrus phase from causing a general increase in the activity level which could have been falsely interpreted as maternal behavior.

METHOD

Three parts are included in the method section: the subjects are identified, the procedures are described, and the tests of maternal behavior are specified. In this study, enough detail is present to allow its replication. For example, in describing the subjects the authors identify the breeding farm, age, sex, cage size and construction, feed, water schedule, and nesting material. There is a practical limit to the amount of methodological detail that can be presented. For this reason some items, such as temperature, humidity, lighting, and so on, are deleted, and one could logically infer that these items were controlled.

Procedures

Blood was withdrawn from the donors within 48 hr after parturition, after it was clearly established that these Ss were performing maternal behavior normally. Blood taken from estrous-cycling donors was withdrawn within 1 hr after the vaginal smear indicated either proestrus or diestrus. Between 6 and 8 cc of blood was withdrawn from the heart. To withdraw the blood, the donors were anesthetized with ether, and the heart was surgically exposed by a chest incision to one side of the midline. The blood was withdrawn using a 40 mm. 16-gauge needle with a 20-cc syringe containing 15–20 units of Heparin Sodium to prevent blood clotting.

About 4 min elapsed from the time the donor was judged to be completely anesthetized until the blood was first transferred from the syringe to a test tube and centrifuging was started. When a zone of clear plasma, free of blood cells, appeared 3–4 cc of plasma was drawn into a 5-cc syringe and injected into a subject with a 20 mm. 27-gauge needle. Plasma injection was completed in 4 min. Each experimental S received all of its plasma from a single donor female.

The S was lightly anesthetized with ether, a small incision was made on the inner surface of the upper thigh, and the right femoral vein was exposed. The needle was inserted into the vein, a small amount of blood was withdrawn, and then the plasma (or saline solution) was injected slowly over a period of 2½ min. The incision was closed with wound clips. In this inbred strain of rats, plasma transfer between any two strain mates does not result in anaphylactic shock.

Maternal Behavior Tests

Each animal was given a 15-min retrieving test 1 hr before blood was withdrawn from the proestrus and diestrus donors and plasma or saline was injected into the recipient Ss. Since no S retrieved during this test it was not necessary to eliminate any of them from the experiment.

Tests following the injection of either plasma or saline were begun after it was judged that Ss were fully recovered from the anesthetic used during the injection. Each S was judged to be recovered if it was able to walk around the edge of a

In the procedure section, Terkel and Rosenblatt demonstrate their considerable talent for unambiguous communication. They explain in detail the method used for blood transfusion. A researcher with some surgical skill could exactly replicate their procedure.

A critical aspect of this study is in the definition of the dependent variable: maternal behavior. The validity of scientific inquiry, to a large extent, rests upon the operational definition of dependent variables and the testing of these variables. The present authors suggest several behavioral characteristics that typify maternal behavior, including latency in retrieval of pups, crouching over young, licking young, and nest building.

bell jar maintaining its balance; if it fell from the edge it was retested at a later time. It was possible for the first postinjection test to begin in all Ss 1 hr after the injection since all Ss were fully recovered from the ether about 5 min after the injection.

Five pups, 5–10 days of age, were placed at the front of each Ss cage. Retrieving was observed for 15 min, following which observations for 1-min periods at 20-min intervals were made over the next 2 hr. During the 1-min period of observation the occurrence of retrieving, crouching over the young, licking the young, nest building, and other maternal and nonmaternal items of behavior were recorded. Nesting material had previously been spread over the floor. At the end of the 2-hr test the pups were left with Ss until the next morning at which time they were removed and replaced by a fresh litter of five pups in the same age range. The test procedure was repeated daily until an S retrieved pups in two consecutive daily tests. Since retrieving is usually the last item of maternal behavior to appear when pups are used to induce it, all Ss had already shown the other main items of maternal behavior (i.e., crouching over young, licking young, nest building) by the termination of testing.

Several Ss that had been injected with maternal plasma were observed continually on the first day following the first test to see if times of maternal behavior would appear between the first and the second test, 22 hr later.

RESULTS

d Mean latencies in days for the onset of retrieving for the various groups, shown in Table 15.1, indicate that plasma taken from a lactating mother within 48 hr after delivery is capable of inducing a more rapid maternal response to pups than saline and either proestrus or diestrus plasma ($F = 9.79$, $df = 3/28$, $p < .01$; data transformed to square roots). Under the combined influence of maternal plasma and stimulation from pups, retrieving appeared in an average of 2 days. This time was significantly shorter than when saline or proestrus plasma was injected or when diestrus plasma was injected (Duncan's New Multiple Range test at the .05 level). Proestrus plasma combined with the proestrus condition of the recipient virgin was similar in its effect on maternal behavior to saline but diestrus plasma given to females in diestrus produced a significant delay in the mean latency for the onset of retrieving (Duncan's New Multiple Range test at the .05 level).

RESULTS

The essence of this research paper is typified in the first sentence (d) of the results section. A main result of this experiment is plainly stated: Plasma of lactating rats is capable of inducing more rapid maternal behavior than other substances. Statistical evidence in the form of the F statistic or analysis of variance is then presented. In addition to the F

Table 15.1 MEAN LATENCIES IN DAYS
FOR THE ONSET OF RETRIEVING

GROUP	N	MEAN	SE
Maternal plasma	8	2.25	.97
Proestrus plasma	8	4.62	1.21
Diestrus plasma	8	7.00	2.96
Saline	8	4.00	1.41
Untreated[a]	14	5.79	2.69

[a]Taken from Rosenblatt (1967).

In a previous study (Rosenblatt, 1967) it was established that maternal be-havior (i.e., retrieving and other items) can be induced in virgins by exposure to pups, without any prior injection, with an average latency of 5.79 ± 2.69 days. The saline-injected proestrus plasma- and diestrus plasma-injected Ss of the present study had average latencies which did not differ significantly from Ss in the earlier study (Mann-Whitney $U = 40$–42, $p > .10$). The average latency of the maternal plasma-injected Ss was, however, signifi-cantly shorter than that of the Ss that were only exposed to pups (Mann-Whitney $U = 20$, $p = .02$).

The onset of retrieving was accompanied in all groups by the occurrence of the three other main items of maternal behavior (i.e., crouching over the young, licking, and nest building). With pups continuously present, the onset of an item of maternal behavior, and particularly the onset of retrieving, was followed by its appearance from then on in each of the subsequent daily tests. Observations of the Ss between the first and second tests led us to believe that the maternal plasma induced maternal behavior more rapidly than our formal test procedure was capable of detecting. Several Ss that were injected with maternal plasma began to show maternal behavior in attenuated fashion within 4–8 hr after the injection and the beginning of exposure to pups, although, several hours earlier, during the first scheduled test of maternal behavior, they were indifferent to the pups. Two of these were fully maternal, according to our criteria for the virgins, by the second test, which was begun 1 day after the injection. Others were not fully ma-ternal until the third test was begun 2 days after the injection.

statistic, the authors employ Duncan's New Multiple Range test and the Mann-Whitney U statistic. Duncan's test is used to ascertain whether significant differences exist between each of the means, while the Mann-Whitney is a special type of analysis for ranked data with two classes of information.

In the second paragraph of the results section the authors introduce an untreated control group from a previous study. This procedure is ob-viously not as desirable as one that would include an untreated control group within the same experiment. There may be differences between

DISCUSSION

e

Our study established for the first time that substances carried in the plasma of the newly maternal rat are capable of increasing the readiness of virgins to respond maternally to pups. We have, therefore, finally found a way of accomplishing what Stone (1925) set out to do when he joined in parabiosis a maternal and nonmaternal rat hoping that blood-borne substances responsible for maternal behavior in the former would induce maternal behavior in the latter. Were it not for the failure of these substances to cross from the maternal to the nonmaternal animal, because of selective transmission across the parabiotic union, Stone would have demonstrated what we have found and perhaps the effect would have been stronger with the continuous exchange of blood that he attempted.

f

The present study does not enable us to identify the substance or substances that are responsible for increasing the maternal responsiveness of the virgins or to determine whether these substances act on the virgin via the endocrine system or directly upon the nervous system. Initially we thought that dividing our plasma control group into two groups, one receiving proestrus plasma

the two experiments in the sample of rats, the handling of the rats, or the experimental procedures. On the other hand, the procedure used in this series of studies is fairly standardized and as such makes a cross-experiment comparison somewhat feasible (although not totally desirable). Data from one experiment should not be analyzed with a second experiment unless the experimenter is quite sure that the subjects and procedures of the two experiments are quite similar.

In the third paragraph of this section the authors introduce some *observational* data that their experimental procedures were not sensitive enough to detect. Observational data of this type are often quite valuable in helping to clarify or further interpret the results. However, it should be noted that data of this type are usually not collected as systematically or precisely as the data collected on the formal testing of the dependent variable. Therefore these data should be treated as supplementary to the results found in the formal testing of the dependent variable, and this is how the authors present them.

DISCUSSION

The discussion of the article begins with a bold statement (e) that is a combination of empirically validated evidence and a logically inferred statement. Terkel and Rosenblatt quickly point out that the hypothesis tested in this study is not a new one and, in addition, suggest a reason for previous failures to validate the hypothesis by other research workers. They also state a limitation (f) by noting that their study did not specif-

and the other diestrus plasma, the virgins themselves being in the corresponding phases of the estrous cycle, would enable us to make a first step in identifying the active substances. To the extent that the diestrus blood plasma combined with the diestrus condition of the recipient virgin produced a delay in the onset of maternal behavior we have been partially successful. However, any identification of ovarian hormones or pituitary secretions as the active substances would be highly speculative and incapable of substantiation at this time. Our findings therefore await further analysis of hormonal secretions during the estrous cycle, pregnancy, and parturition.

An added finding of importance does emerge from this study which was surprising to us. Our previous work indicated that maternal responsiveness increases gradually during pregnancy (Lott & Rosenblatt, 1967), and we interpreted this as indicating that the hormonal conditions during pregnancy gradually sensitized the neural substrate of maternal behavior thereby preparing for the appearance of maternal behavior at parturition. The present study suggests that there need be no prolonged period (i.e., 22 days of pregnancy) of sensitization for substances contained in maternal plasma to have their effect on maternal behavior. It would appear that the gradual increase in maternal responsiveness which we found during pregnancy after Caesarean-section deliveries (Lott & Rosenblatt, 1967) need not be built up by a continual addition of "units" of maternal responsiveness. Rather the level of maternal responsiveness at each period of pregnancy reflects for that particular moment the current capability of the blood to stimulate maternal behavior and this capability presumably undergoes a continuous increase until it is fully established around parturtion. In this respect then our findings agree with those of Moltz and Weiner (1966) and Denenberg, Grota, and Zarrow (1963) that hormonal secretions at parturition are likely to be important for the induction of maternal behavior.

REFERENCES

Beach, F. A., & Wilson, J. R. (1962). Effects of prolactin, progesterone, and estrogen on reactions of nonpregnant rats to foster young. *Psychological Report 13*, 231–239.

Cosnier, J. (1963). Quelques problèmes posés par le "comportement maternel provoqué" chez la ratte. *CR Soc. Biol.*, Paris, *157*, 1611–1613.

Cosnier, J., & Couterier, C. (1966). Comportement maternel provoqué chez les rattes adultes castrées. *CR Soc. Biol.*, Paris, *160*, 789–791.

Denenberg, V. H., Grota, L. J., & Zarrow, M. X. (1963). Maternal behavior in the rat: Analysis of cross-fostering. *Journal of Reproduction and Fertility*, *5*, 133–141.

Koller, G. (1952). Der Nestbau der Weiber Mause und seine hormonale Auslosung. *Verh. dtsch. zool. Ges.*, Freiburg, 160–168.

Koller, G. (1955). Hormonale und psychische Steuerung beim Nestbau Weiber Mause. *Zool. Anz., (Suppl.)*, *19*, 125–132.

Lott, D. F. (1962). The role of progesterone in the maternal behavior of rodents. *Journal of Comparative and Physiological Psychology*, *55*, 610–613.

Lott, D. F., & Fuchs, S. S. (1962). Failure to induce retrieving by sensitization or the injection of prolactin. *Journal of Comparative and Physiological Psychology, 55,* 1111–1113.

Lott, D. F., & Rosenblatt, J. S. (1967). Development of maternal responsiveness during pregnancy in the rat. In B. M. Foss (Ed.), *Determinants of infant behavior IV.* London: Methuen.

Moltz, H., & Weiner, E. (1966). Effects of ovariectomy on maternal behavior of primiparous and multiparous rats. *Journal of Comparative and Physiological Psychology, 62,* 382–387.

Richards, M. P. M. (1965). *Aspects of maternal behaviour in the golden hamster.* Unpublished doctoral dissertation, Cambridge University, 1965.

Rosenblatt, J. S. (1967). Non-hormonal basis of maternal behavior in the rat. *Science, 156,* 1512–1514.

Stone, C. P. (1925). Preliminary note on maternal behavior of rats living in parabiosis. *Endocrinology, 9,* 505–512.

Wiesner, B. P., & Shard, N. M. (1933). *Maternal behaviour in the rat.* London: Oliver and Boyd.

Zarrow, M. X., Sawin, P. B., Ross, S., & Denenberg, V. H. (1962). Maternal behavior and its endocrine bases in the rabbit. In E. L. Bliss (Ed.), *Roots of behavior.* New York: Harper & Row.

This research was supported by National Institute of Mental Health Research Grant MH-08604 to J. S. R. and Biological Medicine Grant FR-7059 to J. T. We wish to thank D. S. Lehrman and B. Sachs for reading the manuscript. Publication No. 50 from the Institute of Animal Behavior, Rutgers University, Newark.

ically identify the substance(s) within the blood responsible for increasing maternal behavior. One can capture a feeling of excitement in this research, and even the newcomer to psychology can anticipate the next development.

QUESTIONS

1. Speculate as to the results if greater quantities of blood had been transfused.
2. Would you care to make a generalization to human transfusion? Why is this generalization warranted or unwarranted? Do you see any "practical" application of this study?
3. Why did the authors include a saline group? A proestrus group? A diestrus group? An untreated group?
4. What significance do you attribute to the results of the diestrus group?
5. What group(s) would you like to add?
6. In addition to the behavioral indices of maternal behavior, suggest several physiological measures of maternal tendencies.

Could these be quantifiable? Would evaluation of these changes be important to this study? Why? Why not?

7. This paper might as readily have been published in a physiological journal. What is the relationship between psychology and physiology?

8. Why are rats used in psychological studies?

CHAPTER
16

Humor

INTRODUCTION

In our daily interactions with other people, we are constantly interpreting people's actions to try to ascertain the "true feelings" or the "intentions" of the actors. Someone may compliment you because he or she genuinely likes what you are doing or because he or she wants to "butter you up." Politicians may publicly endorse a particular program because they actually believe in the program or because they want the votes that they think such endorsement will bring. We know that many times the words and actions of people do not accurately reflect their feelings. Consequently, without our necessarily being aware of it, we use a series of "tests" to determine whether a person's actions are responses to the current situation or if they reflect his or her actual beliefs.

SPECIAL ISSUES

Attribution Theory

Social psychologists are beginning to analyze the tests that people use to determine the motives of others. Many of these tests seem to fall quite nicely into a framework that is generally referred to as *attribution theory*. This theory sets down some of the principles of attributing causes to the behavior of others. There have been many experiments that have tested hypotheses derived from attribution theory. The Suls and Miller article is an interesting application of the theory to humor. In this experiment the subjects read about a person who tells an anti-women's liberation joke to an audience who was either conservative or liberated, and the audience reacts with either laughter or a glare. The subjects were then asked to rate how chauvinistic the joke teller was. Thus we have an interesting case of the subjects trying to ascertain the true feelings of the joke teller based on the characteristics and reactions of the audience.

Humor as an Attributional Index

JERRY M. SULS
STATE UNIVERSITY OF NEW YORK AT ALBANY
RICHARD L. MILLER
GEORGETOWN UNIVERSITY

The present study examined the circumstances under which observers make attributional judgments about an actor on the basis of his humor. Subjects read a description of a group discussion in which a man told an anti-women's liberation joke to a group of women. The variables manipulated were the audience's anticipated reaction (approval or disapproval) and the audience's actual reaction (approval or disapproval). The results indicated that the actor was seen as least chauvinistic when disapproval was anticipated but approval was received. The actor was seen a most chauvinistic when disapproval was both anticipated and received.

a

Although a considerable amount of attention has been paid to the cognitive and motivational aspects of humor (Goldstein & McGhee, 1972; Levine, 1969; Suls, 1972), little is known about the social importance of humor. The present paper reports an exploratory study concerned with what kinds of social perceptions are made about persons who tell jokes under various social circumstances.

b

Based on theories of attribution (Jones & Davis, 1965; Kelley, 1971) it would seem that the degree to which someone's joke is presumed to reflect his true beliefs would depend on the situational constraints present when the joke is told. Two attributional principles may provide a useful framework for examining the social consequences of joke telling. The first, Kelley's (1971) augmentation principle, states that a facilitative cause for a behavior is seen as more effective if the behavior occurred despite the opposing effect of an inhibitory cause. The second, the discounting principle, claims that the role of a given cause in producing a given effect is discounted if other

SOURCE: Reprinted by permission from *Personality and Social Psychology Bulletin*, 1976, *2*, 256–259.

ANALYSIS
REVIEW OF LITERATURE

The first section of this paper nicely points out the purpose of this research as well as the theory behind it. Paragraph (a) outlines the general problem to be explored in the experiment. Paragraph (b) explains the

plausible (facilitative) causes are also present. According to the augmentation principle, if an actor behaves contrary to situational pressures (expected disapproval), his personal belief should more likely be seen as causing his behavior than if the behavior were consistent with situational pressures (expected approval). The discounting principle suggests that when disapproval is expected, the actor's belief is the only plausible (facilitative) cause but when approval is expected, both the actor's belief and social approval are plausible causes and thus render the role of the actor's belief and social approval more ambiguous. Thus, both principles may lead observers to attribute more sincerity to an actor when negative consequences are the expected outcome of his behavior (Jones, Davis, & Gergen, 1961; Mills & Jellison, 1967). **On this basis we would expect that a joke told to a potentially hostile audience would be perceived as more indicative of the joke teller's attitude than would a joke told to a potentially friendly audience.**

Although the actual reaction to any behavior may be distinguished from the anticipated reaction, little attention has been given to the effects of actual reactions on attributions. For example, Jones and Davis' theory stresses the subjective anticipation of decision-outcomes rather than their actual occurrence. There is, however, good reason to consider audience reaction to humor. If an audience reacts with disapproval or hostility, it may indicate that they took the joke seriously (i.e., that they perceived the joke teller as expressing his opinion which is contrary to theirs). If, on the other hand, the audience reacts favorably, this may reflect either their agreement with the joke's content or their perception that the joke teller is being playful. Whether the audience reacts with approval or disapproval may be an important cue for the observer's judgment of the joke as indicative of the actor's true beliefs. The present exploratory study attempted to determine the effects of anticipated and actual reactions to humor on the perception of an actor's true beliefs.

theory behind the research, which in turn leads to a logical derivation of the first hypothesis (**c**).

In the next paragraph (**d**) the authors discuss another problem to be explored in the experiment—that of the audience's reaction to humor. The authors offer no hypothesis for this variable; rather they point out that audience reaction may be an important cue in judging an actor's true beliefs.

METHOD

This type of research is fairly easy to execute. The equipment costs are minimal, as the experimental manipulation consists of different booklets of mimeographed material which are given to different subjects. It is

METHOD

Subjects were 86 students enrolled in introductory psychology classes. There were approximately an equal number of males and females. The subjects were asked to participate in a study on social perception and were given one-page descriptions of a group interaction to read.

The description indicated that a group of four people—three women and one man—participated in a discussion of several social issues: welfare, busing, and so on. All four discussants were described as knowing one another socially through their local civic association and as being reasonably well acquainted thus giving the actor presumably a basis upon which to form expectations about the consequences of his behavior. To manipulate the anticipated audience reaction, one-half of the subjects read information indicating that the three women were politically and socially very conservative; the other half of the subjects received information indicating the women were politically and socially "very liberated." During the course of the conversation the man in the group, John, told a joke which "put down" certain aspects of the women's liberation movement. At this point in the conversation the women either "laughed heartily with John" or "glared

sometimes called "pencil-and-paper" research, as this is the only equipment needed. The subjects are usually tested in large groups, which allows for the data to be collected quickly as well as at little cost. The availability of equipment, the cost of the research, the availability of subjects, and the time involved are all important concerns in doing research. This type of research seems to solve most of these problems.

In (e) we learn that subjects were from introductory psychology classes and there were an almost equal number of males and females. The experiment was described as one on "social perception," and the "stimulus material" was a single-page description of a group interaction.

The group interaction is explained in (f). Note that there are two independent variables: (1) audience attitudes and (2) audience reaction. And there are two levels of each of the independent variables. For the first variable, the audience was either very conservative or very liberal. For the second independent variable, the audience reaction was either laughter or a glare. Thus we have a 2 × 2 factorial design similar to that explained in Chapter 2. In a 2 × 2 design there are four treatment groups (2 × 2 = 4), which are diagrammed below:

	AUDIENCE REACTION Laughter	Glare
AUDIENCE ATTITUDE Conservative	1	2
Liberal	3	4

In the description John tells an anti-women's liberation joke. In experimental group 1 the audience is conservative and responds with

silently at him." This served as the manipulation of actual audience reaction. One-half of the subjects read the first reaction; the other half read the second. After more information about the group interaction, i.e., other topics covered, subjects were asked a number of questions about the discussion including an estimate of John's opinion about the women's liberation movement. Subjects were also asked to rate how likable they found John and to indicate their own opinion of women's liberation. The subjects made their responses on 7-point Likert-type scales with smaller numbers representing greater "chauvinism," "likableness," and "unfavorableness."

laughter; in experimental group 2 the audience is conservative and responds with a glare; and so on.

After the subjects had read the description, they checked several Likert-type scales (g) concerning how "chauvinistic" and how likable John was, as well as other questions. A typical 7-point Likert scale is shown below. Subjects are asked to check the space that best represents how chauvinistic they think John is.

_____ extremely chauvinistic
_____ somewhat chauvinistic
_____ slightly chauvinistic
_____ neutral or undecided
_____ slightly liberated
_____ somewhat liberated
_____ extremely liberated

Subjects are given a numerical score from 1 to 7 based on the space they checked. Low scores indicate greater chauvinism.

RESULTS

Table 16.1 presents the mean liking score and the mean chauvinism score for each of the four treatment groups.

Since the study deals with women's liberation, it is reasonable to assume that the women subjects may respond differently from the men subjects to the various experimental treatments. To test this assumption, the means of the men subjects across the four treatments were compared to that of the women subjects to see if they differ (it becomes a 2 × 2

Table 16.1   THE EFFECTS OF ANTICIPATED AND ACTUAL
AUDIENCE REACTIONS ON ATTRIBUTIONS OF BELIEF AND LIKING

AUDIENCE	REACTION	ATTRIBUTIONS OF BELIEF	LIKING
Liberated	Laugh ($n = 22$)	4.46	2.13
	Glare ($n = 20$)	2.76	3.48
Conservative	Laugh ($n = 21$)	3.53	3.26
	Glare ($n = 21$)	4.09	2.86

RESULTS

Since a preliminary analysis of the data indicated no sex differences across conditions, the data were collapsed across sex. **A 2 (anticipated reaction) × 2 (actual reaction) analysis of variance was performed on the subjects' estimates of John's attitude toward women's liberation.** The analysis of variance indicated a significant main effect of actual audience reaction ($F = 4.72$, $df = 1/82$, $p < .03$) and a significant anticipated reaction × actual reaction interaction ($F = 15.72$, $df = 1/82$, $p < .001$). Post-hoc comparisons indicated that the actor was seen as most chauvinistic when he addressed a potentially hostile audience and received disapproval as a result of telling his joke. The actor was seen as least chauvinistic when he addressed a potentially hostile audience that reacted favorably. In contrast, when the actor addressed a potentially friendly audience, attributions were less extreme and did not differ across conditions of actual reaction.

The liking ratings were also subjected to a 2 × 2 analysis of variance, which indicated a main effect of actual reaction ($F = 4.18$, $df = 1/82$, $p < .04$) and a significant two-way interaction ($F = 11.96$, $df = 1/82$, $p < .01$). Post-hoc comparisons indicated that the actor was seen as most likable when disapproval was anticipated but approval was received; the actor was seen as least likable when disapproval was anticipated and received.

× 2 design with male-female as one factor). In (**h**) we learn that there were no such sex differences.

A 2 × 2 analysis of variance of John's attitude toward women's liberation indicated a significant main effect of actual audience reaction and a significant interaction effect. The authors describe this in (**i**). However, a graphic representation would be helpful here. In Figure 16.1 the two types of audiences are represented on the horizontal axis, and the two types of audience reactions are represented by separate lines. The analysis of variance (**i**) indicated a significant main effect of audience reaction, which indicates that, in general, John was seen to have a more liberated attitude when he received a laugh reaction to his joke. (The student should reread the section on factorial designs in Chapter 2 for a discussion of main and interaction effects in the analysis of variance.)

However, the interaction effect described in (**i**) qualifies the main effect. The post-hoc comparisons (i.e., statistical tests done on all *pairs* of means) indicate that there was no statistically significant difference between the means when the audience was conservative (potentially friendly) (X in Figure 16.1); however, there was a significant difference when the audience was liberated (potentially hostile) (Y in Figure 16.1). The authors base their interpretation of the results on this analysis.

In (**j**) the authors report the analysis for the liking ratings. Since they are very similar to the ratings for chauvinism, the authors explain why this is so.

An examination of the attribution means and the liking means shows that when the actor was seen as least chauvinistic he was seen as most likable and when most chauvinistic least likable. This pattern is explicable since the subjects' mean opinion of women's liberation was 4.75 indicating that they were favorable toward women's liberation. Not surprisingly, subjects liked those who they believed held similar beliefs (i.e., others who were in favor of women's liberation).

DISCUSSION

The results indicated that anticipated and actual reaction to the joke interacted such that the most extreme attributions were made where the anticipated reaction was negative. For the condition where disapproval was both anticipated and received, the observer was reasonably certain that the actor's joke reflected his beliefs. This was probably true since (a) there was no apparent external cause for telling the joke and (b) the audience took it seriously. Conversely, where it was anticipated that a joke would receive a hostile reaction but instead was greeted with approval, observers attributed the least belief-humor consistency. It is reasonable to assume that a group

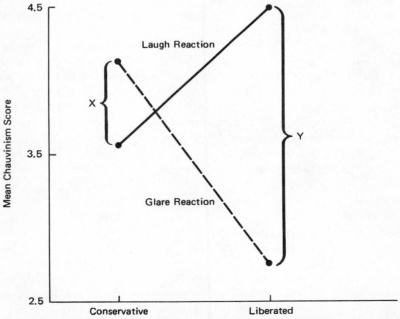

Figure 16.1   The interaction effect of the antiwomen's liberation when the audience was conservative (X) and when the audience was liberated (Y).

would only laugh at a joke that was hostile to their views if the group knew that the joke teller was "teasing" and did not really agree with the content of the joke at all. In fact in order to laugh the audience must have assumed that the actor was in agreement with them. This means that observers could see the actor's beliefs as directly opposite from those expressed in the joke.

The results indicated that less extreme attributions occurred when anticipated audience reaction was favorable. In the case where approval was anticipated and received, the actor may have told the joke simply to get approval or the audience didn't take the joke seriously, or they agreed with the joke. Due to the many possible causes for the actor's behavior, this is an ambiguous situation, so it is reasonable that the observers made less certain inferences about the actor's beliefs. Also, when approval was expected but not received a problematic situation for the observer was created. Why would a group not laugh at a joke which was consistent with their beliefs? The answer could be that either the joke was perceived to be in bad taste or the audience thought the joke teller was being facetious by ridiculing their views. For either reason observers did not rate the actor as holding the view expressed by the joke.

k

To summarize, the results suggest that anticipated and actual audience reaction interact such that the actor was seen as *most* in agreement with the beliefs expressed in his joke when disapproval was anticipated but approval was received. Less extreme attributions occurred when approval was anticipated regardless of the actual reaction since the possible external cause (expected approval) seemed to render ambiguous the actor's real belief.

The present results suggest that under certain social circumstances humor is used as an attributional index. In addition the findings suggest that the resulting attributions may differ depending on whether actual and anticipated social reactions are consistent or inconsistent, positive or negative. It remains to be determined what other situational elements increase or decrease the information value of humor and whether the actual $\times$ anticipated reaction interaction holds for nonhumorous attitudinal statements.

DISCUSSION

The discussion section is fairly brief and consists of an interpretation of the results. Here the authors attempt to explain the pattern of results. Some of this explanation is necessarily speculation and you may be able to think of other reasons for the results obtained.

Finally, the authors end with a summary of the results (k) which can be viewed as a statement of the conclusions drawn from this experiment.

REFERENCES

Goldstein, J. H., & McGhee, P. E. (Eds.). (1972). *The psychology of humor.* New York: Academic Press.

Jones, E. E., & Davis, K. E. (1965). From acts to dispositions: The attribution process in person perception. In L. Berkowitz (Ed.), *Advances in experimental social psychology* (Vol. 2). New York: Academic Press.

Jones, E. E., Davis, K. E., & Gergen, K. E. (1961). Role playing variations and their informational value for person perception. *Journal of Abnormal and Social Psychology, 63,* 302–310.

Kelley, H. (1971). *Attribution in social interaction.* Morristown, N.J.: General Learning Press.

Levine, J. (Ed.) (1969). *Motivation in humor.* New York: Atherton.

Mills, J., & Jellison, J. (1967). Effect on opinion change of how desirable the communication is to the audience addressed. *Journal of Personality and Social Psychology, 6,* 98–101.

Suls, J. (1972). A two-stage model for the appreciation of jokes and cartoons. In J. H. Goldstein & P. E. McGhee (Eds.), *The psychology of humor.* New York: Academic Press.

QUESTIONS

1. Explain how the first hypothesis (c) is a logical derivation of the theory (b).
2. Could some hypotheses about the effect of audience reaction (d) be derived from the theory (b)? If so, what would they be?
3. Draw a figure illustrating the results of the liking ratings and interpret the results.
4. How would you assign subjects to treatments in this experiment?
5. An obvious advantage of this type of "pencil-and-paper" research is that complex social situations can be studied under controlled conditions rather easily. What are the disadvantages of this type of research?
6. Design an experiment to test the effect of audience attitudes and audience reaction on attitude attribution in a situation that does not involve humor.

CHAPTER
17

Alcohol and Perception

INTRODUCTION

There is little doubt that excessive consumption of alcohol has an adverse effect on hand–eye performance: One simply has to look at the carnage on our highways due to alcohol abuse to confirm that statement. But it may be that alcohol does not affect all psychological processes equally or that different levels of alcohol ingestion may affect different processes. To untangle some of the complex relationships between alcohol ingestion and the perception of motion, Roger MacArthur and Robert Sekuler examined the effect of various levels of alcohol ingestion on three motion perception skills.

This article is to be analyzed by the student; however, there are several distinct features to which we would like to direct your attention. Notice that the research is theoretically motivated, but also may have some practical application. The authors also are very specific about the conditions of alcohol ingestion. As you study this section, note the many careful procedures used by the authors. Also, look at the way MacArthur and Sekuler treat their data. Although some of the statistical concepts may be challenging to you, they deserve your close analysis as we think the treatment is very sophisticated.

As you read the article we suggest that you try to answer the following questions:

1. What is the general theme of this paper?
2. How has past research addressed the issues?

3. What is the problem? The hypothesis?
4. What did the experimenters do? What was asked of the subjects? (See *Method* section)
5. What was the independent variable? The dependent variable?
6. How were the results collected? How were they analyzed?
7. Briefly summarize what was done and what the results were.
8. How did the authors discuss their experimental results?

Alcohol and Motion Perception

ROGER D. MACARTHUR and ROBERT SEKULER
NORTHWESTERN UNIVERSITY, EVANSTON, ILLINOIS

Three motion perception skills were measured under different levels of alcohol ingestion. Our method for detecting decrements in visual information processing proved sensitive to blood alcohol levels as low as .02%. Alcohol in small doses increased reaction times to the onset of motion, particularly to slow speeds, but did not reduce the ability to allocate attention effectively. In view of these findings, certain motion perception tests may be valuable assays for detecting impaired performance with low blood alcohol levels.

Among the serious consequences of alcohol abuse is reduced ability to process visual information. Diminished vision from excessive consumption of alcohol has been implicated in industrial, home, and automobile accidents (Ryan, Salter, Cox, & McDermott, 1976). However, impaired performance associated with low levels of alcohol intoxication has often been difficult to measure and quantify. This has been particularly true for tests involving temporally modulated stimuli.

The present research examined the effects of alcohol ingestion upon three measures of visual function. We wished to determine which, if any, of these three visual capacities might be diminished by ingestion of alcohol. This determination could establish more precisely the mechanisms by which alcohol ingestion actually contributes to risk of accidents. In one test, we measured the ability of an observer to pay attention to two different directions of movement. A related test has been used to work on selective auditory attention designed to predict road accident frequency (Kahneman, Ben-Ishai, & Lotan, 1973). In fact, it has been claimed that "alcohol reduces our ability to effectively allocate our attention" (Shinar, 1978). We sought to put the generality of such a claim to test.

We used two additional tasks specifically designed to measure possible impairment of response to motion with alcohol ingestion—the precision with which an observer could judge the direction of a target's movement and his speed of response to targets moving at different velocities. Performance on all three tasks was compared with and without alcohol ingestion.

SOURCE: Reprinted by permission from *Perception & Psychophysics*, 1982, 31(5), 502–505. Some portions of the original article have been omitted.

METHOD

Apparatus

In all of the experiments reported here, stimuli were random dot patterns presented under computer control on the face of a cathode ray tube (CRT). At any given moment, between 400 and 500 random dots were visible to the observer. The dots were centered within an 8-deg circular aperture. The computer controlled precisely both direction and speed with which the dots moved across the CRT (see Ball & Sekuler, 1979, for further details of the display). Dots in all three tasks were presented at a luminance sufficient to make them approximately 50 times their own detection thresholds. Thus, the observer was asked either to judge the direction, in the first task, or to make a response to motion onset, in the second and third tasks, to dots that were highly visible. Reaction times were defined by the elapsed time between onset of motion and the observer's response. Observers viewed the CRT display monocularly from a distance of 57 cm. A small black dot in the center of the CRT provided a fixation point upon which observers were to hold their gazes.

Subjects

We tested nine male observers, averaging 24 years of age. Each had visual acuity of 20/20 or better (or corrected to 20/20 or better). The mean acuity, measured just prior to the experiment by means of a Bausch & Lomb Orthorater, was 20/18. In addition, all observers considered themselves to be light to moderate drinkers, consuming an average of four drinks per week (range: one to eight).

Procedure

Conditions of Alcohol Ingestion

We tested observers under three conditions of alcohol ingestion. All doses were administered in a drink that had a total of 3 ml liquid/kg body weight. The highest dose consisted of 1 g/kg body weight of 190-proof ethanol, diluted 1:2 with Hawaiian Punch to produce the required total volume of liquid. The second dose consisted of .5 g/kg body weight of alcohol diluted 1:5 with Hawaiian Punch. We also administered a placebo consisting of Hawaiian Punch alone. Thus, in the high-dose condition, observers received between 64 and 89 ml of 190-proof ethanol, depending on body weight. In every case, though, the high-dose drink was 32% ethanol. The amount and percentage of ethanol in the low-dose condition were exactly half those in the high-dose condition.

To minimize olfactory and visual cues, each drink was presented in an opaque, lidded cup through which a straw protruded. Two ice cubes were put into each drink. Observers were allowed 20 min in which to consume the drink through the straw. Each observer received all three doses, with the order of the dose individually randomized. Only one dose was given per day. Test periods began immediately before ingestion and at 30, 75, 120, and 180 min after the 20 min allowed for ingestion of the drink had expired. During each test period, all three tasks were run in a single, 15- to 20-min session. The order of tasks within a test period was randomized.

The experiment always began at 1:00 P.M. and ended at approximately 5:00 P.M. Since we wished to minimize the chances that observers would become nauseated by the alcohol, they were instructed to eat lunch before coming to the laboratory. The importance of keeping time and amount of intake constant was stressed to them. Observers gave an oral description of the time and content of meals prior to the start of the experiment, and we found that they consistently had followed instructions. These steps were taken to try to keep each observer's rate of alcohol absorption as constant as possible from day to day.

Practice Sessions

Initially, each observer had three practice sessions on all the experimental tasks. These practice sessions, designed to familiarize the observers with the three tasks and to stabilize performance, were conducted over either 1 or 2 days. Practice was completed between 1 and 7 days prior to the collection of actual experimental data.

Breath Analysis

All blood alcohol levels were estimated by using a Breathalyzer (Alco-Tector, Decatur Electronics). Measurements were made immediately before and immediately after each period, and the mean of the two was calculated. Mean blood alcohol levels were somewhat lower than would have been the case if subjects had fasted before the experiment began. Furthermore, the peak of intoxication is likely to have occurred sometime between the first two test periods after drink consumption, and sampling when we did yielded points on either side of the peak (see Table 17.1). The peak mean readings were .02% and .06% for the .5-g/kg and 1.0-g/kg doses. Standard deviations associated with these peak scores were .008 and .017, respectively.

Tasks

One task, "direction judgment," required the observer to estimate the direction in which dots moved across the CRT. Five trials were presented at each of 10 possible directions ranging from 9 deg right to 9 deg left of vertical (at 2-deg intervals). Directions were presented in a random order. Each presentation lasted 1.2 sec, with an intertrial interval of 2.2 sec. The speed of movement was 4 deg/sec. Observers were instructed to hold their gazes on the fixation point as long as the dots were visible. After the dots had been extinguished, observers moved their gazes to a protractor encircling the CRT and read off the perceived direction of the dots' movement. These numerical reports constituted the data for this task.

Table 17.1   MEAN PERCENT BLOOD ALCOHOL LEVELS AT VARIOUS TIMES

DOSE*	TIME SINCE DRINK (IN MINUTES)				
	0	30	75	120	180
.5	.00	.02	.02	.01	.00
1.0	.00	.06	.05	.05	.04

*In milliliters per kilogram of body weight.

In the second task, "reaction time to movement," the observer was required to react to the onset of dot movement across the CRT. In each sitting, the observer was tested on a total of 96 trials, 12 at each of 8 speeds: .125, .25, .5, 1, 2, 4, 8, and 16 deg/sec. On each trial, the dots first appeared on the screen as a stationary pattern. Then, after a random foreperiod lasting between 1 and 2 sec, the dots instantly accelerated to the required speed and moved rightward across the screen. The interval between trials was 1.5 sec. Note that the observer could not predict from one trial to the next the speed of movement to which he would have to respond. The response was a buttonpush that observers were to make as rapidly as possible to the onset of movement, ignoring the speed insofar as possible. Although the dots actually did move on each trial, observers were told that sometimes a trial with no movement might be presented. This misinformation was given to insure that, especially at the slowest speeds, observers would not respond until they actually had seen motion.

The third experimental task, "direction uncertainty," compared reaction times to onset of movement under two basic conditions—one in which the observer could be certain from trial to trial of the direction the movement would take, and another in which from trial to trial the observer was uncertain as to the direction the movement would take. Certainty blocks had 24 trials, all with rightward movement; uncertainty blocks had 24 trials, with movement either to the right (12 trials) or to the left (12 trials), in random order. Before each block, the observer was told in which condition he would be tested. Two blocks of certainty alternated with two blocks of uncertainty in an ABBA fashion, with a total of 96 trials run in any test period. On each trial, the dots first appeared as a stationary pattern and then, after a random period of between 2 and 3 sec, instantly accelerated to 4 deg/sec. The observers were instructed to push a button as soon as the dots began to move, ignoring the movement's direction. The intertrial interval was approximately 3 sec.[1]

RESULTS

Before the results are described, a word is in order about the processing of data from each experimental task. For judgments of direction, we calculated the mean judgment for each of the 10 possible directions presented in any block of trials. For the reaction time to various speeds, we identified the highest and lowest reaction times to each of the eight speeds and discarded them; we also discarded reaction times less than 100 msec or greater than 900 msec (see below for justification). From the remaining reaction times, we computed the geometric mean for each speed. Because distributions of reaction times tend to be skewed, the geometric mean more validly reflects the central tendency of the reaction time distributions than does the arithmetic mean (Woodworth & Schlosberg, 1954). With the uncertainty procedure,

[1]Note that the length of the foreperiods and intertrial intervals differed between Experiments 2 and 3. The difference was necessary to allow the computer to perform the calculations required between trials. However, we did manage to keep constant the ratio of foreperiod to intertrial intervals between the two experiments.

we also identified the highest and lowest reaction times for each direction within each block and discarded them prior to data analysis. From the remaining reaction times, we calculated the geometric means.

Direction Judgments

The data were converted to error scores defined by the absolute value of the difference between the observer's judgment of direction of the movement and the actual direction. The absolute (or unsigned) error was chosen as the response measure because the symmetrical distribution of our observers' errors would have given mean errors of approximately 0 deg if the signed error had been used instead.

Since only five judgments had been obtained for each direction, we collapsed all directions before data analysis began. Error scores were entered in an analysis of variance (ANOVA) with two factors—dose of alcohol (three levels) and time of testing (five times). This ANOVA revealed one source of variance, that of alcohol dose, to be statistically significant $[F(2,16) = 4.50, p < .03]$. Surprisingly, with alcohol the errors actually decreased slightly. The mean error was 3.0 deg for the placebo condition, 2.6 deg for the half-dose condition, and 2.7 deg for the full-dose condition.

Reaction Time to Movement

Again, an ANOVA was done. Here the design included velocity (eight levels), dose (three levels), and time relative to dose (five times). Significant sources of variance were the main effect of velocity $[F(7,56) = 128.00, p < .001]$, the main effect of the linear component of dose $[F(1,8) = 11.53, p < .01]$, the interaction of the linear components of dose and velocity $[F(1,8) = 6.07, p < .04]$.

As velocity increased from .125 deg/sec through 16 deg/sec, mean reaction times decreased from 613, through 447, 362, 310, 283, 264, and 257, to 254 msec. Increasing doses of alcohol produced an overall elevation in reaction times. From placebo through 1.0 g/kg, the times increased from 335 through 351 to 360 msec. Not unexpectedly, then, alcohol seems to increase reaction time to onset of movement. But the patterns of statistically significant interactions require that this overall effect of alcohol be considered in more detail. For instance, the interaction between dose and velocity reflects the fact that alcohol primarily elevated reaction times to the slower speeds of movement. Thus, reaction times to the slowest speed increased by nearly 50 msec as one went from the placebo to the 1.0-g/kg dose, while reaction times to the most rapid speed increased by only 16 msec over the same range of conditions. The triple interaction among velocity, dose, and time reflects the fact that alcohol had its primary effect on measurements between 30 and 75 min after completion of drink. In other words, as one would expect, the effect of alcohol was a quadratic function of time, first increasing and then decreasing. These relationships are shown in Figure 17.1 for three of the times of testing. Note that the reaction times at 180 min for the .5-g dose are equal to or slower than the times for the 1.0-g dose, despite the fact that the blood alcohol level is 0 in low-dose subjects and .04 in high-dose subjects. This unexpected result might indicate that other factors, such as fatigue, are exerting a large effect in the later stages of the experiment, causing the reaction times to cluster and masking any smaller effect of alcohol.

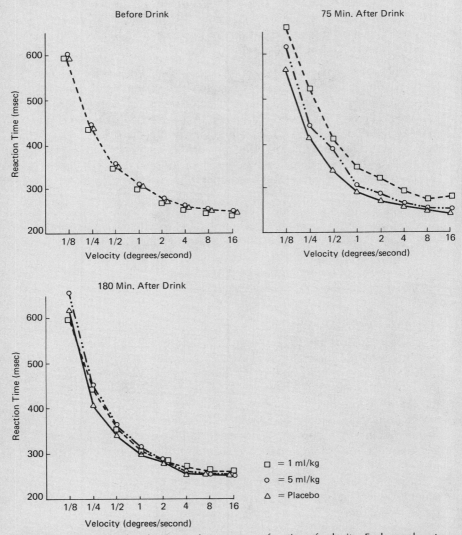

Figure 17.1   Mean reaction time to motion onset as a function of velocity. Each panel portrays data at one time relative to the consumption of the drink (before, 75 minutes after, and 180 minutes after). Within each panel, one curve is shown for each dose of alcohol.

An ANOVA on the intraobserver standard error of the mean reaction times showed that the effect of velocity was significant, with higher standard errors being associated with the slower speeds of movement [$F(7,56) = 178.80$, $p < .001$], suggesting some heterogeneity of variance. However, the validity of inferences about alcohol's effect on RT is not comprised by that heterogeneity of variance because no term involving alcohol even approached significance in the ANOVA on standard errors.

Two classes of responses, premature and tardy, were examined. A premature reaction time was defined as less than 100 msec. A genuine response to movement would have required 150 msec at least (Woodworth & Schlosberg, 1954), so premature reaction times indicate that the oberver was anticipating movement rather than actually responding to its onset. A tardy reaction time was defined as more than 900 msec. Such a response time is more than three standard deviations greater than the mean of any of our reaction time distributions and shows that the observer had great difficulty in detecting the motion. Premature responses were rare (2%), and their frequency was unaffected by any other variables in the experiment. In other words, alcohol did not seem to increase impulsive behavior (see also Sekuler & MacArthur, 1977). The overall frequency of tardy responses for all but the slowest speed also was low (.3%); at that slowest speed, the incidence of tardy responses rose to 20%. However, alcohol did not have any obvious effect on the number or distribution of tardy responses.

Direction Uncertainty

As before, an ANOVA was done. The design included certainty vs. uncertainty blocks (2 levels), dose (3 levels), and time relative to dose (5 times). The mean reaction time to rightward movement in conditions of direction certainty was 253 msec, whereas the mean reaction time in conditions of direction uncertainty was 266 msec. The ANOVA showed this difference to be statistically significant [$F(1,8) = 27.59$, $p < .001$]. Thus, this 13-msec elevation in reaction time, although small, proved to be highly significant, confirming the results of Sekuler and Ball (1977). No effect involving dose of alcohol accounted for a statistically significant portion of the variance. Although observers were generally handicapped by having to pay attention to two different directions of movement, this handicap seemed not to be modulated by the dose of alcohol that they had ingested.

DISCUSSION

The following summarizes the main findings and implications of the present work. Alcohol, in low doses, significantly lengthens reaction time to visual movement. This effect is particularly strong with slower motion. Therefore, we feel this particular visual perception task may be a valuable technique by which to measure impaired performance with low blood alcohol levels.

We should note that the results of Experiment 2 do not demonstrate an effect of alcohol that is specific to *motion* perception. In fact, it may be that the results reflect a general decrement in RT to stimulus change.

Contrary to previous claims, alcohol does not necessarily reduce our ability to allocate attention effectively (Kahneman et al., 1973; Moskowitz & Sharma, 1974; Shinar, 1978). In particular, alcohol, at least at low doses, did not affect observers' ability to pay attention to two possible directions of motion. Our results require that earlier claims about alcohol's effect on attention be restated in less general form.

Surprisingly, low doses of alcohol made observers' judgments of direction of motion more, rather than less, precise. Furthermore, the effect was not dose dependent, since the low and high doses of alcohol were almost equally effective in improving accuracy. These results may reflect an alcohol-induced potentiation of involuntary, tracking eye movements (Sekuler & MacArthur, 1977; Wilkinson, Kime, & Purnell, 1974). Although a fixation point always was present in the center of our screen, if the observer had traced by eye the movement from the fixation point to the protractor around the CRT, he could have virtually perfectly identified the direction of movement from the path. Observers did in fact comment that, with any dose of alcohol, they had considerable difficulties in maintaining fixation. However, in most other tasks, if alcohol diminished stability of fixation and rendered the observer more susceptible to involuntary tracking eye movements, this same effect would be deleterious to visual performance and could play a major role in reducing visual stabilization of posture.

REFERENCES

Ball, K., & Sekuler, R. (1979). Masking of motion by broadband and filtered directional noise. *Perception & Psychophysics, 26*, 206–214.

Kahneman, D., Ben-Ishai, R., & Lotan, M. (1973). Relation of a test of attention to road accidents. *Journal of Applied Psychology, 58*, 113–115.

Moskowitz, H., & Sharman, S. (1974). Effects of alcohol on peripheral vision as a function of attention. *Human Factors, 16*, 174–180.

Ryan, G. A., Salter, W. E., Cox, C. J., & McDermott, F. T. (1976). Blood alcohol and road trauma survey. *Medical Journal of Australia, 2*, 129–131.

Sekuler, R., & Ball, K. (1977). Mental set alters visibility of moving targets. *Science, 198*, 60–62.

Sekuler, R., & MacArthur, R. D. (1977). Alcohol retards vision recovery from glare by hampering target acquisition. *Nature, 270*, 428–429.

Shinar, D. (1978). *Psychology on the road: The human factor in traffic safety.* New York: Wiley.

Wilkinson, I. M. S., Kime, R., & Purnell, M. (1974). Alcohol and human eye movement. *Brain, 97*, 785–792.

Woodworth, R. S., & Schlosberg, H. (1954). *Experimental psychology.* New York: Harper & Row.

We wish to thank Angelo Corella of the Illinois Department of Public Health for advice, Jeff Levitt and Kimball Laden of Rush-Presbyterian-St. Luke's Medical Center for the generous loan of breathalyzers, and the Evanston, Illinois, Police Department for verifying the accuracy of the breathalyzers. This research was supported by a grant from the Scientific Advisory Board of the Distilled Spirits Council of the United States and a fellowship from the National Science Foundation. The first author is now at the University of Illinois College of Medicine, Chicago, Illinois. Address reprint requests to the second author at Cresap Neuroscience Laboratory, Department of Psychology, Northwestern University, Evanston, Illinois 60201.

ADDITIONAL QUESTIONS

1. What general theoretical issue is addressed by this research? What practical problems may be answered by this research?
2. How did the experimenters control for subject variables (e.g., body weight)? Why did they control for these variables?
3. Why did the experimenters use practice sessions?
4. In what way were the subjects misled? Is this justified? Is this acceptable from an ethical standpoint? Why or why not?
5. Why were some data discarded? What is the justification for discarding these data?
6. What practical application might be made of the conclusions of this research?

Russian Vocabulary

INTRODUCTION

Learning a foreign language can be a time-consuming and tedious process. Students of foreign languages have not only to acquire a different sentence structure for the foreign language but they must also learn thousands of vocabulary words in order to become fluent in that language. This difficult task has been approached from many angles, including the utilization of mental imagery to aid in remembering words.

SPECIAL ISSUES

Practical and Theoretical Problems

Many research projects grow out of practical problems, while other projects are inspired by theoretical issues; and still a third class of research develops out of both practical and theoretical problems. In selecting cases to represent the diverse types of problems in experimental psychology, we have tried to draw examples of practical problems (e.g., identification of cola beverages), theoretical problems (e.g., picture memory), and the current case of practical and theoretical problems (e.g., the keyword method in learning Russian vocabulary).

Practical problems in psychology are an important source of experimentation as they are often addressed to specific questions, to which the answers are of direct use. A researcher in an industrial setting, for example may want to know

often used the keyword method in the control condition, thus diminishing the true differences between conditions. Moreover, many subjects had studied at least one Romance language and were able to learn some words in the control condition by recognizing them as cognates. The results suggested that it would be useful to evaluate the keyword method using a between-subject design and a foreign language that was less obviously related to languages previously studied by the subjects.

Russian was selected for the work reported here. In addition to being a non-Romance language Russian posed a special challenge to the keyword method because it involves a number of frequently recurring phonemes that do not occur in English. Also, from a practical viewpoint, for many students the Russian vocabulary is more difficult to learn than is the vocabulary of, say, German, French, or Spanish; it would be useful if the keyword method proved to be an effective means of teaching Russian vocabulary.

A 120-word Russian test vocabulary was divided into three comparable 40-word subvocabularies for presentation on separate days. The subjects were run under computer control. They received instructions from a cathode-ray display scope, listened to recorded foreign language words through headphones, and typed responses into the computer by means of a console keyboard. The experiment began with an introductory session (Day 0), during **d** the first part of which **subjects were familiarized with the equipment;** during the second part they were assigned to the keyword and control groups and given instructions on the appropriate learning method. On each of the following three days (Day 1, Day 2, and Day 3) one of the test subvocabularies was presented for study and testing. On each of these days three study/test trials were given. The study part of a study/test trial consisted of a run through the subvocabulary; each foreign word was pronounced and, depending upon the treatment group, either the keyboard and English translation were displayed (keyword group), or the English translation alone was displayed (control group). A test trial consisted of a run through the subvocabulary in which each foreign word was pronounced and 15 sec were allowed for subject to type the English translation. A comprehensive test covering all 120 items of the vocabulary was given the day after the pre-

The experimenters gave the subjects sufficient time to familiarize themselves with the equipment and the procedure (DAY 0). It is important that subjects in an experiment be sufficiently acquainted with the procedure and equipment that they must manipulate. Unless proper practice is allowed, a subject may make more errors at the beginning, until he or she becomes familiar with the procedure (**d**). The 120-word vocabulary was divided into three 40-word lists to be presented to the subjects on three successive days. The dependent variables in this experiment are the scores on the nine daily tests, the scores on a comprehensive test on Day 4, and the scores on the delayed comprehension test

e   sentation of the last subvocabulary (Day 4). **A similar test was given approximately 6 weeks later.**

METHOD

Subjects

Fifty-two Stanford University undergraduates served as subjects (26 males and 26 females). All were native speakers of English, none had studied Russian, and none had participated in prior experiments using the keyword method.

Stimulus Material

A test vocabulary of 120 Russian nouns with associated keywords was selected; a sample of 20 items is presented in Table 18.1. The test vocabulary represents a cross section of vocabulary items typically presented in the first-year Russian curriculum at Stanford University. English translations of the Russian vocabulary were ranked according to imageability as determined both by the judgment of the experimenter and the Paivio (Note 1) image values for those English words for which values were available. The average Paivio value for the 15 most imageable words was 6.72, and the average for the 15 least imageable words was 2.51. The keywords were selected by a four-person committee whose members were familiar with the keyword method. For some items, the committee chose keyword phrases rather than single keywords; a total of 38 keyword phrases were used in the test vocabulary. The test vocabulary was divided into three subvocabularies of 40 words each, matched in abstractness and imageability.

Procedures

f   **During the first session** (Day 0) **the experimenter showed each subject how to start the computer program that conducted the experiment. The program itself explained all of the remaining procedures.** After giving instructions on the use of the keyboard and audio headset, the program introduced keywords as a means of focusing attention on the sound of a Russian word. In order to provide all subjects with experience in the procedures, practice was given on a randomized list of 30 words (not included in the test vocabulary); a Russian word was spoken and its keyword was given in which each Russian word was spoken and its keyword was displayed in brackets for 5 sec. Afterwards, a test (randomized for each subject)

six weeks later (**e**). The independent variable is the method of presentation (keyword or control method).

COMPUTER-BASED PROCEDURE

The procedure used by Atkinson and Raugh introduces an advanced technology in experimental psychology (**d, f**). Rather than have an experimenter physically present stimuli to subjects, the stimulus infor-

Table 18.1   A SAMPLE OF 20 ITEMS FROM THE
RUSSIAN VOCABULARY WITH RELATED KEYWORDS

RUSSIAN	KEYWORD	TRANSLATION
Vnimánie	[pneumonia]	Attention
Délo	[jello]	Affair
Západ	[zap it]	West
Straná	[strawman]	Country
Tolpá	[tell pa]	Crowd
Linkór	[Lincoln]	Battleship
Rot	[rut]	Mouth
Gorá	[garage]	Mountain
Durák	[two rocks]	Fool
Ósen´	[ocean]	Autumn
Séver	[savior]	North
Dym	[dim]	Smoke
Seló	[seal law]	Village
Golová	[Gulliver]	Head
Uslóvie	[Yugoslavia]	Condition
Dévushka	[dear vooshka]	Girl
Tjótja	[Churchill]	Aunt
Póezd	[poised]	Train
Krovát´	[cravat]	Bed
Chelovék	[chilly back]	Person

was given in which each Russian word was spoken, and 10 sec were allowed to start typing the keyword. If a response was begun within 10 sec, the time period was extended from 10 to 15 sec; otherwise, the program advanced to the next item. A second randomized study of the 30 practice words was given, followed by a newly randomized test. Throughout the experiment, the same training and randomized presentation procedures were followed.

g

After the keyword practice, **subjects were randomly assigned to the experimental and control groups with the constraint that both groups were to contain an equal number of males and females.** They were given the appropriate written

mation was stored in a computer program. By using a computer program and computer hardware, the experimenters reduce the influence of subject-experimenter bias. In addition, the use of a computer-based procedure means that all subjects receive the same stimuli. Because computer-based experiments afford a high degree of experimental reliability (in addition to increasing the range of potential stimuli), they are rapidly becoming an integral part of the standard equipment in experimental psychology.

RANDOM-GROUP DESIGN VERSUS WITHIN-SUBJECT DESIGN

In the present experiment Atkinson and Raugh chose a random-group (**g**) design (**c** and **d**), which reduced the possibility that control subjects

instructions on the method for associating Russian words and English translations. The experimental instructions were like the keyword instructions for Experiment III presented in Raugh and Atkinson (in press). They explain that while a Russian word was being pronounced, a keyword or keyword phrase would be displayed in brackets at the left-hand margin of the screen and the English translation would appear to the right. Experimental subjects were instructed to learn the keyword first and then picture an imagery interaction between the keyword and the English translation; the experimental instructions also stated that if no such image came to mind, they could generate a phrase or sentence incorporating the keyword and translation in some meaningful way. The control instructions explained that while each Russian word was pronounced, the English translation would be displayed near the center of the screen. Control subjects were told to learn in whatever manner they wished; they were not given instructions on the use of keywords or mental imagery.

After the instructions were given, a practice series of 10 Russian words was presented in which each Russian word was spoken while the English translation was displayed; for subjects in the experimental group the approximate keyword was also displayed with each English translation. Following this a test trial was given in which each Russian word was spoken and the subjects attempted to type the English translation. A second study trial was given and was followed by a second test trial, concluding Day 0. The subjects were told the practice on the 10-word list was like the procedure for the remainder of the experiment.

They returned the following day for the Day 1 session. For each subject the computer program randomly selected one of the three-40 word subvocabularies for presentation. Day 1 consisted of three successive study-test trials. The study trial was exactly like the study trial at the end of Day 0: Each Russian word was spoken while, depending upon the group, the cathode-ray tube displayed either the keyword and English translation, or the English translation alone. For both groups the presentation was timed for 10 sec per item. The test trials were identical for both groups: Each Russian word was spoken and the subject had 10 sec to begin a response. No feedback was given; an incomplete or misspelled response was scored as incorrect.

Day 1, Day 2, and Day 3 (which fell on consecutive days) followed identical formats. The only difference was that each day involved a different randomly assigned subvocabulary.

---

would use the keyboard method (see Chapter 5). It is important that the distinction between these two techniques be clearly understood. In a random-group design two groups of subjects are used—an experimental group and a control group. Comparison *between* the two groups are made. The cases of the Spanish-keyword and the Russian-keyword experiments provide us with lucid examples of within-subject and random-subject designs. The design models are shown in Figure 18.2.

The comprehensive test followed on Day 4. The comprehensive test was exactly like a daily test trial, except that it covered the entire 120-word test vocabulary. For the sixth and final session (the delayed comprehensive test), subjects were called back about 30 to 60 days (average 43 days) from Day 0 to take a randomized repeat of the comprehensive test. They had not been forewarned that they would be tested at a later date.

RESULTS

The Day 0 keyboard practice phase of the experiment was identical for both the experimental and control groups. The results of the keyword tests averaged over trials were 51% for male keyword subjects and 53% for male control subjects; the comparable scores for females were 59% and 58% respectively. The results indicated that the average for keyword subjects was 55% and the corresponding average for control subjects was 56%. The results indicate that the keyword and control groups were evenly matched so far as performance on the pretest was concerned.

RESULTS

The means for the practice phase of the experiment showed no statistically significant differences between the experimental group and the control group during the practice session. It is common for experimenters

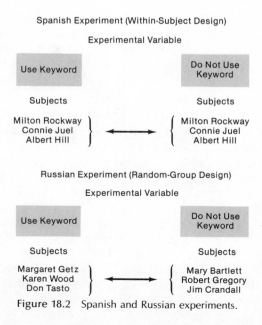

Figure 18.2   Spanish and Russian experiments.

Table 18.2   PROBABILITY OF A CORRECT RESPONSE ON THE COMPREHENSIVE
TEST AS A FUNCTION OF TREATMENT GROUP, SEX, AND STUDY DAY

STUDY	KEYWORD			CONTROL		
DAY	MALE	FEMALE	M	MALE	FEMALE	M
Day 1	.55	.73	.64	.27	.40	.33
Day 2	.63	.76	.70	.38	.47	.43
Day 3	.80	.82	.81	.60	.67	.63
M	.66	.77	.72	.42	.51	.46

Table 18.2 presents results of the comprehensive test in which the probability
of a correct response is given as a function of sex, treatment group, and day
on which the word was studied; for example, the table shows that on the
comprehensive test females in the keyword group responded correctly to
76% of the words that they had studied on Day 2, whereas males responded
correctly to 63% of the words studied on Day 2. A Sex × Treatment analysis
of the comprehensive test data was made in which performances on the
Day 1, Day 2, and Day 3 subvocabularies were viewed as repeated trials.
It was found that keyword subjects were superior to the control subjects, F
$(1, 48) = 35.8$, $p < .001$; moreover, the females performed significantly
better than the males, $F (1, 48) = 5.9$, $p < .025$. No interactions between
sex and treatment were found; the error mean square used in these tests was
.023.[2] Because the subjects were volunteers we cannot say whether the sex
differences reflect a sampling error or an actual difference between males
and females. In any case, the results suggest that for vocabulary-learning
experiments of this sort, care should be taken to insure that males and females
are evently divided among treatment groups.

[2]An inspection of frequency histograms indicated unimodal distributions for both the
keyword and control groups. There was no evidence to suggest that some subjects in the
keyword group performed unusually well, whereas the others were comparable to control
subjects.

to test the difference between means of the experimental group and the
control group in order to determine if random assignment has succeeded
in making the groups approximately equal. In this case, since the experimental and control groups did not differ significantly during the
practice session, the groups can be considered equivalent regarding their
ability to learn the vocabulary words without the keyword method. A 2
× 2 analysis of variance (sex × treatment) was performed, and it was
found that the keyword subjects performed significantly better than the
control subjects ($p < 0.001$) and the female subjects performed significantly better than the males ($p < 0.025$) (h). No interactions between sex
and treatment were found. Figure 18.3 shows that keyword subjects were
superior on all three test trials on all days. The authors stress the importance of insuring that males and females are equally represented in
the treatment groups.

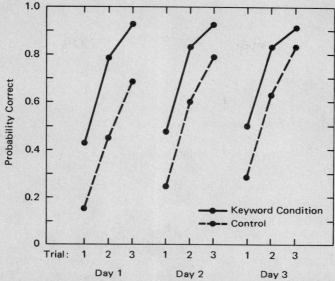

Figure 18.3   Probability of a correct response over test trial on Day 1, Day 2, and Day 3.

Figure 18.3 presents the probability of a correct response on each of the three test trials for Day 1, Day 2, and Day 3. The keyword group in all cases obtained superior scores; in fact, on each day the keyword group learned at least as many words in two study trials as the control group learned in three trials.

A question of some interest is whether keyword phrases facilitate learning as much as single keywords do. Our data cannot answer the question because we did not systematically vary the number of keywords used for each Russian item. Nevertheless, the data are suggestive. In the experimental condition 38 items involved the use of keyword phrases instead of a single keyword. For example, the keyword phrase "narrow road" was associated with the word *naród,* and "tell pa" was associated with *tolpá.* The average performance of the keyword phrase items on the comprehensive test was .74 in the keyword condition and .44 in the control condition. The corresponding

Atkinson and Raugh address the question of whether keyword phrases were less effective than single keywords. However, a direct answer to this question is impossible, because the experiment was not designed to test this hypothesis. From the results of the current experiment, the authors infer that there probably is no difference in effectiveness between the keyword phrase and the single keyword in effectiveness. Further research is needed to confirm this hypothesis.

averages for single keyword items were .71 and .45, respectively. Thus, the probability of learning a keyword-phrase item was about the same as the probability of learning a single keyword item.

DISCUSSION

It should be kept in mind that our results are for subjects who have not had previous training in Russian. It may well be that supplying the keywords is most helpful to the beginner, and becomes less useful as the subject gains familiarity with the language and the method. We have run an experiment using a Spanish vocabulary where subjects were instructed in the keyword method, but during study of an item received a keyword only if they requested it by pressing an appropriate key on their computer console (Raugh & Atkinson, in press). We call this variant of the keyword method the free-choice procedure. When an item was initially presented for study a keyword was requested 89% of the time; on subsequent presentations of the item the subject's likelihood of requesting the keyword depended upon whether or not he missed the item on the preceding test trial. If he missed it, his likelihood of requesting the keyword was much higher than if he had been able to supply the correct translation. Otherwise, however, the likelihood of requesting a keyword was remarkably constant from one day of the experiment to the next; that is, there was no decrease in keyword requests over the three study days, where on each day the subject learned a new vocabulary. It is interesting to note that performance on the comprehensive test for the free-choice group was virtually identical to the performance of a group that was automatically given a keyword on all trials. Not much of a diffence would be expected between the two groups since the free-choice subjects had such a high likelihood of requesting keywords. Nevertheless, these findings suggest that the free-choice mode may be the preferred one. In the free-

DISCUSSION

Practical Issue

In this section the authors discuss various implications of their experimental findings. The first question they consider is: Should the experimenter provide the keyword to the subject, or should he or she allow the subject to generate his or her own keyword? Based on other experimental evidence the authors state that the subjects tended to do better when the experimenters supplied the subjects with keywords than when the subjects made up their own keywords (i). The difference in performance between words with keywords supplied and words without keywords supplied, however, was not very great, so this finding must be interpreted cautiously.

choice procedure subjects report that they generally wanted a keyword, but that there were occasional items that seemed to stand out and could be mastered.

Results using the keyword method raise a number of issues; some of these issues are discussed elsewhere (Raugh & Atkinson, in press) and will not be reviewed in this paper. Of special interest to the experiment reported here is the question: Should the experimenter supply the keyword, as we have done, or can the subject generate his own more effectively? The answer to this question is somewhat complicated. In an unpublished experiment similar to the one described here, all subjects were given instruction in the keyword method. During the actual experiment half of the items were presented for study with a keyword, whereas no keyword was provided for the other items. The subjects were instructed to use the keyword method throughout. When a keyword was provided they were to use that word, when no keyword was provided they were to generate their own. On the comprehensive test the subjects did better on the keyword supplied items than on the others, but the size of the difference was small in comparison to the difference between groups reported in this paper. Instruction in the keyword method was helpful, and somewhat more so if the experimenter also supplied the keywords. In summary, the answer to our question is that subjects appear to be somewhat less effective when they must generate their own keywords; but results from the free-choice procedure indicate that keywords need only be supplied when requested by the subject.

i

A theoretical framework for interpreting these results is provided by Atkinson and Wescourt (1974). According to their theory, early in the learning process the memory structure for a given item involves only two independent links (what we have called the acoustic and imagery links). However, with continued practice a third link is formed directly associating the foreign word with its English translation. It is this direct link that sustains performance once an item is highly practiced. The subject may still have access to the keyword but the retrieval process based on the direct association is so rapid that the subject only recalls the keyword under special circum-

j

Theoretical Issue

Atkinson and Raugh then tie their experiment to a theoretical framework by Atkinson and Wescourt (j), which maintains that two independent links (imagery and acoustic) are formed in the learning process of an item but that with continued practice a third link, between the item and its English translation, is formed. Thus, the indirect link between acoustic and imagery aspects facilitates the learning of a direct link between the actual item and the translation. The indirect link is discarded as the direct link is learned.

stances, such as when he is consciously trying to do so or has a retrieval failure in the primary process. But the less direct chain of the acoustic and imagery links has the advantage that it is easily learned and provides a crutch for the subject as he learns the direct association; it facilitates the learning of the direct association by insuring that the subject is able to recall items early in the learning process.

k   **Our experimental findings indicate that the keyword method should be evaluated in an actual teaching situation.** Starting this year, we will be running a computerized vocabulary-learning program designed to supplement a college course in Russian. The program will operate much like our experiments. When a word is presented for study it will be pronounced by the computer and simultaneously the English translation will be displayed on a cathode ray tube. The student will be free to study the item in any way he pleases, but he may request that a keyword be displayed by pressing an appropriate button on his console. Students will be exposed to about 800 words per quarter using the computer program, which in conjunction with their normal classroom work should enable them to develop a substantial vocabulary. We, in turn, will be able to answer a number of questions about the keyword method when it is used over an extended period of time. Many foreign language instructors believe that the major obstacle to successful instruction is not learning the grammar of a language, but acquiring a vocabulary sufficient to engage in spontaneous conversation and read materials other than the textbook.

If the instructional application proves successful, then the keyword method and variants of it deserve a role in language-learning curricula. The keyword method may prove useful only in the early stages of learning a language and more so for some classes of words than others. The method may not

CONCLUDING REMARKS

The authors conclude the article by discussing the application of their findings. They state that there is evidence that students use mediating strategies in vocabulary learning and cite the similarity between these mediating strategies and the keyword method. They cite the applicability of the method in foreign language instruction and mention plans to set up a vocabulary learning program using the method in conjunction with a college Russian course (k).

This discussion section provides a model for the student learning technical writing. It includes a discussion of the findings, implications of these findings which are suggestive of future research, and practical applications of the findings. The authors also fit the experiment into a theoretical framework, all of which makes this section an example of good technical writing.

be appropriate for all learners, but there is the possibility that some, especially those who have difficulty with foreign languages, will receive particular benefits.

REFERENCE NOTE

1. Paivio, A. (1973). *Imagery and familiarity ratings for 2448 words: Unpublished norms.* Unpublished manuscript. Available from Allan Paivio, Department of Psychology, University of Western Ontario, London, Ontario, Canada.

REFERENCES

Atkinson, R. C., & Wescourt, K. T. (1974). Some remarks on a theory of memory. In P. Rabbit & S. Dornic (Eds.), *Attention and performance V.* London: Academic Press. (1972).

Bower, G. (1972). Mental imagery and associative learning. In L. Gregg (Ed.), *Cognition in learning and memory.* New York: Wiley.

Bugelski, B. R. (1968). Images as mediators in one-trial paired-associate learning. II: Self-timing in successive lists. *Journal of Experimental Psychology, 77,* 328–334.

Ott, C. E., Butler, D. C., Blake, R. S., & Ball, J. P. (1973). The effect of interactive-image elaboration on the acquision of foreign language vocabulary. *Language Learning, 23,* 197–206.

Paivio. A. (1971). *Imagery and verbal processes.* New York: Holt, Rinehart and Winston.

Raugh, M. R., & Atkinson, R. C. (in press). A mnemonic method for the learning of a second language vocabulary. *Journal of Educational Psychology.*

Schnorr, J. A., & Atkinson, R. C. (1970). Study position and item differences in the short- and long-term retention of paired associates learned by imagery. *Journal of Verbal Learning and Verbal Behavior, 9,* 614–622.

Yates, F. (1972). *The art of memory.* Chicago: University of Chicago Press.

This research was supported by the Office of Naval Research, Contract No. N00014-67-A-0012-0054, and by Grant MN-21747 from the National Institute of Mental Health. The authors wish to thank Richard D. Schupbach and Joseph A. Van Campen of the Department of Slavic Languages and Literatures at Stanford University for assistance in preparing the vocabulary used in the work reported here and for advice on problems of vocabulary acquisition in second-language learning.

QUESTIONS

1. Why did Atkinson and Raugh find it necessary to test the keyword method on a Russian vocabulary when it had already proved successful in aiding the subjects in the acquisition of a Spanish vocabulary in a previous experiment?
2. Why did the experimenters assign equal numbers of males and females to the control and experimental groups?
3. State two hypotheses that could be tested that were suggested by this study.

4. Identify a type of experiment that directly fits into a theoretical framework in psychology.
5. Under what conditions should an experimenter use a within-subject design and under what conditions should an experimenter use a between-subjects design?
6. In what fields could the keyword method be applied other than the learning of a foreign language?

19

Therapy for Anger

INTRODUCTION

In our contemporary society the occurrence of stress and angry reaction among individuals is apparent. The normal stressors encountered in daily living seem to cause irritations, which frequently are manifest in anger. In spite of the frequency of such reactions and a need for greater understanding of the treatment of such reactions, it is somewhat surprising that the phenomenon has not been more thoroughly studied. In the present article by Raymond Novaco, "anger disorders" and their treatment are studied.

Several important topics related to the special problem encountered in research in clinical psychology are introduced in "Stress Inoculation: A Cognitive Therapy for Anger and Its Application to a Case of Depression."

SPECIAL ISSUES:

Single-Subject Design in Clinical Studies

In this study by Novaco, only one subject was used. In other parts of this book we have discussed small *n* designs (Chapter 14, "Creative Porpoise"). Studies using a single subject present special problems in that the contrast effects due to

263

the intervention of treatment (or independent variable) are usually made between behavior measures collected during one time period as compared to another period. During the initial period, careful measures are made of responses in order to establish a base rate (or baseline) of the behavioral characteristics which exist before the introduction of the independent variable. These behavioral characteristics may be measured by multidimensional scales, that is, measures on several behavioral traits. If a researcher is interested in the emotional changes that occur after the introduction of an independent variable—say, the injection of a mood-altering drug—the baseline might include a description of several measures of affect, e.g., expression of happiness, spontaneity, sociability. After the introduction of the independent variable, in this case the mood-altering drug, subsequent evaluations on the same multidimensional characteristics are made. These measures are contrasted with baseline data and constitute the main effect of the study.

Because individual differences plays a critical role in single-subject designs, a thorough description of the subject is necessary. In research involving psychotherapy, this description of the client includes a "case history"; a special feature of the study by Novaco is the case history. The author reviews the clinical background of the subject, which includes a history of the subject's work situation, symptoms, hospitalization, reaction to stressful situations, and previous treatment. The identification of these characteristics is particularly important in clinical studies as they provide the reader with a background upon which to base further generalization of the treatment.

Stress Inoculation: A Cognitive Therapy for Anger and Its Application to a Case of Depression

RAYMOND W. NOVACO
UNIVERSITY OF CALIFORNIA, IRVINE

Clinical interventions for anger disorders have been scarcely addressed in both theory and research in psychotherapy. The continued development of a cognitive behavior therapy approach to anger management is presented along with the results of its application to a hospitalized depressive with severe anger problems. The treatment approach follows a procedure called "stress inoculation," which consists of three basic stages: cognitive preparation, skill acquisition and rehearsal, and application practice. The relationship between anger and depression is discussed.

a   The treatment of anger-based disorders has escaped the attention of both psychotherapy theory and research. This is a puzzling state of affairs when one considers the abundance of laboratory research on aggression. An approach to the treatment of chronic anger and its experimental analysis has been presented by Novaco (1975). Although scattered reports of circumscribed interventions with anger exist (Herrell, 1971; Kaufmann & Wagner, 1972; Rimm, deGroot, Boord, Reiman, & Dillow, 1971), there is a distinct need for concerted work in this area. In an effort to promote a more detailed conception of anger problems and of therapeutic interventions for anger, the present article describes the further development of a cognitive behavior therapy and presents the results of its application to a hospitalized depressive with severe anger problems.

SOURCE: Article is abridged and reprinted by permission from *Journal of Clinical Psychology,* 1977, *45,* 600–608, and from the author.

ANALYSIS

REVIEW OF LITERATURE

The first paragraph (a) is an excellent example of an introductory statement. The author creates a need for the present type of experiment and a hint as to the treatment, and suggests how such results may be applied to similar clients. Note that in his paper a formal hypothesis is not stated. In general, this represents a departure from previous studies analyzed in this book, but the practice is not erroneous. The author has provided sufficient detail for the reader to infer the hypothesis without formally stating it.

Persons having serious problems with anger control have been successfully treated by a cognitive behavior therapy approach (Novaco, 1975, 1976b). In the present article, this treatment approach has been further developed to follow a procedure called *stress inoculation* that has been applied to problems of anxiety (Meichenbaum, 1975; Meichenbaum & Cameron, Note 1) and pain (Turk, Note 2). The stress inoculation therapy consists of developing the client's cognitive, affective, and behavioral-coping skills and then providing for the practice of these skills with exposure to regulated doses of stressors that arouse but not overwhelm the client's defenses. The stress inoculation approach involves three basic steps or phases: (a) cognitive preparation, (b) skill acquisition and rehearsal, and (c) application practice. This approach to anger problems was first developed by me with regard to the training of police officers (Novaco, in press). The present article describes the further development of this approach as a clinical procedure and its application to the anger problems of a depressed patient on an acute psychiatric ward.

METHOD

Treatment Procedure

b   ### Cognitive Preparation

The initial phase educates clients about the functions of anger and about their personal anger patterns, provides for a shared language system between client and therapist, and introduces the rationale of the treatment approach. This process is facilitated through an instructional manual for clients.[1] The text of the manual describes the nature and functions of anger, when anger becomes a problem, what causes anger, and how anger can be regulated.

The components of the cognitive preparation phase are (a) identifying the persons and situations that trigger anger; (b) fostering recognition of the difference between anger and aggression; (c) understanding the cognitive, somatic, and behavioral determinants of anger, with particular emphasis on anger-instigating self-statements; (d) discriminating justified from unnecessary anger; (e) recognizing the signs of tension and arousal early in a provocation sequence; and (f) introducing the anger management concepts as coping strategies.

c   ### Skill Acquisition and Rehearsal

The process of skill acquisition involves a familiarization with the three sets of coping techniques, modeling of the techniques by the therapist, and then rehearsal by the client. At the cognitive level, the client is taught how to alternatively view situations of provocation by changing personal constructs (Kelly, 1955) and by modifying the exaggerated importance often attached to events (Ellis, 1973).

Two central cognitive devices for anger regulation are a task orientation to provocation and coping self-statements. A task orientation involves attending to desired outcomes and implementing a behavioral strategy directed at producing these outcomes. Self-instructions are used to modify the appraisal of provocation and

[1]The client instructional manual and the therapist treatment procedure manual may be obtained from the author.

to guide coping behavior. Pragmatically, the self-instructions are designed to apply to various stages of a provocation sequence: (a) preparing for a provocation, (b) impact and confrontation, (c) coping with arousal, and (d) subsequent reflection in which the conflict is either unresolved or resolved. Stages (c) and (d) provide for the possible failure of self-regulation and for the prolongation or reinstigation of arousal by ruminations.

At the affective level, the client is taught relaxation skills and is also encouraged to maintain a sense of humor as a response that competes with anger.

Relaxation training was conducted via the Jacobsen method of tensing and relaxing sequential sets of muscle groups. Mental as well as physical relaxation is emphasized, as relaxation imagery is used to induce a light somnambulistic state. Deep breathing exercises supplement the muscle tension procedures. Previous research (Novaco, 1975) has found that the cognitive-coping component is comparatively more effective than the relaxation- training component. However, relaxation training enables the client to become aware of the tension and agitation that lead to anger and provides an important sense of mastery over troublesome internal states.

The behavioral goals of treatment are to promote the effective communication of feelings, assertive behavior, and the implementation of task-oriented, problem-solving action. Generally speaking, the client is induced to use anger in a way that maximizes its adaptive functions and minimizes its maladaptive functions (Novaco, 1976a). Among the positive functions of anger are its energizing effects, the sense of control that it potentiates, and its value as a cue to cope with the problem situation.

The therapeutic procedure enables clients to recognize anger and its source in the environment and to then communicate that anger in a nonhostile form. This controls the accumulation of anger, prevents an aggressive overreaction, and provides a basis for changing the situation that caused the anger.

Inducing a task-oriented response set facilitates the occurrence of problem-solving behavior. It is emphasized that anger is an emotional reaction to stress or conflict that is due to external events being at variance with one's liking. Poor behavioral adjustment is linked to the inability to provide problem-solving responses that are effective in achieving desired goals (Platt & Spivack, 1972; Shure & Spivack, 1972). Anger management involves a strategic confrontation whereby the person learns to focus on issues and objectives. The execution of such behavior is explicitly prompted and progressively directed by the use of coping self-statements.

d   Application Practice

The value of application practice has been emphasized by a variety of skills-training approaches (D'Zurilla & Goldfried, 1971; Meichenbaum, 1975; Rich-

METHOD

In the introduction, the author defines the "stress inoculation" approach and in sections (b), (c), and (d) gives a verbal description of the stages of this approach. Of special interest is the attempt to operationally define the concepts, thus allowing subsequent investigators to replicate the

ardson, 1973; Suinn & Richardson, 1971). The regulation of anger is a function of one's ability to manage a provocative situation, and in this treatment phase, clients are given an opportunity to test their proficiency. The client is exposed to manageable doses of anger stimuli by means of imaginal and role-playing inductions. The content of the simulations is constructed from a hierarchy of anger situations that the client is likely to encounter in real life. By progressively working on each hierarchy scene, first imaginally and then by role playing, the client is enabled to sharpen anger management skills that had been rehearsed with the therapist.

Case History

The client was a 38-year-old male who had been admitted to the psychiatric ward of a community hospital with the diagnosis of depressive neurosis. Upon admission he was judged to be grossly depressed, having suicical ruminations and progressive beliefs of worthlessness and inadequacy. He was a credit manager for a national business firm and had been under considerable job pressure. Quite routinely, he developed headaches by midafternoon at work. He had recurrent left anterior chest pain that was diagnosed by a treadmill stress test procedure as due to muscle tension.

There client had been hospitalized for 3 weeks when the attending psychiatrist referred him to me for the treatment of anger problems. At that time I had initiated a staff training program for the treatment of anger. The principal behavior setting in which problems with anger control emerged were at work, at home, and at church. The client was married and had six children, one of which was hyperactive.

Circumstances at work had progressively generated anger and hostility for this man's superiors, colleagues, and supervisees. His anger at work was typically overcontrolled. He would actively suppress his anger and would then periodically explode with a verbal barrage of epithets, curses, and castigations when a conflict arose. At home, he was more impulsively aggressive. The accumulated tensions and frustrations at work resulted in his being highly prone to provocation at home, particularly in response to the disruptive behavior of the children. Noise, disorders,

e

---

study. Direct your attention to the components of cognitive preparation in (a) through (f) in section (b) and the various stages of a provocation sequence [(a) through (d) in section (c)].

Case History

Single-subject studies, especially of a clinical nature, typically involve a comprehensive case history in which the salient psychological characteristics of the subject are documented (e). The case history is designed to reveal the basic descriptive facts of the subject with little interpretation of these facts. The writing style is lucid and without embellishment.

and the frequent fights among the children were high anger elicitors. Unlike his behavior at work, he would quickly express his anger in verbal and physical outbursts. Although not an abusive parent, he would readily resort to physical means and threats of force (e.g., "I'll knock your goddamn head off") as a way to control the behavior of his children. The children's unruly behavior often became a problem during church services. The client's former training in a seminary disposed him to value family attendance at church, but serious conflict was often the result. In an incident just prior to hospitalization, the client abruptly removed two of his boys from church for creating a disturbance and threatened them to the extent that one ran away. At this point he had begun to realize that he was reacting "out of proportion," but he felt helpless about instituting the desired changes in behavior.

During hospitalization, treatment sessions were conducted three times per week for 3½ weeks. Following discharge, follow-up sessions were conducted biweekly for a 2-month period. During these sessions, anger diary incidents were discussed, and there was continued modeling, rehearsal, and practice of coping procedures.

Dependent Measures

f   Proneness to provocation was first assessed by means of an inventory of 80 situation descriptions for which the respondent rates anger on a 5-point scale. This is a revised version of the instrument reported in Novaco (1975). The current scale has been found to have a high degree of internal consistency across various subject populations.

g   **Behavior ratings were obtained by a clinical psychologist with 9 years of postdoctoral experience.** Dichotomous ratings were obtained on 14 behavior descriptions indexing the expression of anger (e.g., impatient with others, expresses resentment toward others, threatens to harm someone, quick to react with antagonism) and 8 behavioral descriptions indexing constructive coping and improvement (e.g., humorous and good natured, optimistic about future, tolerant and considerate of others, deals with conflict constructively, is positively assertive). Scaled ratings were obtained on three behavioral dimensions: (a) "The person is relaxed and at ease," (b) "the person demonstrates hostility or anger (expressed anger)," and (c) "the person is restraining anger and resentment (suppressed anger)." These items were rated on 8-point scales ranging from "not at all" to "exactly so" verbally anchored at each interval point.

h   **The reliability of the behavior ratings for the dichotomous items was examined for behavior in group therapy, occupational therapy, and recreation for a total**

Dependent Measures

Measurement of a client's response in a clinical study frequently poses a problem of objectivity. What indices of reduced anger could be used which would both reflect a valid measure of emotional change and yet could be used by subsequent researchers in a replication of the study? The author is very specific in identifying the special tests (**f**) and behavioral ratings (**g**) and (**i**); see page 270. Furthermore, in (**h**), the reli-

of nine observation periods with an average of eight patients observed in each setting. Four psychiatric nurses with an average of 6 years of experience served as the raters along with the clinical psychologist. The mean percentage of agreement was 84.6% for each pair of raters on the 22 items. For the 3 scaled items, the ratings of the psychologist were correlated with those of three psychiatric nurses who, respectively, observed the patient in three group therapy sessions. The average correlations were .7581 for Item 1, .8593 for Item 2, and .6621 for Item 3. Thus, greatest consistency was obtained for ratings of "expressed anger" and least consistency for ratings of "suppressed anger." The psychologist's ratings were also correlated with the patient's ratings of himself for behavior in group therapy, which resulted in correlations of .9242, .6770, and .7521 for the respective items. The judgments of this observer were concluded to be sufficiently reliable for the present case analysis.

Self-monitoring of anger reactions by the patient is integral to the treatment procedure, as in other self-control therapies (Mahoney & Thoresen, 1974). The patient was asked to maintain a diary of anger experiences for which two ratings on a 7-point scale were obtained—(a) the degree of anger arousal experienced

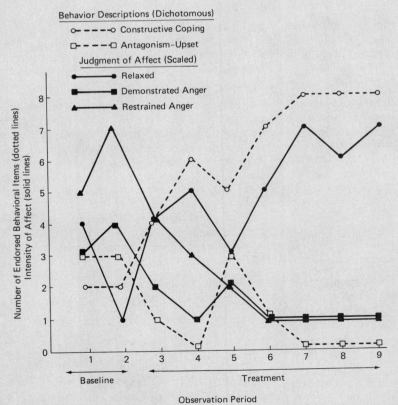

Figure 19.1 Clinician's ratings of patient's behavior for anger and anger management during hospitalization.

and (b) the degree of anger management achieved. Although the reduction in frequency and intensity of anger is a central goal of treatment, it is assumed that in certain instances, the arousal of anger is justified and appropriate. Therefore, it is also important to assess the degree to which anger regulation resulted in a constructive response to provocation. This patient's anger experiences were concentrated at home and at work. Only rately was he provoked while in the hospital. Hence, the self-ratings were obtained for weekend home visits while on pass from the hospital and then on a daily basis following discharge and return to work.

RESULTS

k

Pretreatment assessment of the patient's proneness to provocation by means of the anger inventory resulted in a total inventory score of 301. At discharge, the inventory was readministered and a total score of 258 was obtained. This decrease of 43 points represents a decrease of greater than 1 standard deviation based on the recent random college sample and on the chronic anger clients of previous research. It also is comparable to the change obtained on this instrument for subjects in a controlled treatment design (Novaco, 1975).

l

Continuous measures of change were obtained from the behavior rating scales and from the client's self-ratings. Behavior ratings of anger were obtained during group therapy. Since the patient had already been in the hospital for 3 weeks prior to the referral for treatment, only a 1-week baseline could be obtained. **Results for the behavior description measures and for the scaled ratings of affect are contained in Figure 19.1.** It can be seen that over the

ability of the ratings was assessed. Finally, self-monitoring measures by the patient were made (j).

One may argue that the above dependent measures lack the objectivity of measures made in some nonclinical experiments. For example, the dependent measure of a rat pressing a bar in a "Skinner box" leaves little room for ambiguity. Even though there is little doubt about the objectivity of bar-pressing responses, judgments about emotional states in clinical research can also achieve a high degree of objectivity; but one needs to specify, as completely as possible, the measurement instruments and their validity.

RESULTS

The overall results of the inventory scale given during pretreatment are contrasted with the results of the scale given at discharge (k). The results are also presented in Figure 19.1 [see (l)]. The researcher has shown five different measures of the patient's behavior for anger and anger

course of treatment, the patient underwent improvement on each index. The number of positive behavior descriptions endorsed and the scaled judgments of relaxed appearance increased over time, whereas antagonistic behavior and judgments of demonstrated and suppressed anger decreased. **After the third week of treatment, the attending psychiatrist, the staff, and the patient concurred in the decision for discharge.**

m

Convergent evidence of the patient's improvement was obtained from the self-monitored ratings of anger and anger management. Data were recorded on a weekend pass during the hospitalization period and on a continuous daily basis after discharge. **Results for the 60 days of recorded self-monitoring for frequency, degree of anger, and degree of control are contained in Figure 19.2.** The first 15 days of observation represent 5 weekends during the hospitalization period. As previously indicated, little anger was experienced with regard to events in the hospital environment, and these data are not plotted.

n

In the absence of a more satisfactory baseline period (which could not be extended due to the patient's length of stay on an acute facility prior to referral for treatment), changes in behavior must be examined as treatment progresses. On the weekend prior to treatment, the mean frequency was 4.0 per day, and by the patient's report this was quite standard. It can be seen that the mean frequency of anger incidents decreased from 1.55 per day during the first 20-day period to 1.10 per day for the second 20-day interval, and to .40 per day for the third interval. The mean anger magnitude ratings for the intervals were 4.30, 4.62, and 4.25, whereas the mean ratings of anger management were 4.30, 4.97, and 4.50, respectively. This indicates that when he did become angry, the degree of regulation was commensurate with or greater than the degree of arousal. It should also be noted that incidents of high achievement in self-control were occasionally not recorded by the client, who, like many depressives, was reluctant to "make too much" of his accomplishments. Hence, he periodically neglected to record an incident of high control and low anger. Although ratings were unfortunately not obtained from his wife, she consistently remarked that he was showing progressive improvement, in marked contrast to his previous behavior.

o

Among the most apparent and important behavioral changes achieved was the reduction of impulsive aggression as a punitive response to disruptive

---

management in this figure. Study the relationship between the judgment of "relaxed" behavior and "restrained" anger in this curve.

In section (m), it was the judgment of the psychiatrist, the staff, and the patient that the patient be discharged.

The long-term effects of the treatment are shown in (n) and in Figure 19.2.

In section (o) the author gives a very personal insight into the success of the treatment and into the severity of the original problem.

behavior by the children. During the treatment period numerous incidents of *physical* fighting among the boys were recorded in the diary as having aroused anger. There were 10 recorded incidents in the home, 5 in the car, and 2 in church. Considering the latter two settings, one can appreciate the magnitude of the disturbance and its potential impact. The incidents described as having occurred in the car, for example, were indeed unnerving. Fist fights between the boys resulted in bleeding mouths, black eyes, and lumps on the head. In one instance, a weight-lifting bar was used as a weapon. Despite the magnitude of these disturbances, the client only struck his children three times during the 3-month treatment period. Prior to treatment it had been quite routine for him to hit the boys in response to fighting or prephysical quarrels.

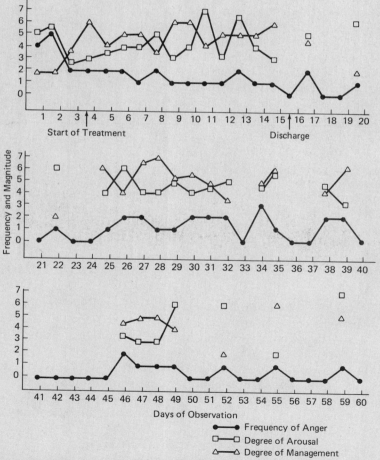

Figure 19.2 Self-monitored ratings of frequency of anger, magnitude of anger, and degree of anger management.

DISCUSSION

The self-regulation of anger is an important facet of the ability to cope with stress. The present article has reported the continued development of a cognitive behavior therapy approach to the treatment of anger problems and the extension of these procedures to a hospitalized depressed patient. **The stress inoculation model, first developed by Meichenbaum (1975), has been extended to the domain of anger and has been shown to be effective with someone having a severe psychological disturbance.**

p

The severity of the clinical problem is a nontrivial factor. That this factor is not to be overlooked is demonstrated by comparing the length of treatment in the present case with that of previous controlled research. In the clinical sample investigated by Novaco (1975), extensive pretreatment assessment eliminated all but those subjects having serious anger problems. Novaco (1975), however, used an outpatient sample, and significant results were achieved in 6 treatment sessions. Yet in the present case, comparable results required 11 sessions in the hospital and 4 as an outpatient. In addition, the clinical procedures had improved, an instructional manual had been incorporated, and relaxation therapy was available on a daily basis by tape. In the present case the severity of the anger problem was interlocked and compounded with the psychological deficits associated with depression.

DISCUSSION

In keeping with the format outlined in this book, the author discusses his research in the context of a theoretical issue raised in the introduction (**p**).

QUESTIONS

1. Discuss the special problems in developing objective dependent measures in clinical research.
2. The results are not discussed in terms of statistical probability. Why not?
3. Why were pretreatment measures made?
4. Why did the investigator choose to use multidimensional measures of behavior?
5. What precautions were undertaken to assure the objectivity of behavioral ratings?
6. Why is a detailed case history part of this study?
7. What special ethical issues are raised by this study?

REFERENCE NOTES

1. Meichenbaum, D., & Cameron, R. (1973). *Stress inoculation: A skills training approach to anxiety management.* Unpublished manuscript, University of Waterloo, Ontario, Canada.
2. Turk, D. (1974). *Cognitive control of pain: A skills- training approach.* Unpublished manuscript, University of Waterloo, Ontario, Canada.

REFERENCES

D'Zurilla, T., & Goldfried, M. (1971). Problem solving and behavior modification. *Journal of Abnormal Psychology, 78,* 107–126.

Ellis, A. (1973). *Humanistic psychology: The rational-emotive approach.* New York: Julian Press.

Feshbach, N. D. (1975). Empathy in children: Some theoretical and empirical considerations. *Counseling Psychologist 5,* 25–30.

Herrell, J. M. (1971). Use of systematic desensitization to eliminate inappropriate anger. *Proceedings of the 79th Annual Convention of the American Psychological Association, 6,* 431–432. (Summary)

Kaufmann, L., & Wagner, B. (1973). Barb: A systematic treatment technology for temper control disorders. *Behavior Therapy, 3,* 84–90.

Kelly, G. A. (1955). *The psychology of personal constructs* (Vol. 2). New York: Norton.

Meichenbaum, D. A. (1975). Self-instructional approach to stress management: A proposal for stress inoculation training. In C. Spielberger & I. Sarason (Eds.), *Stress and anxiety* (Vol. 2). New York: Wiley.

Novaco, R. W. (1974). The treatment of anger through cognitive and relaxation controls. *Journal of Consulting and Clinical Psychology, 44,* 681.

Novaco, R. W. (1975). *Anger control: The development and evaluation of an experimental treatment.* Lexington, Mass.: Lexington Books.

Novaco, R. W. (1976). The functions and regulation of the arousal of anger. *American Journal of Psychiatry, 133,* 1124–1128.

Novaco, R. W. (In press). A stress inoculation approach to anger management in the training of law enforcement officers. *American Journal of Community Psychology.*

Platt, J. J., & Spivack, G. (1972). Problem-solving thinking of psychiatric patients. *Journal of Consulting and Clinical Psychology, 39,* 148–151.

Richardson, F. (1973). A self-study manual for students on coping with test-taking anxiety. (Tech. Rep. No. 25) Austin: University of Texas Press.

Rimm, D. C., deGroot, J. C., Boord, P., Reiman, J., and Dillow, P. V. (1971). Systemic desensitization of an anger response. *Behavior Research and Therapy, 9,* 273–280.

Shure, M. B., & Spivack, G. (1972). Means-ends thinking, adjustment, and social class among elementary-school-aged children. *Journal of Consulting and Clinical Psychology, 38,* 348–353.

Suinn, R., & Richardson, F. (1971). Anxiety management training: A nonspecific behavior therapy program for anxiety control. *Behavior Therapy, 2,* 498–510.

Weissman, M. M., Klerman, G. L., & Paykel, E. S. (1971). Clinical evaluation of hostility in depression. *American Journal of Psychiatry, 128,* 261–266.

Requests for reprints should be sent to Raymond W. Novaco, Program in Social Ecology, University of California, Irvine, California 92717.

CHAPTER
20

Office Environments

INTRODUCTION

An experimental psychologist should not be content to work on parochial issues with standard solutions, but should be open to practice science on an expanded scale. In addition, the fully informed professional psychologist is aware of international findings and theories.

In this selection we present an article by Heidmets and Niit of Tallinn, Estonia, which originally appeared in a periodical of Tartu University called *Problems of Perception and Social Interaction*. The article "An Activity Analysis of Office Environments: Reality and Preferences" is included in this section for two reasons: (1) It illustrates a type of study published in foreign journals and (2) it addresses an important theoretical issue examined in a "nontraditional" (i.e., non-Western) way.

As you read the article pay particular attention to the careful development of a theoretical position that has definite applied consequences. Also, you will note that attention to details—both procedural and statistical—is not as evident as in the other examples in this book. Nevertheless, the authors address an important practical issue, collect data on the topic, represent the data in graphic and statistical form, and form conclusions based on their observations. All these components are critical ingredients of good experimental design. Although the study may lack some of the sophisticated techniques and impressive statistical analysis evident in many of our examples, it is important to recognize that valid research can be done in different ways.

An Activity Analysis of Office Environments: Reality and Preferences*

M. HEIDMETS AND T. NIIT

Abstract. The paper describes the methodology used for studying the activity structure of administrative personnel in office buildings. The incongruences between the actual and desired situation are outlined along with the possibilities of application of these findings for planning and design of office buildings. Some cross-cultural differences in the use of office space are also discussed.

1. INTRODUCTION

What should the effective office environment look like? In answering this question it is assumed that the needs and requirements of office users toward their environment arise from the *activities* that they carry out daily at their workplace (i.e., office building). Thus, to design an ideal office one has to describe rather precisely both the actual and desired *structure of activities* of the personnel and their environmental context. The measurement of activity structure and finding out environmental conditions most congruent with it was the main task of this research project.

a

Due to the limits of space it is not possible to review the relevant literature on studies of office environment. We can only state that our approach differs from the majority of published studies in the following ways:

—We proceed from *real activities* (and not from evaluations, experiences, etc.). From the viewpoint of design it seems to be much more useful to know

*This is a slightly revised version of a paper presented at the symposium "Future Developments in Office Environments," 20th International Congress of Applied Psychology, Edinburgh, July 25–31, 1982. We would like to thank Dr. Alan Hedge for presenting the paper in our absence. Funding for the research reported in this paper was provided by Central Research and Experimental Planning Institute of Spectacle Buildings, Sporting Complexes and Administrative Buildings in Moscow. Reprint requests should be sent to Mati Heidmets or Toomas Niit, Environmental Psychology Research Unit, Tallinn Pedagogic Institute, Narva mnt. 27, Tallinn, Estonia, USSR.

ANALYSIS

In many foreign publications the major sections (e.g., abstract, introduction, method, etc.) are labeled and sometimes numbered.

(a) As the authors note, no literature review is included because of space limitations. The absence of citations of previous work is unusual,

what a person does and wants to do rather than to know his age, education, personality characteristics, etc.,

—We deal with *all* main activities that are carried out in an office building (and do not concentrate on work-related activities as often has been done in the past). This approach derives from a hypothesis that sometimes the workplace functions as a "club," where socializing plays a major role along with work activities.

—*All locations* of activities beginning with one's desk and ending with the town outside the office building are included.

b

So we can say that it is a kind of systems approach that has guided our research. In this sense our approach, which analyzes environment as a set of objects and places that have become a part of activities, is in opposition to the approach where environment is an object of perception and evaluation. In the latter case it is never clear how a verbal evaluation is connected with real behavior; that is, we can state that a place or thing is beautiful but still may act with hostility toward it.

2. METHOD

Subjects

c

The study was carried out in three different cities (Tallinn, Moscow, and Pskov) in 12 different office buildings altogether, which represented three different categories of offices: offices of industrial plants, architectural offices, and offices of cultural institutions. The respondents ($N = 260$) represented all possible categories of administrative and office personnel, and more than 15 percent of the employees in every institution participated in the study.

---

but the authors adroitly lay the foundation for the study in the next few paragraphs.

(b) The language in this section may seem peculiar. Bear in mind that the authors' principal language is Estonian. Nevertheless, they do convey their meaning accurately and, to some, the writing is less stuffy and formal than is found in some Western publications.

(c) The subjects are identified. However, less attention is given to the characteristics of the sample than one would ordinarily find in Western publications. No description is given of the age, gender, level of education, training, and the like. Are these factors important?

The study was conducted with subjects from three different cities— Tallinn, Moscow, and Pskov. The location of workers was a principal independent variable in this study. All three cities are in what is now generally considered to be the USSR.

Procedure

A game board like a standard office plan was used (see Figure 20.1). It differentiates between five locations where activity could take place (one's own workplace, workroom, place or special room on the floor used by floor personnel only, place in the building for common use, and the outside of the office building) and the number of people in the company of or with whom the activities were carried out (alone, in the company of one or two other persons, in a group of 4–6 persons, and in a larger group). The respondents had to lay the labels describing 15 activities typical to the office work and personnel in appropriate places on the game board first to describe their actual arrangement of work and thereafter the desired one.

These activities were: quiet work; talking about work with coworkers, talking about work with visitors; telephone calls; receiving private visitors; talking about nonwork matters with coworkers, drinking tea or coffee; smoking; toilet and grooming; mental relaxation (reading newspapers and magazines, playing chess or board games, solving crosswords, etc.); physical recreation; meetings, committees; having lunch; putting coats away; parties and banquets.

The disturbing factors for each activity were identified for the actual situation and the preferred environmental conditions for the desired situation.

3. RESULTS

The spatial and social organization of administrative and office work varies markedly. On the average, 50 percent of all activities are performed alone, the other half in groups of various size. The relative share of group activities is bigger among recreational activities (about 70 percent). Nevertheless about 30 percent of work activities are also carried out in groups.

We can trace strong cultural differences on this dimension. In Estonia, the frequency of acting alone is twice as high as in Moscow; the frequency of acting in large group is about three times lower in Estonia. It is interesting to note that these differences in a real situation do not influence the preferences. As can be seen from Figure 20.2, the desired situation (an "ideal office") is rather similar for both regions on this dimension.

On the basis of this research project it appears that administrative personnel are quite mobile: about 50 percent of work activities are performed in the

---

(d) The distribution of 15 activities typical to the type of work and personnel on the game board was the main dependent variable.

(e) The main finding is reported verbally and graphically in this section. Note that no statistical tests (e.g., F, t, or χ^2 are used). Non-Western periodicals do employ statistics, but not as rigorously as in U.S., British, Canadian, and Australian publications; however, in spite of the use of descriptive statistics, the results of the study are clearly portrayed.

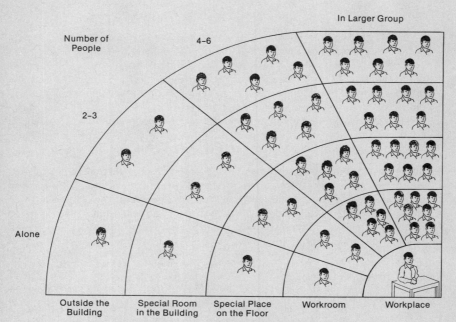

Figure 20.1   The game board used in the study.

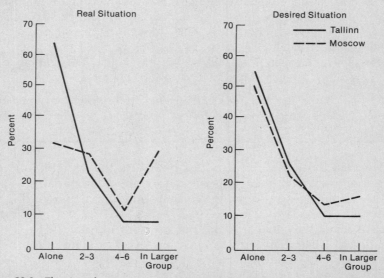

Figure 20.2   The size of groups engaged in various activities in administrative buildings.

workroom, the other half out of it, and about 15 percent even out of the administrative building. Every employee goes about once during the day out of the building ("to the city"). The in-house mobility is also quite remarkable (see Figure 20.3).

As one can see from the figure, actually most of the activities are concentrated in the workroom, whereas desired location for almost all of the nonwork activities and for the part of work activities is given to some kind of special room on the floor or in the building. The main reason for this is that the behavioral overload of the workroom inhibits the performance of essential work activities. The main disturbing factor is noise from talking, walking, telephone calls, visitors, and so on. It has often become impossible to work in the workroom, and not rarely "quiet work" is done at home. The workplace has frequently turned into a club due to the lack of a needed separate structure.

The "ideal" or "preferred activity" structure is usually the case when the workroom has been "cleared" of work activities. Other activities are localized in specific rooms on the floor (visiting, smoking, etc.) or in the building (relaxation, lunch, meetings, etc.). We could present the appropriate spatial structure of activities for different kinds of administrative buildings. One example is presented in Figure 20.4. To achieve the depicted model there must be fewer people in one room (the average actual number in this research was between 5 and 15 in different kinds of office buildings; the perference ranges from 2 to 4), and we have to create a set of functionally specialized rooms in the building.

f

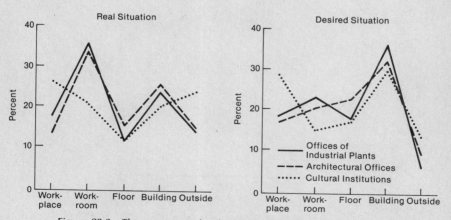

Figure 20.3   The percentage distribution between various locations.

(f) The results are discussed and then the authors present a model of the relationship between the number of people, activities, and special rooms needed for the people and activities.

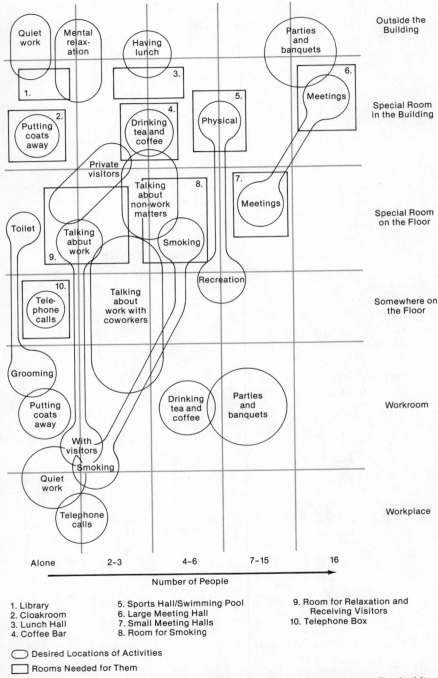

Figure 20.4   Desired locations of activities and special rooms needed for them in office buildings.

4. CONCLUSION

g   The method used seems appropriate given the desire to make practical proposals to the architects. It was really possible to make proposals about the optimal spatial location of activities since many of the everyday activities in offices are not located where they should be and could be.

This "projective" method seems to be a proper one for extracting the information about behavior from the respondents, as they had a complete picture of all the possibilities and the locations of activities.

Because this project was actually the first psychological study of office environments in the Soviet Union, it was more important for us to get a picture of the problems that exist relating to office buildings and to find a method of research rather than to get vast amounts of every kind of data.

(g) The authors make several practical suggestions in this section.

QUESTIONS

1. What differences exist between this study and the other experiments in this section?
2. Are the authors' intentions made clear?
3. Could you replicate this experiment?
4. Do you think that the results would hold up if the study was done in San Francisco, Omaha, and Detroit? Why? Why not?
5. Comment on the independent and dependent variables. Are they clearly specified?
6. Go to your library and select one or two foreign publications, preferably non-Western publications. Read and report on them to your class, noting differences in style and content.

CHAPTER
21

Prosocial Behavior

INTRODUCTION

One of the most hotly debated subjects in recent years has been the effect of television viewing on children's behavior. The average American child spends more hours per week watching television than attending school, and several educators have suggested that television may have replaced school as the primary "socializing" influence outside of the family.

The effects of television on children's behavior has been the subject of numerous psychological experiments, governmental studies, and congressional investigation. Most of this effort has centered on the effects of television violence and has asked whether television violence leads to violence in society. Less often studied is whether television can lead to (or cause) *prosocial* behavior. Prosocial behaviors are those behaviors that are believed to benefit others, such as helping people in distress.

The following experiment is an experimental study of the effect of television viewing on prosocial behavior. However, we have not analyzed the case for you. Instead, you are to write an analysis of the experiment. In your analysis the following outline is suggested:

1. What is the general topic of the research?
2. What pertinent literature is reviewed?
3. What is the research problem?

4. What are the authors testing that has not been done in previous research; that is, what is unique about this experiment?
5. What is the hypothesis?
6. Name the independent variable(s).
7. Name the dependent variable(s).
8. How were subjects assigned to treatments?
9. What controls were used to prevent extraneous variables from confounding the results?
10. How were the data collected?
11. How were the data analyzed?
12. Summarize the results. Draw a bar graph illustrating the results.
13. Summarize the discussion.

Effects of a Prosocial Televised Example on Children's Helping

JOYCE N. SPRAFKIN, ROBERT M. LIEBERT, and RITA WICKS POULOS
STATE UNIVERSITY OF NEW YORK AT STONY BROOK

The possibility that regularly broadcast entertainment television programs can facilitate prosocial behavior in children was investigated. Thirty first-grade children, 15 boys and 15 girls, were individually exposed to one of three half-hour television programs: a program from the *Lassie* series which included a dramatic example of a boy helping a dog, a program from the *Lassie* series devoid of such an example, or a program from the family situation comedy series *The Brady Bunch*. The effects of the programming were assessed by presenting each child with a situation that required him to choose between continuing to play a game for self-gain and helping puppies in distress. Children exposed to the *Lassie* program with the helping scene helped for significantly more time than those exposed to either of the other programs.

Since its spectacular rise as an entertainment medium in the 1950s, television's possible effects on the behavior of the young has captured the interest of many investigators. Throughout the years the most compelling question has remained the possible detrimental effects of violent programming, resulting in considerable knowledge of the influence of televised aggression (e.g., Liebert, Neale, & Davidson, 1973). Remarkably little is known, on the other hand, about the degree to which the medium might also inculcate prosocial attitudes and behaviors. The present report is concerned with this latter issue.

That certain commercially broadcast programs may serve as positive socializers is suggested by numerous investigations employing specially prepared programming presented in a television format. Such studies have shown that modeling cues provided by a television set can increase the generosity of young observers (Bryan, 1970), foster adherence to rules (Stein & Bryan, 1972; Wolf & Cheyne, 1972), augment delay of gratification (Yates, 1974), and facilitate positive interpersonal behaviors (O'Connor, 1969). Only one study, so far as we know, endeavored to determine the potential of *broadcast* televised materials for instigating prosocial forms of behavior in children. Over the course of 4 weeks, Stein and Friedrich (1972) exposed 3- to 5-year-olds to one of three television diets: aggressive (Batman and Superman cartoons), prosocial (episodes taken from "Mister Rogers Neighborhood"),

SOURCE: Reprinted by permission from *Journal of Experimental Child Psychology*, 1975, *20*, 119–126.

or neutral (such scenes as children working on a farm). Generally confirming earlier work, certain types of prosocial behavior were found to increase after exposure to the "Mister Rogers" program, relative to the other two types of input. In this important demonstration, as the investigators themselves indicated (Friedrich & Stein, 1973), it was not feasible to control for factors such as the format of the programs, the characters displayed, and the relative balance of entertainment vs. instructional components in each diet. Equally important, "Mister Rogers" is not a commercially produced network program and its direct, exhortative format also might serve to limit the size of its child audience.

In the experiment reported, we wished to determine whether the presence of a specific act of helping that appeared in a highly successful commercial series would induce similar behavior in young viewers when they later found themselves in a comparable situation. To control for various broad characteristics of the program, children in the two principal groups viewed one of two programs from the series *Lassie*, complete with commercial messages. The *Lassie* programs both featured the same canine and human main characters, took place in the same setting, apparently were written for the same audience, and presumably were motivated by the same entertainment and commercial interests. Whereas the programs necessarily differed somewhat in story line, the critical difference was that one of them had woven into it the particular prosocial example while the other did not. This combination of similarities and differences, it was reasoned, would optimize detecting the potential effect of specific modeling cues from entertainment television upon youngsters' later behavior. As a further comparison, a third group was also included, in which a generally pleasant situation comedy program, *The Brady Bunch*, was viewed.

METHOD

Design and Participants

A 3 × 2 factorial design was used, employing the television experience (observation of prosocial *Lassie*, neutral *Lassie*, or *Brady Bunch* program) and the sex differentiation of the subjects. The participants were 30 children, 15 boys and 15 girls, from four first-grade classes in middle-class, predominantly white suburban neighborhood. All prospective subjects were given forms to be signed by their parents authorizing their participation. The letter described the nature of the project, but requested parents to withhold this information from their children. Subjects were selected randomly from the pool of youngsters whose parents returned the consent forms; approximately 80% did so. Within each class and sex, the children were assigned randomly to the treatment conditions.

One experimenter escorted the child from the classroom to the television viewing room, conducted the screening of the videotape, and then brought the child to the second experimenter in an adjacent room. The second experimenter, who was blind to the treatment condition, administered the dependent measure. The former role was assumed by four graduate students over the course of the study; the latter role was always filled by the same 24-yr-old white, female graduate student.

Apparatus and Materials

The television viewing room contained a black and white television monitor with a 9-in. screen, programmed by a hidden Sony video recorder. The experimental room, in which assessment was conducted, contained two pieces of apparatus: a Point Game and a Help button, separated by a distance of 5 ft and each placed on a child-size desk.

The Point Game consisted of a response key and a display box. The response key was a 1½ by ¼ in. rectangular blue button mounted on a 4 by 6½ in. inclined platform which served as a handrest. The display box was a 6 by 8 in. gray metal box on which a blue 15-W light bulb and a Cramer digital timer were mounted. The 10's and 1's columns of the timer were covered with black tape. When the button was pressed, the bulb lit and the timer was activated.

The Help button was a small, green, circular button mounted on a 7 by 5 in. gray metal box with the word "Help" printed in black letters directly below it. The Help button activated a second Cramer digital timer which was located outside the experimental room.

Earphones were located next to the game on the desk. They were connected via an extension cable to a tape recorder located outside the experimental room. The tape contained 30 sec of silence followed by 120 sec of dogs barking.

Procedure

Introduction to the Treatment Condition

The first experimenter escorted the child from the classroom to the television viewing room, which was a small classroom containing desks, chairs, and the videotape monitor. She told the child that he was going to play a game shortly, but that Miss [the second experimenter] was not ready for him yet. She then turned on the video recorder surreptitiously, turned on the monitor openly, and suggested that the child watch while she did some work in the back of the classroom. All children appeared quite interested in the half-hour programs and the experimenter worked busily to discourage the child from initiating conversation.

Experimental and Control Conditions

Children in the experimental condition viewed the prosocial Lassie program. The episode dealt largely with Lassie's efforts to hide her runt puppy so that it would not be given away. The story's climax occurred when the puppy slipped into a mining shaft and fell onto the ledge below. Unable to help, Lassie brought her master, Jeff, and Jeff's grandfather to the scene. Jeff then risked his life by hanging over the edge of the shaft to save the puppy. The selection of the program was based on the inclusion of this dramatic helping scene.

In one control condition, the neutral *Lassie* program was shown. Featuring the same major characters as the prosocial program, it dramatized Jeff's attempts to avoid taking violin lessons. It was devoid of any example of a human helping a dog, but necessarily featured the animal in a positive light.

In the second control condition, the youngsters viewed *The Brady Bunch,* a non-aggressive, nonanimated children's program which did not feature a dog. The program depicted the youngest Brady children's efforts to be important by trying to set a record for time spent on a seesaw. It provided a measure of children's willingness to help after exposure to a popular program that showed positive interpersonal family encounters but no cues pertinent to either human or canine heroism.

Assessment of Willingness to Help Puppies in Distress

Assessment consisted of placing the child in a situation that required him to choose between continuing to play a game for self-gain and trying to get help for puppies in distress.

1. *Introduction of the game.* After the television program was viewed, the first experimenter told the child that Miss *[the second experimenter]* was ready for him, escorted the child to the experimental room, performed appropriate introductions, and departed. The child was invited to play the Point Game in which he could earn points by pressing a button that lit a bulb. The number of points obtained was displayed on a timer visible to him. The experimenter showed the child a display of prizes, situated on two desks, that varied in size and attractiveness—from small, unappealing erasers to large, colorful board games. It was noted that "some prizes are bigger and better than others," and that "the more points you get in the game, the better your prize will be." The experimenter then told the child that she was going to let him play the game alone so that she could finish some work down the hall.

2. *Introduction of distress situation.* The experimenter started to leave, then returned and exclaimed:

Oh, there's something else I'd like you to do while I'm gone. You see, I'm in charge of a dog kennel a few miles from here—that's a place that we keep puppies until a home can be found for them. I had to leave the puppies alone so that I could come here today, but I know whether the dogs are safe or not by listening through these earphones [indicating] which are connected to the kennel by wires. When I don't hear any noises through the earphones, I know the dogs are OK—they're either playing quietly or sleeping. But if I hear them barking, I know something is wrong. I know that they are in trouble and need help. If I hear barking, I press this button [indicating Help button], which signals my helper who lives near the kennel, and when he hears the signal, he goes over to the kennel to make sure nothing bad happened to the dogs.

While I'm gone, I'd like you to wear the earphones for me while you play your game. If you don't hear any barking, everything is OK at the kennel; but if you hear barking, you can help the puppies if you want by pressing the Help button. You might have to press the button for a long time before my helper hears the signal, and there's a better chance he'll hear it if you press it a lot. You can tell when he has heard the signal when you hear the dogs stop barking.

Remember, this game is over as soon as I come back—you won't be able to get any points after that. Try to get as many points as you can because the more points you get, the better your prize will be. You know, if the puppies start barking, you'll have to choose between helping the puppies by pressing the Help button and getting more points for yourself by pressing the blue button. It's up to you.

After being assured that the child understood the situation, the experimenter placed the earphones on the child and left the room. When outside the experimental room, she immediately turned on the tape recorder and a stopwatch. The child first heard 30 sec of silence followed by 120 sec of increasingly frantic barking. The experimenter recorded the latency of helping, reentered the experimental room 60 sec after the cessation of barking, and awarded a prize that was loosely based on the number of points earned. After the child was escorted to his classroom, the total seconds of helping was recorded from the timer located outside the experimental room.

RESULTS AND DISCUSSION

The primary dependent measure was the number of seconds the Help button was pressed. The mean helping scores for all cells of the design, and the standard deviations for each, are shown in Table 21.1. A 3 (Treatment) × 2 (Sex) analysis of variance of these scores yielded only one significant effect, for treatment conditions ($F \times 4.97$, $p < .025$); the main effect for Sex was not significant ($F = 2.18$). Dunnett's t tests revealed that subjects who saw the prosocial *Lassie* program helped significantly more than those in the neutral *Lassie* ($t = 2.08$, $p < .05$) or the *Brady Bunch* condition ($t = 2.79$, $p < .01$). The latter groups did not differ from one another ($t = .66$).[1]

There was also a marginal tendency for sex and treatment condition to interact ($F = 2.80$, $p < .10$). As is apparent descriptively in Table 21.1, the effect appeared because the impact of the prosocial *Lassie* program, relative to the neutral one, tended to be greater for girls than for boys; moreover, the girls tended to respond differently to the neutral *Lassie* and *Brady Bunch* programs whereas the boys did not. Both of these tendencies may be due, in part, to a possible ceiling effect for the boys; in the two control conditions they tended to help more than their female counterparts. Further, nine out of the ten boys in the two control conditions rendered some help, whereas only five out of ten girls in the control conditions did so. (Four of the five boys who saw the prosocial *Lassie* program, and all five of the girls, rendered some help to the distressed puppies.)

[1]A parallel 3 × 2 analysis of variance of latency to help scores revealed no significant differences, but the tendency was consistent with amount of helping: 24.6 for the prosocial *Lassie*, 36.1 for the neutral *Lassie*, and 55.4 for the family situation comedy.

Table 21.1   MEANS [AND *SD*] OF HELPING
SCORES IN SECONDS FOR ALL GROUPS

	SEX OF SUBJECT		
	FEMALE	MALE	COMBINED
Prosocial Lassie	105.48	79.80	92.64
	[20.37]	[47.38]	[36.94]
Neutral Lassie	34.80	68.20	51.50
	[46.30]	[53.36]	[50.28]
Brady Brunch	8.50	66.50	37.50
	[14.18]	[45.55]	[44.13]

Despite these different tendencies, the overall results of this experiment disclose clearly the validity of the basic demonstration we sought to produce: At least under some circumstances, a televised example can increase a child's willingness to engage in helping behavior. This finding extends the earlier work with simulated television materials and the experimental field study reported by Stein and Friedrich (1972) by demonstrating the influence of commercial broadcast programming designed primarily as entertainment. The present design also permitted us to isolate the particular aspect of the program, a prosocial example by the protagonist, as the central ingredient necessary for such an effect to occur; the alternative *Lassie* show and the generally warm *Brady Bunch* program produce significantly less helping. Inasmuch as the former control program featured the same major characters, setting, and general dramatic style as the prosocial *Lassie* show, it is unlikely that factors other than the specific modeled example influenced helping. This conclusion is further bolstered by the comparison of helping responses between children who saw either the neutral *Lassie* or *Brady Bunch* shows; the mere presence of a canine hero in the former did not itself significantly facilitate helping of puppies in distress. Theoretically, then, the present results support the social learning view of Bandura (1969, 1971) and Liebert (1972, 1973; Liebert, Neale, & Davidson, 1973) that the effects of television on behavior are mediated by specific modeling cues and the interpretation of these cues by the child, rather than general format and other global considerations. What is more, the dependent measure involved helping the experimenter as well as helping the puppies; to the extent that the children responded to this aspect of the situation, the results represent even more generalization from the specific altruistic behavior modeled in the prosocial show.

The practical implications of the present demonstration are also clear: It is possible to produce television programming that features action and adventure, appeals to child and family audiences, and still has a salutary rather than negative social influence on observers. A detailed understanding of the modeling processes involved in such influences may contribute to the production of other socially desirable programs in the future.

REFERENCES

Bandura, A. (1969). *Principles of behavior modification.* New York: Holt, Rinehart & Winston.

Bandura, A. (1971). Analysis of modeling processes. In A. Bandura (Ed.), *Psychological modeling.* New York: Aldine-Atherton.

Bryan, J. H. (1970). Children's reactions to helpers: Their money isn't where their mouths are. In J. Macauley & L. Berkowitz (Eds.), *Altruism and helping behavior.* New York: Academic Press.

Friedrich, L. K., & Stein, A. H. (1973). Aggressive and prosocial television programs and the natural behavior of preschool children. *Society for Research in Child Development, 38* (whole Monogr.).

Liebert, R. M. (1970). Television and social learning: Some relationships between viewing violence and behaving aggressively (overview). In J. P. Murray, E. A. Rubinstein, & G. A. Comstock (Eds.), *Television and social behavior. Vol. II: Television and social learning.* Washington, D.C.: U.S. Government Printing Office.

Liebert, R. M., (1973). Observational learning: Some social applications. In P. J. Elich (Ed.), *Fourth western symposium on learning.* Bellingham, Wash.: Western Washington State College.

Liebert, R. M., Neale, J. M., & Davidson, E. S. (1973). *The early window.* New York: Pergamon Press.

O'Connor, R. D. (1969). Modification of social withdrawal through symbolic modeling. *Journal of Applied Behavior Analysis, 2,* 15–22.

Stein, A. H., & Friedrich, L. K. (1973). Television content and young children's behavior. In J. P. Murray, E. A. Rubinstein, & G. A. Comstock (Eds.), *Television and social behavior. Vol. II: Television and social learning.* Washington, D.C.: U.S. Government Printing Office.

Stein, G. M., & Bryan, J. H. (1972). The effect of a television model upon rule adoption behavior of children. *Child Development, 43,* 268–273.

Wolf, T., & Cheyne, J. (1972). Persistence of effects of live behavioral, televised behavioral, and live verbal models on resistance to deviation. *Child Development, 43,* 1429–1436.

Yates, G. C. P. (1974). Influence of televised modeling and verbalisation on children's delay of gratification. *Journal of Experimental Child Psychology, 18,* 333–339.

This study was supported, in part, by grants from General Foods Corporation and General Mills, Inc. to Eli A. Rubinstein, John M. Neale, and the second and third authors for the investigation of television's prosocial influence. Grateful acknowledgment is due the Wrather Corporation for providing the *Lassie* programs, and to the principal, Mrs. Ann Littlefield, and staff of Smithtown Elementary School, Smithtown, New York, for valuable collaboration in this work. Special thanks also are due Elaine Brimer, Ann Covitz, Francine Hay, David Morgenstern, and Steven Schuetz for their capable assistance as experimenters.

ADDITIONAL QUESTIONS

1. What design principles are illustrated in this experiment? Is the sample potentially biased by this constraint? Explain.
2. Think of another experiment that would test some of the hypotheses or issues raised by this experiment.

3. Why did the experimenter get parental consent? When do you need the consent of parents or of the subjects themselves before the experiment?
4. Was the experimental treatment realistic? That is, do you think that the children believed that the situation was real or contrived? Explain.
5. The experiment involved deception in that the children were led to believe they were monitoring a dog kennel. Discuss the issue of deception in psychological research. Under what conditions is it permissible?
6. The experimenters apparently did not "debrief" the subjects, that is, tell them the true nature of the experiment after it was over. When is debriefing necessary or desirable, and when can it be ignored?
7. Compare your answers to questions 3, 5, and 6 with the ethical principles in Chapter 7.

22

Birdsong Learning

INTRODUCTION

Growing up in the rolling hills of Nebraska, I remember my first introduction to the identification of birds through their song. "A bob-white can be easily recognized by its song because he says his name," my father told me, and then he whistled, " ♩♩♪ ," and said, "bob-bob-white." My enthusiasm for bird identification—yea, all sorts of nature studies—soared. I was convinced that this perfect mnenomic device was but one example of the divine harmony between verbal labels and nature's phenomena—in this case, the name of a bird and its song. My early enthusiasm was quickly dashed when I tried to apply the same technique to the sound of a robin or a wren. And the complex melody of the meadowlark was nothing like the word "meadowlark" (however, the "caw" of the blackbird did seem to fit the animal).

It was somewhat gratifying to learn that serious researchers are also concerned with the problems of labeling birdsongs, and in the article just presented, the authors look at the identification of birdsongs by means of a clever technique. In the article by Richard Walk and Michael Schwartz, the authors study birdsong identification in which visual stimuli are used as mnemonic devices. You will find that the experiments deal with an interesting theoretical problem as well as illustrating a unique design problem.

This article was selected because it contains several important design features. As you study this experiment, pay particular attention to these elements: (1) This is a multiple-experiment study. The researchers present two separate experiments that address a common problem but deal with separate aspects of the problem. (2) The experimenters "pretested" subjects to establish a base rate of birdsong identification. (3) The design is primarily a random group design with a visual code group vs. a "no-code" group, but also two measures were taken on both groups which is a "repeated measure design." (4) The experimenters report the loss of detailed records for some subjects. Notice how they handle this problem.

SPECIAL ISSUES

Multiple Experiments

Psychological research in the past several years has become increasingly complex. Because of the expanded nature of the subject, researchers have been inclined to undertake studies which examine several aspects of a common problem. Some journals of the APA encourage multiple-experiment manuscripts. Even a casual reading of the current literature in psychology will confirm this observation.

The practice of multiple-experiment articles has several strengths. In a multiple-experiment article, it is possible to approach a psychological issue from various experimental viewpoints, while a single experiment may tell only a portion of the answer. Several experiments may reveal a more complete picture of the issue. Another aspect of multiple-experiment studies is that "programmatic" research can be reported in a single experiment. Many contemporary researchers develop a long-range, progressive scheme of research effort. This scheme may involve a series of experiments in which later experiments are predicated on the outcome of earlier experiments. This type of research is particularly amenable to multiple-experiment articles.

Birdsong Learning and Intersensory Processing

RICHARD D. WALK
GEORGE WASHINGTON UNIVERSITY, WASHINGTON, D.C.
MICHAEL L. SCHWARTZ
YALE UNIVERSITY SCHOOL OF MEDICINE, NEW HAVEN, CONNECTICUT

Two experiments were performed in which subjects learned to attach names to birdsongs. In the first experiment, subjects who were instructed to generate their own visual codes were much superior to those not given any instructions except those of learning the birdsongs. In the second experiment, both those given a model code for half of the birdsongs and those who made their own visual codes were superior to controls without visual codes. The experiments show the way in which learning in the auditory modality can benefit from visual symbols.

a   Eleanor Gibson, in her influential book *Perceptual Learning and Development,* reports that W. H. Thorpe used visual spectrograms to help his students learn to identify birdsongs (Gibson, 1969, p. 211). This is an example of the use of a visual model or crutch to help to learn a complex auditory pattern. **It has implications for intersensory processing, the way in which one modality can influence learning in another one.** Birdsongs are readily available in our environment and on commercial phonograph records, yet few of us can identify with certainty any bird sounds except the most obvious, such as the caw of the crow or the cock-a-doodle-doo of the rooster. But these common bird sounds are not songs. Byron once characterized, with poetic insight, a birdsong as "the sweetest song ear ever heard."

The present series of experiments concerned the learning of birdsongs. The first experiment investigated the effect on such learning of having the subjects

SOURCE: Reprinted from *Bulletin of the Psychonomic Society*, 1982, *19*, 101–104, by permission of the Psychonomic Society and the authors. The original article has been abridged.

ANALYSIS

REVIEW OF LITERATURE

The review of literature is unusually brief and proselike (the authors even quote a poet!). However, the essential qualities are represented: the basic theory that motivates the study (a), the problem (b), and the hypothesis (c), the latter two in the form of a question.

construct their own visual code of the birdsong as compared with those given instructions just to learn the songs. The second experiment had an additional group that had model visual patterns of half of the birdsongs.

b
c   **Do the visual codes help learning? What are the theoretical implications of such auditory-visual studies?**

GENERAL METHOD

Stimulus Materials

Birdsongs

A sample of 10 common birdsongs was selected from a record that contained a large number of birdsongs. The songs chosen were: Baltimore oriole, cardinal, common goldfinch, mockingbird, purple finch, robin, song sparrow, towhee, tufted titmouse, and winter wren.

d   **The songs were placed in random order on a cassette tape.** A total of six random orders, one for each phase, was constructed as follows: Phase 1 (pretest)—birdsongs followed by 8-sec pause; Phases 2–5 (training)—birdsongs identified at end of 8-sec pause; Phase 6 (posttest)—birdsongs followed by 8-sec pause. Each phase played all 10 birdsongs, with the trial identified for a phase so it could be matched with the answer sheet, for example: *"Trial one.* Birdsong [e.g., cardinal]. *Trial two.* Birdsong [e.g., winter wren] . . . ,"* and so on. With six phases and 10 trials for each phase, the experiment had a total of 60 trials. In the training section, as described, the name of the song came at the end of the 8 sec, whereas no feedback was given for pretest and posttest. The tape lasted about 25 min because of the taped instructions, the birdsongs, pauses for writing the answers, and 30-sec pauses
e   between phases.[1]

[1]Note that the length of the foreperiods and intertrial intervals differed between Experiments 2 and 3. The difference was necessary to allow the computer to perform the calculations required between trials. Hoowever, we did manage to keep constant the ratio of foreperiod to intertrial between the two experiments.

GENERAL METHOD

The experiment is introduced by a general method section in which the stimulus material and procedure common to both experiments is described.

The birdsongs used in the experiment are identified and then (d) the authors tell us that six random orders of the songs were used. Presumably this was done to guard against any systematic errors due to presentation order. (We know that first and last items of a series of items are learned better than middle items and that effect may interact with the independent variable.) The exact makeup of the stimulus material is then disclosed. The authors place in a footnote (e) the specific phonograph record used in obtaining the stimulus material.

Procedure

All subjects were told that they would hear a tape with birdsongs on it and that they would be asked to match the songs they heard with the names of the birds listed on the response sheet. The taped instructions reiterated this. On the pretest, the taped instructions said that no identification would be given of a song, but it asked that the subjects guess on every trial. The training phases (Phases 2–5) told the subjects to guess before the song was identified at the end of 8 sec. The posttest instructions told the subjects the birdsongs would not be identified and asked them to answer every trial.

EXPERIMENT 1

In this experiment, some of the subjects were asked to construct their own visual code to help identify the birdsongs.

Method

Subjects

The subjects were 87 students in a laboratory course in psychology at George Washington University. They were from five different laboratory sections.

Procedure

Two laboratory sections were given no additional instructions. The other three sections were told that a way to assist learning would be to take notes on the characteristics of each birdsong. For example, high and low marks might illustrate high and low notes in the songs.

Analysis

Not all subjects requested to use notes actually did so. Those requested to use notes, but who used none at all, compose a second "no-code" group, referred to in the results as "No-Code 2," meaning a second group of subjects without a visual code.

The General Procedure, i.e., the procedure used in both experiments, is described in (f).

EXPERIMENT 1

After the general method the experimenters turn to the specific method used in Experiment 1. This section is easy to understand and requires no comment except the section identified as "analysis" (g). This is a departure from normal procedure and indicates that not all subjects used techniques suggested by the experimenters. Nevertheless, data generated by these subjects are of interest and Walk and Schwartz tell how these data were analyzed.

RESULTS

The results in terms of accuracy by group are shown in Table 22.1. An analysis of variance showed that the groups did not differ on the pretest; all were at the chance level of about 10% correct. On the posttest, however, all groups were more accurate. Pretest-posttest difference scores showed that both the first no-code group ($p < .01$) and the second no-code group ($p < .05$) had significantly improved scores and the visual code group was markedly improved ($p < .001$). An analysis of variance of posttest scores of the three groups found a significant group effect [$F(2,84) = 17.13$, $p < .01$], and the student's t range statistic demonstrated that the visual code group was significantly superior ($p < .01$) to the other two groups.

The performance of the three groups on each individual birdsong is shown in Table 22.2. The use of a visual code helped performance on all songs, except that the superiority of the visual code group was slight for the common goldfinch. The use of a visual code particularly helped in discriminating the songs of the mockingbird, robin, towhee, and tufted titmouse. The high performance of the visual code group on the Baltimore oriole, cardinal, towhee, and winter wren should also be noted. Of course, contextual factors, such as the place of a song on the list and the confusability of competing songs, is a factor in the task.

h

Table 22.1   AVERAGE PERCENT CORRECT IDENTIFICATIONS
FOR THE THREE GROUPS IN EXPERIMENT 1

TIME	GROUP		
	NO-CODE 1	NO-CODE 2	VISUAL CODE
Pretest	9	11	10
Posttest	24	25	53
N	38	17	32

RESULTS

The Results section of the first experiment is clearly presented. The authors use a standard statistical test (an analysis of variance) and report the level of significance. In addition to the statistical analysis, the data are shown in two tables: the first contains the average correct identification for the three groups and the second contains the percent correct identification for each birdsong. The authors comment on the degree of improvement for some species (h). The section is concluded with some brief comments, but the major discussion is postponed until the results of Experiment 2 are presented.

Table 22.2   AVERAGE PERCENT CORRECT IDENTIFICATIONS ON
THE POSTTEST FOR EACH BIRDSONG, BY GROUP, IN EXPERIMENT 1

| | GROUP | | |
BIRDSONG	NO-CODE 1	NO-CODE 2	VISUAL CODE
Baltimore Oriole	37	29	63
Cardinal	47	47	81
Common Goldfinch	18	6	25
Mockingbird	11	12	47
Purple Finch	13	12	34
Robin	24	0	47
Song Sparrow	24	47	56
Towhee	21	35	72
Tufted Titmouse	11	24	44
Winter Wren	32	35	66

To conclude Experiment 1, the use of a visual code markedly improved
performance on this auditory task. But the visual code must be used. Those
given instructions to use the code who ignored the instructions performed
no better than the control group that did not receive such instructions. Further
discussion of these results is deferred until after Experiment 2 is presented.

EXPERIMENT 2

Method

Subjects
The main subjects were 37 students from three laboratory sections. Detailed
records were lost for an additional 34 subjects from two laboratory sections.

Procedure
Two of the laboratory sections had the same procedure as Experiment 1.
The third laboratory section had a model code pattern made for five of the
birdsongs (see Figure 22.1). These birdsongs were chosen to have equal
performance on Experiment 1 by those who used a visual code. The songs

EXPERIMENT 2

In this article two experiments are presented: a practice which is becoming increasingly common in the psychological literature.

The method section of the second experiment does not repeat the information in the first experiment but adds information that is specific to
the new procedure. Because the model given to subjects could not be
described verbally, the authors present a sample of the model in Figure
22.1.

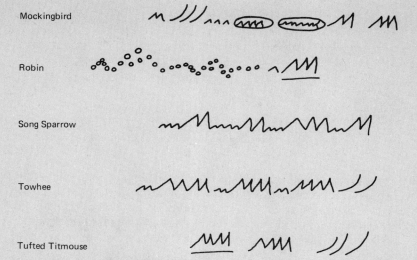

Mockingbird

Robin

Song Sparrow

Towhee

Tufted Titmouse

Figure 22.1   The model or pattern furnished subjects in the model code group of Experiment 2.

chosen, mockingbird, robin, song sparrow, towhee, and tufted titmouse, were correctly identified 52.8% of the time on the posttest by the visual code group of Experiment 1. The other birdsongs had 53.4% correct identifications on the posttest by the visual code group.

The modeled birdsong group was given a dittoed sheet with models of the songs (Figure 22.1). These subjects were not allowed to turn the sheet over until after the pretest phase of the experiment.

Analysis

One laboratory section ($n = 18$) was a modeled code group; the other ($n = 16$) was a group that constructed its own visual codes. All subjects requested to use a visual code actually did so.

RESULTS

The main results are the three sections for which complete data were available. The other two sections will serve to fill in and help give perspective to the experiment.

Several issues are raised by Experiment 2. The first is the loss of detailed records for part of the subjects. One may argue that a professional article should reflect the purest techniques available and thus, if the detailed records were lost, that another group should have been run in which complete records were available. On the other hand, consider how the experimenter used the data generated by the group whose records

Table 22.3   AVERAGE PERCENT CORRECT IDENTIFICATIONS FOR
EACH GROUP IN EXPERIMENT 2, SHOWING POSTTEST PERFORMANCE
ON MODELED AND NONMODELED SONGS

			POSTTEST		
GROUP	N	PRETEST	OVER-ALL	MODEL SONGS*	NO MODEL
No Code	14	10	24	21	26
Own Code	11	7	59	55	64
Model Code	12	11	50	57	43

*Only the "model code" group actually had a model or pattern of the five birdsongs.

The results are shown in Table 22.3. They are very comparable to those of the first experiment. Pretest scores were around the chance level of 10%; the no-code group had 24% on the posttest and the visual code group had over 50%. The subjects furnished model patterns of the birdsongs were similar in their performance to those who made their own visual code.

Statistical analysis confirmed this. The groups did not differ on the pretest. A posttest showed a significant difference [$F(2,34) = 8.61, p < .01$] among groups, and the two code groups were significantly superior on the student's t range statistic ($p < .01$) to the group not instructed to use a visual code.

Two further questions are relevant to this analysis. Are those with a pattern model superior on the modeled birdsongs to those who construct their own visual code? Does the use of a patterned model for half of the songs generalize to improve performance on the other songs?

This is also shown in Table 22.3. Both the modeled group and the own-code group were superior to the no-code group on the five modeled songs [$F(2,34) = 7.13, p < .01$]. The two code groups were significantly different from the no-code group but not from each other. The answer to the first question is negative. The second question is more complex. On songs without a model, the analysis of variance among groups was again significant [$F(2,34) = 6.05, p < .01$], but the significant difference in the means was for the own-code group compared with the no-code group, and the difference among means almost showed the own-code group (64%) significantly superior to the patterned model group (43%). The modeled song group seemed

were lost (i). These data were not considered to be the primary findings but were presented as supplementary information. Despite the cautious use of data from their subjects by the authors, we suggest that your research should be restricted to groups on which complete records are maintained. There may be some exceptions to this rule, as in the case of field-based experiments where a nonrepeatable situation occurs (e.g., a death of a national figure, a riot, an accident). Because of the capricious nature of the event and subject sample, some data and/or characteristics of the sample may be unavailable or lost in these special situations.

somewhat superior on nonmodeled songs to the no-code group (43% vs. 26% accuracy), but the difference was not significant and the trend may have been due to an artifact. The artifact is that the song models raise the chance level on the nonmodeled songs. Thus, to answer the second question, we have no evidence that the use of models generalizes (perhaps by increasing attention to identifying features) to improve performance on songs without a model.

i

Results from those for whom statistical analysis could not be performed were similar. The own-code subjects had 9% correct on the pretest and 47% on the posttest, with 40% correct on the other group's modeled songs and 54% on the nonmodeled songs. The modeled group had 11% correct on the pretest and 46% on the posttest, with 54% accuracy on the songs with a model and 37% on the songs without one. Compared with the controls, with 26% accuracy, the 37% correct on the songs without a model for the modeled group would seem to show that the modeled experience does not generalize to help performance on the nonmodeled songs. The slight superiority could be an artifact.

Thus, those for whom no analysis could be performed reinforced the conclusions from the other subjects. The modeled songs are a definite aid to learning, but the use of the models does not seem to generalize beyond them.

DISCUSSION

j

These experiments show that visual codes, whether generated by the subjects themselves or from models provided by the experimenters, help subjects to learn auditory songs. The results seem to be fairly specific: Those subjects instructed to use a code who used none did not benefit, and learning of the nonmodeled songs in Experiment 2 was not helped by the models provided for some of the songs. Explanation of the effect in terms of the rehearsal of, participation in, or attention to the task provided by the suggestion of or presence of the codes does not seem to be appropriate because of the specificity. We seem to have specific intersensory facilitation based on vision.

Birdsongs are hard to describe in words, but complex visual patterns can help in their identification. Neither the subjects' own codes or the models

DISCUSSION

In section (j) the authors repeat the major findings of their two experiments and an abbreviated interpretation of their data. The remainder of the discussion section is devoted to a description of their results and the research of others; additionally, they discuss the theoretical implication of their experiments. Finally, the Discussion section is ended by raising

shown in Figure 22.1 are obvious sound spectrograms, but they do contain intermodal elements that aid learning. Some of the self-generated visual codes from Experiment 1 are reproduced in Walk (1978). They have few obvious intersubject similarities, and yet they did aid learning.

The broader theoretical question concerns high-level similarities and differences in the coding of different modalities. One thinks of Gibson's (1969) concept of "higher order invariances" or similarities, such as sharpness or smoothness, that interconnect modalities. Even 6- to 8-month-old infants can identify as similar or different patterns in the visual and auditory modalities (Allen, Walker, Symonds, & Marcell, 1977).

Birdsongs have no obvious matching codes like those the culture has provided for reading and music. One could teach oneself musical notation, of course. Sound spectrograms may help, but our subjects did not need them.

k

We need more research on how the modalities are related to each other. Where should the present research lead? Is it best to ask what "distinctive features" best interconnect the two modalities and how we can characterize the intersensory principles of perceptual organization? Or will any symbol help? We doubt that. If we named the birdsongs (the "squeaky one," "the busy one," etc.), we might get just as much facilitation as we do with visual models, but we could not randomly connect the names (or the models) to the songs.

several questions that remain unanswered (k) and may serve as the springboard for further experimental work.

QUESTIONS

1. How were subject responses gathered? Suggest an alternative way that may be more "naturalistic."
2. Suggest and design an experiment that uses verbal labels as a mnemonic device in birdsong learning, in place of visual models.
3. Why did the authors divide the subjects who were instructed to use the notes (Experiment 1) into two groups: those who actually used the notes and those who did not use the notes? Why did they use these two groups in their statistical analysis?
4. In Experiment 2, the authors compare some models that were given to subjects as a means to encoding the birdsongs–birds. What are the implications for these subjectively drawn models, especially with regard to developing comparative experiments?
5. What purpose did the pretest serve (Experiment 1)? Was it necessary? Why? Why not?
6. Comment on the loss of subjects' data.

REFERENCE NOTE

1. Birdsongs were taken from the phonograph record "A field guide to bird songs of Eastern and Central North America" (revised) (HM-4671). Boston: Houghton Mifflin, 1971. A copy of the tape used in this experiment can be obtained from the first author at cost.

REFERENCES

Allen, T. W., Walker, K., Symonds, L., & Marcell, M. (1977). Intrasensory and inter-sensory perception of temporal sequences during infancy. *Developmental Psychology, 13,* 225–229.

Gibson, E. J. (1969). *Principles of perceptual learning and development.* New York: Appleton-Century-Crofts.

Walk, R. D. (1978). Perceptual learning. In E. C. Carterette & M. P. Friedman (Eds.), *Handbook of perception (Vol. 9). Perceptual processing.* New York: Academic Press.

Reprints may be obtained from Richard D. Walk, Department of Psychology, George Washington University, Washington, D.C. 20052.

CHAPTER
23

Note Taking

INTRODUCTION

The second most widespread habit among college students is probably note taking. From courses in physics, to anatomy, to philosophy, to sociology, to history, to biology, to English, to football, to anthropology, students and notebooks are ubiquitously intertwined. And woe to the unfortunate student who misplaces his or her treasured notes before an examination. But what is the effect of note taking on recall? This very practical question has not escaped the eye of educational psychologists, who have investigated the question with the hard precision of experimentation. This experiment by Judith Fisher and Mary Harris is an example of how a very practical educational problem can be brought under experimental control. The specific question Fisher and Harris ask is: "Is note taking a beneficial activity for the student, and if so, does it operate as a method that aids the student in recalling the material simply by the process of note taking itself or by the fact of having an external record of the lecture to refer to after the lecture?" Of particular interest in this experiment are the conflicting results regarding the effect of note taking on recall (a), with some experiments reporting a significantly lower recall for note takers than for students who did not take notes.

ATTRIBUTES OF FIELD EXPERIMENTS

Field experiments generally lack the control of laboratory experiments. They are not devoid of control, however. In many experiments (for example, the article on note taking) the researcher can specify the nature of the subjects, the independent variable, the dependent variable, and many situational stimuli that may influence the results. Before considering some specific issues raised by the article on note taking, a few of the positive attributes of the field technique will be mentioned. The value of field techniques are that they can be conducted in the subject's natural setting, with little disturbance in his or her normal behavior. Subjects are not required to report to the laboratory in field experiments. This means that the research can be conducted in a realistic setting which tends to enhance the credibility of the study. In some forms of field research, the subject is unaware that he or she is a subject. For example, a social psychologist may be interested in the ratio of Bush to Dukakis bumper stickers on a college campus and in a church parking lot. The subject is not even present—except by proxy.

SPECIAL ISSUES

Subject Variables

Fisher and Harris are mindful of the possible influence of the sex of the subject on their results and test its influence (**k**). Although a sex influence was found (males did more poorly than females), the influence of sex did not significantly interact with the experimental conditions. These sex results are in themselves interesting, and, as the researchers are quick to point out in their discussion (**p**), further investigation into the effect is warranted. Generally, these results are contrary to other experiments which show that subjects do better on learning-memory tasks when the experimenter is of the opposite sex. The sex effect in the present experiment may be only slightly affected by the lecturer being a woman and more greatly affected by the sample of male students who were enrolled in a course in human growth and development.

Effect of Note Taking and Review on Recall

JUDITH L. FISHER and MARY B. HARRIS
UNIVERSITY OF NEW MEXICO

Male and female college students were randomly assigned to five treatment groups combining different note-taking and review combinations. Recall was measured immediately and three weeks later. The results showed that a combination of taking notes and reviewing one's own notes produced the most recall, while not taking notes and reviewing the lecture "mentally" produced the least recall. Females recalled significantly more data than males, but opinions concerning note taking and efficiency of notes were not related to recall outcome. Quality of notes was positively correlated with free-recall score for two of the three note-taking groups and with short-term objective test score for two of the three note-taking groups.

Although the practice of taking notes and subsequent review of those notes for study purposes is widespread, the experimental evidence with respect to the efficacy of note taking is not clear. DiVesta and Gray (1972) have shown that note taking leads to an increase in the number of ideas recalled from a prose passage. Crawford (1925b) found that note-takers had significantly higher immediate and long-term recall scores than non–note-takers, and Howe (1970b) showed that the probability of recalling an item that occurred in the subject's notes was about seven times that of items not in his notes. **Other studies, however, have shown no significant differences between different note-taking strategies and not taking notes on a measure of recall** (Howe, 1970a; McClendon, 1958) **and even significantly lower recall scores for note-takers than non–note-takers during a film presentation** (Ash & Carlton, 1953).

a

SOURCE: Reprinted with permission from *Journal of Educational Psychology*, 1973, *65*, 321–325.

ANALYSIS

REVIEW OF LITERATURE

In the review of literature (**a**) the authors point out conflicting results and the problem.

METHOD

The problems presented in the present study are to discover if note taking and review of notes aid students in recall, to investigate the nature of note taking (encoding versus an external memory function), and to assess

There were five types of packets compiled according to experimental treatment condition. Each packet consisted of a cover sheet, which contained a questionnaire concerning note-taking habits, and instructions to either take or not take notes. Paper was provided for those in the three note-taking groups. Later pages instructed the subject about review procedures, providing subjects in the NN-RLN and N-RLN conditions with a copy of the lecturer's notes. The last three pages of each packet consisted of a free-recall page where the subject was instructed to put down all that he could remember of the lecture and a two-page objective test consisting of 15 multiple-choice and four short-answer items. These packets were put in random order before being distributed to the classes. The posttest consisted of nine multiple-choice items, which had been asked on the short-term objective test, four new multiple-choice items, and one three-part short-answer question; assignment of questions to the posttest and short-term objective test was done randomly.

Procedure

After the subjects assumed their regular seats, the packets were distributed. Subjects were instructed to fill out the questionnaire and read the instructions at the bottom of the page. These instructed the subject either to take notes on the following

---

notes but could not review them would do better than non–note-takers allowed to review the lecturer's notes, if encoding were indeed important. The design of the experiment can be conceptualized as follows:

	REVIEW CONDITION		
Note-takers	Review lecturer's notes	Mental review	Review own notes
Non–note-takers	Review lecturer's notes	Mental review	———

In (**e**) the time sequence of the test session is specified.

Subjects were assigned to five groups (**f**), which are shown below:

Note Condition	Review Condition	Subject Assignment			
Non–note-takers	Review lecturer's notes	S_{A1}	S_{A2}	$S_{A3} \ldots$	S_{An}
	Mental review	S_{B1}	S_{B2}	$S_{B3} \ldots$	S_{Bn}
Note-takers	Review lecturer's notes	S_{C1}	S_{C2}	$S_{C3} \ldots$	S_{Cn}
	Mental Review	S_{D1}	S_{D2}	$S_{D3} \ldots$	S_{Dn}
	Review own notes	S_{E1}	S_{E2}	$S_{E3} \ldots$	S_{En}

In (**g**) Fisher and Harris tell us that the lecturer was a female graduate student and that female subjects outnumbered male subjects by a ratio of nearly 3:1; more will be made of this later.

In the description of "materials" (**h**) the authors state that "every effort was made to keep experimental conditions as close to actual classroom

blank pages or to listen to the lecture without taking notes or turning the first page of the packet. The lecture was then delivered. At the conclusion of the lecture, the subjects were told to turn to page A and follow the instructions. The instructions on this page varied with experimental treatment, so that subjects in the N-MR and NN-MR groups were instructed to review mentally what they heard in the lecture. Subjects in the N-RON group were instructed to review the notes they had taken, while those in the N-RLN and NN-RLN groups were instructed to review the set of the lecturer's notes, which was provided in the packet. At the end of 10 minutes, the subjects were told to turn to page B, where they were instructed to write down as much of the lecture as they could remember. They were also instructed to continue to the last two pages when they had completed the free-recall section. These last two pages contained multiple-choice and short-answer questions. At the end of 30 minutes all papers which had not been turned in were collected. No mention of a posttest was made. Three weeks after the lecture was given, the experimenter brought the posttest of long-term recall to class and gave it to those subjects who were present and who had heard the original lecture. Since this was the last week of classes of the regular year, only 71 of the original subjects were present to take the posttest.

conditions as possible." Thus, the experimenters hoped to maximize the "natural" field in which data are collected and to minimize the possible "artificial" influence of a laboratory setting.

Each subject received a packet that contained everything needed in the initial part of the experiment. It is especially important to note that the dependent variable in this experiment was evaluated by several tests: a free-recall test, a multiple-choice test, and a short-answer test. The posttest conditions also contained several types of tests, with some of the items being new items and some old. These multiple measures are separately analyzed in the results section. In an effort to specify direct cause-and-effect relationships between experimental conditions and results, many times it is necessary to make complex measures of the subjects' responses. Such is the case in this experiment. Note taking may influence test results, but Fisher and Harris go beyond this global proposition and direct their research toward finding out exactly what type of test results are most influenced.

The experimenters were careful to set up an objective scoring system so that a reliable score could be assigned to the ambiguous data.

It should also be noted that the free-recall scores and the objective-test scores correlated quite highly, which would suggest that both tests are a measure of similar information or abilities.

The results were measured in three dependent measures (i). Prior to the main statistical analysis, a test for homogeneity of data was run (j). Had the data not been homogeneous, it would have required special treatment.

RESULTS

i **The three basic dependent measures were those of short-term free-recall scores, short-term objective test scores, and long-term objective posttest scores.** Since t tests revealed no significant differences between the two classes on any of the measures and since such differences would not have been of theoretical interest anyway, data from the two classes were pooled in all further analyses. Hartley's F_{max} tests revealed that variances of scores on both the free-recall test ($F_{max} = 3.29$, $df = 5/25$) and the short-

j term objective test ($F_{max} = 1.09$, $df = 5/25$) were homogeneous. The 5 × 2 unweighted means analyses of variance of treatment and sex **revealed significant treatment effects and significant sex differences, indicating that while women in all treatment conditions showed a higher mean score on the free-recall test ($\overline{X}_F = 18.58$, $\overline{X}_M = 12.17$; $F = 7.295$, $df = 1/102$, $p < .01$) and short-term objective test ($\overline{X}_F = 15.19$, $\overline{X}_M = 11.56$; $F = 9.850$,**

k **$df = 1/102$, $p < .01$) than did men, no sex by treatment interactions approaching statistical significance were found.** Women did not score significantly higher than men on the posttest ($F = 3.267$, $df = 1/61$). Therefore, in order to increase the n in each group, data from men and women were pooled in all further analyses.

Scores on the short-term free-recall and objective tests were significantly positively correlated ($r = .70$, $p < .01$), as were the free-recall and posttest scores ($r = .66$, $p < .01$) and the short-term objective test and posttest scores ($r = .75$, $p < .01$).

The free-recall tests were scored by giving 1 point for each idea recalled from the lecture, according to a master list of the 101 points contained in the lecture, 91 of which were also contained in the lecturer's notes given to the subjects for review. All tests were scored by one judge who was unaware of the subject's experimental condition. A second judge was then trained in the scoring procedure using 20 protocols, and then both raters scored a second 20 randomly selected protocols, producing an interscorer reliability coefficient of .98. The mean scores for the five experimental groups on the free-recall test are presented in Table 23.1. **A one-way analysis of**

l **variance revealed a significant treatment effect ($F = 3.06$, $df = 4/107$, $p < .05$),** and a post hoc comparison using **Scheffé's method indicated that**

m **the groups could be ordered in the following fashion:** N-RON (weight = + 3) > NN-RLN (+ 2) > N-RLN (+ 1) > N-MR (− 1) > NN-MR (− 5) ($F = 11.95$, $p < .05$).

In (**k**) the authors are mindful of gender differences (see *Special Issues* at the beginning of this chapter).

A principal finding is in (**l**) and then the data are further processed by means of Scheffé's test (**m**), which specifically orders the significant

Table 23.1   MEAN SCORES OF EXPERIMENTAL GROUPS

DEPENDENT MEASURE	EXPERIMENTAL GROUP				
	N-RON	NN-RLN	N-RLN	N-MR	NN-MR
Free recall	21.8	19.2	17.8	15.3	12.4
Short-term objective	17.5	15.9	13.7	13.7	11.5
Posttest	13.6	13.9	10.7	9.1	8.1

Note: Abbreviations: N-RON = took notes—reviewed own notes; NN-RLN = took no notes—reviewed the lecturer's notes; N-RLN = took notes—reviewed the lecturer's notes; N-MR = took notes—mental review; and NN-MR = took no notes—mental review.

The mean scores for the short-term objective test are also presented in Table 23.1. A one-way analysis of variance revealed a significant treatment effect ($F = 2.89$, $df = 4/107$, $p < .05$), and a post hoc comparison using Scheffé's method indicated that the groups could be ranked in the following order: N-RON (weight = $+3$) > NN-RLN ($+2$) > N-RLN (-1) = N-MR (-1) > NN-MR (-3) ($F = 11.18$, $p < .05$).

The mean scores for the posttest are also presented in Table 23.1. Although they are in approximately the same order as the other two measures, and of course not independent of them, an analysis of variance showed no significant differences among them ($F = 1.84$, $df = 4/64$).

To determine if there was a relationship between subjects' opinions about note taking and subsequent recall, the answers to the question "How do you feel about taking notes?" on the first page of the packet were analyzed. The three "no comment" remarks received were discarded, and the remainder were classified as either negative remarks or positive (wholly or conditionally) remarks. The t tests between these two groups yielded no significant differences on any of the three dependent measures, the free-recall test ($\overline{X}_N = 14.39$, $n = 31$, $\overline{X}_P = 17.87$, $n = 78$; $t = 1.707$, $df = 107$), the objective test ($\overline{X}_N = 12.71$, $n = 31$, $\overline{X}_P = 14.78$, $n = 78$; $t = 1.54$, $df = 107$), or the posttest measure ($\overline{X}_N = 10.05$, $n = 20$, $\overline{X}_P = 10.47$, $n = 49$; $t = .24$, $df = 67$).

The quality of the notes taken was defined as the number of ideas from the lecture that were included in the notes. A scorer, blind as to the subject's experimental condition, scored all the notes, and a randomly selected 20 sets were scored by a second blind judge. The correlation between the two

groups from each other. This procedure is again used for the short-term objective test (n).

The discussion section of the experiment leads off with a succinct statement of what the experimenters found (o), and throughout the remainder of the discussion, that finding is fitted both into a large con-

Table 23.2   CORRELATIONS OF QUALITY AND EFFICIENCY OF
NOTES WITH SHORT-TERM OBJECTIVE TEST AND FREE-RECALL SCORES

GROUP	n	QUALITY		EFFICIENCY	
		OBJECTIVE	FREE RECALL	OBJECTIVE	FREE RECALL
N-RON	20	.28	.53*	−.23	−.39
N-RLN	26	.42*	.51**	.15	−.03
N-MR	25	.44*	.37	.14	.07

Note: Abbreviations: N-RON = took notes—reviewed own notes; N-RLN = took no
notes—reviewed the lecturer's notes; and N-MR = took notes—mental review.
*$p > .05$.
**$p > .01$.

sets of ratings was .98. The efficiency of the notes was defined, as Howe
(1970b) suggested, by dividing the quality of the notes (i.e., number of ideas)
by the total number of words. The correlations of the quality and efficiency
of notes with the objective and free-recall tests are presented in Table 23.2.
Although all six correlations between quality of the notes taken and the
other measures are positive, only the correlations with free recall for the N-
RON and N-RLN groups and with objective scores for the N-RLN and N-
MR groups are statistically significant. None of the correlations involving
efficiency of note taking were significant.

DISCUSSION

The results of the present study appear to indicate that **note taking serves
both an encoding function and an external memory function, with the latter
function being the more important.** The group showing the best performance,
the N-RON group, was able to utilize both functions, whereas the group
with the lowest recall scores, the NN-MR group, could neither encode the
data nor use notes as an external storage mechanism. Subjects in the second
highest group, the NN-RLN group, were able to use notes as a memory aid
but were unable to encode the data directly; those in the second lowest
group, the N-MR group, were able to take advantage of the encoding but
not external storage function of note taking. It thus appears that of the two
functions, the one of serving as an external memory device has the greater
facilitating effect on recall.

The fact that subjects in the NN-RLN group performed better than those in
the N-RLN group was not predicted, since it was thought that those in the
latter group would benefit from both functions of note taking, whereas those
in the former group would use notes only as an external memory storage
device. A possible explanation for this result is that the substitution of the
lecturer's notes for the subject's own notes in some way interfered with the
consolidation process, thus decreasing recall.

The analyses of the posttest data were not significant, although the means were in approximately the same direction as the short-term free-recall and objective test means, which suggests that the same processes were operating here. An experimental mortality rate of 38% might have affected significance, as might the fact that subjects did not expect to be retested on the material.

Since a significant relationship was not found between opinions about note taking and subsequent recall, it would appear from these data that liking, disliking, or other opinions connected with note taking do not have an effect on recall.

The correlations between quality or efficiency of notes and objective or free-recall data for the three note-taking groups indicated that subjects with notes of good quality generally recalled more data than those with notes of poor quality, but that efficiency of notes did not correlate significantly with other measures. Crawford too found that there was a significant positive correlation between the number of points recorded in his subjects' lecture notes and the number recalled at the time of a quiz he gave (Crawford, 1925a). The present data would appear to confirm this finding, but are in opposition to Howe's report of a significant positive correlation between efficiency of notes and a measure of free recall (Howe, 1970b).

A further finding of the present study was that females had significantly higher mean recall scores than males for all treatment conditions on both the short-term free-recall and the objective tests. This result is in line with the finding of Todd and Kessler (1971), who found that women recalled significantly more words than did men. **Further investigation into the reason for this effect is needed, in order to determine whether the subject matter, sex of the instructor, or a true sex difference in memory is responsible for the results.**

p

In conclusion, the findings of this study indicate that in a typical college lecture setting, the taking of notes of good quality and the subsequent review of these notes are associated with more efficient recall.

q

---

ceptual framework and to other empirical findings (e.g., **p**) The final conclusion (**q**) is direct and to the point.

QUESTIONS

1. The authors note that the failure to find statistically significant differences between the experimental treatments in the three-week follow-up test may have been due to the 38 percent subject mortality. Why would this make a difference? Can you think of a way in which the authors could have eliminated this attrition?
2. Search the current literature for a subsequent article on note taking and recall and prepare a report of it.

REFERENCES

Ash, P., & Carlton, B. J. (1953). The value of note-taking during film learning. *British Journal of Educational Psychology, 23*, 121–125.

Crawford, C. C. (1925a). The correlation between lecture notes and quiz papers. *Journal of Educational Research, 12*, 282–291.

Crawford, C. C. (1925b). Some experimental studies on the results of college note-taking. *Journal of Educational Research, 12*, 379–386.

Deese, J. (1958). *The psychology of learning.* New York: McGraw-Hill.

DiVesta, F. J., & Gray, G. S. (1972). Listening and note taking. *Journal of Educational Psychology, 63*, 8–14.

Howe, M. J. A. (1970a). Note-taking strategy, review, and long term retention of verbal information. *Journal of Educational Research, 63*, 100.

Howe, M. J. A. (1970b). Using students' notes to examine the role of the individual learner in acquiring meaningful subject matter. *Journal of Educational Research, 64*, 61–63.

McClendon, P. (1958). An experimental study of the relationship between the note-taking practices and listening comprehension of college freshmen during expository lectures. *Speech Monographs, 25*, 222–228.

Miller, G. A., Galanter, E., & Pribram, K. (1960). *Plans and the structure of behavior.* New York: Holt, Rinehart and Winston.

Todd, W. B., & Kessler, C. C. (1971). Influence of response mode, sex, reading ability, and level of difficulty on four measures of recall of meaningful written material. *Journal of Educational Psychology, 62*, 229–234.

3. Design a factorial experiment on note taking and recall in which the sex of the subjects and lecturers are integral components.

4. Draw a graph using the data presented in Table 23.1.

5. Based on this research, what practical application would you suggest to a college student in his or her note taking?

6. The students who served as subjects in this experiment were part of a captive audience. Apparently they had no choice as to whether they wanted to participate in the study or not. Also, class time was apparently taken to conduct the study. Do these procedures meet the ethical standards presented in Chapter 7?

7. What control did the experiment exert to be certain that the subjects followed the procedures of the condition to which they were assigned? Why are such controls necessary?

8. The experimenter distributed the entire packet of materials before the start of the study. What potential problems exist? What controls are necessary? Why?

CHAPTER
24

Weight Loss

INTRODUCTION

A major concern in America is obesity. Millions of Americans are overweight. The causes of this condition have been defined as either *endogenous*—originating from some abnormal condition within the body—or *exogenous*—stemming from the excessive ingestion of food. It is within the latter category that most obese people fall. People may eat too much for a variety of reasons: habit, social pressure, boredom, tension, fear, happiness, loneliness . . . the list is seemingly endless.

One "cure" for obesity is simple: eat less. Its execution, however, is nearly impossible for some people. Throughout our country, countless millions pursue the shapely, idealized form of a slender Hollywood starlet or a svelte athlete through an equal number of fad diets, self-help courses, group therapy, fat farms, diet books, and a whole warehouse full of gymnastic paraphernalia, all promising the faithful user a trimmer physique. Fat is big business.

Psychologists have been interested in obese behavior for a long time. Experimental psychologists have studied obesity from the standpoint of its origins and the factors that cause people to ingest more food than their body can metabolize. Practicing clinical psychologists have been more interested in the development of therapeutic procedures that effectively reduce body weight while at the same time embodying an element of psychotherapy that facilitates the client's capacity to cope with the causes of his or her problem. The effectiveness of psychotherapeutic techniques is always a difficult thing to evaluate (How do you define "less schizophrenic"?). In experimental terms, the dependent variable is poorly defined. However, in research on the efficacy of weight-loss techniques, the obvious dependent variable is the number of pounds lost as an immediate result of therapy and as a long-term result of therapy.

Of all the therapeutic techniques developed, one has great intuitive appeal. It is the self-help strategy, which may involve self-reward, self-punishment, or a combination of both. Yet in spite of the wide use of these techniques in popular obesity clinics and clinical settings, the empirical evaluation of their efficacy for both the immediate term and the long term has been woefully lacking. This experiment by Mahoney, Moura, and Wade was selected because it effectively demonstrates how a topic in clinical psychology can be brought under experimental control. You will note that the types of problems it involves—for example, definitions of dependent and independent variables, and the subject sample—are significantly more difficult than in other examples in this book. You should carefully read how these factors were attended to by the researchers and model your own experiments after this one.

SPECIAL ISSUES

Research in Clinical Psychology

Experimental research in psychology is flexible enough to study many different forms of behavior. Yet no matter how diverse the behavior being studied, most forms of experimentation follow a fundamental model: Subjects are treated to experimental variables, and the consequence of the treatment is evaluated. In brief, that is the simple method used by Mahoney, Moura, and Wade in their article on weight loss and weight-loss techniques. Subjects who responded to a newspaper and were treated to experimental variables (weight-loss techniques), and the consequence (weight loss) were evaluated.

Although the experimental model (or experimental paradigm) is the same as that of most research reported in the psychological literature, this paper was selected as a case example because it represents a type of clinical research in which subject characteristics are an integral part of the design. An additional feature of this experiment which has not been introduced before is the "clinical" nature of the design. Clinical research does not differ from other research in psychology in form, but it does frequently differ in the nature of the subject, the treatment, and the dependent variable. Let's consider each of these topics in the present experiment. In the present experiment, a strikingly different group of subjects is used—those who responded to a newspaper ad on weight loss. The selection of subjects in this experiment was a critical part of the design.

Subject Characteristics in Clinical Research. In some clinical research the subjects are, of course, selected from a markedly abnormal population. For example, an investigator may direct his or her efforts toward paranoid subjects, or homicidal schizophrenic subjects, or manic-depressive subjects. No matter what the clinical symptoms of the subjects, care must be taken to define unambiguously their characteristics. And that very issue raises a substantial problem in clinical research. "Abnormal" subjects by definition differ from "normal" subjects. By using a diagnostic description of subjects, an additional variable is introduced to the experiment which may affect the reliability of the results. The variable is the definition of subjects. However, the researcher need not be thwarted in his or her desire to do research with abnormal subjects because of definitional problems. He or she can reduce the variability introduced by subject diagnosis by carefully delineating the precise characteristics of the experimental subjects. It is of no use to define abnormal subjects as "schizophrenics" or "obese" or "phobic." A more precise definition is called for. Great care must be exercised in operationally defining abnormal subjects. In the method section of the paper by Mahoney, Moura, and Wade, you will see an example of how "obesity" was carefully defined (percentage of body weight overweight, height/weight ratio). In addition the researchers apply an eligibility criterion that further delineated the subject sample.

The Independent Variable in Clinical Research. A second characteristic of clinical research is the nature of the treatment, or independent variable. Sometimes this variable is similar to other research discussed in this book, as in the case in which an experimenter studies the effect of a specific drug on behavior. Frequently, however, the independent variable involves the introduction of a psychotherapeutic technique. Here again the researcher needs to worry about the ambiguity that may be introduced. To simply say that the subjects underwent psychotherapy not only belies the complexity of the psychotherapeutic process but also is a poor research technique. Therefore, great care must also be exercised in defining the type of psychotherapy in research, using it as an independent variable. In this present experiment, the researchers spell out in detail the psychotherapeutic methods applied to their subjects.

The Dependent Variable in Clinical Research. The final factor in our skeleton of design principles is the dependent variable, or the subjects' response to the treatment. Clinical research frequently results in complex behavioral changes. When this is the case, the principal changes, or those under investigation, need to be clearly identified and evaluated. In the present experiment, the dependent variable is weight loss, which is an uncomplicated factor. However, many clinical studies yield complex data that require sophisticated analysis. (See Chapter 19.)

It is obvious that treatment research is valuable in the practice of clinical psychology and in psychological theory. But it is also frequently compounded by the addition of definitional problems which must be solved in order for the research to be reliable. We selected this article on weight loss because it represents an example of a type of clinical research and its unique problems. Furthermore, the researchers demonstrated how the definitional problems in clinical research can be brought under control by accurate and detailed delineation of the variables.

Relative Efficacy of Self-Reward, Self-Punishment, and Self-Monitoring Techniques for Weight Loss

MICHAEL J. MAHONEY, NANCY G. M. MOURA, and TERRY C. WADE
STANFORD UNIVERSITY

Obese adults ($n = 53$) were randomly assigned to five groups: (a) self-reward, (b) self-punishment, (c) self-reward and self-punishment, (d) self-monitoring, and (e) information control. All Ss were given information on effective stimulus control techniques for weight loss. This constituted the sole treatment for control Ss. Self-monitoring Ss were asked to weigh in twice per week for four weeks and to record their daily weight and eating habits. Self-reward and self-punishment Ss, in addition to receiving self-monitoring instructions, were asked to reward or fine themselves a portion of their own deposit contingent on changes in their weight and eating habits. After four weeks of treatment, self-reward Ss lost significantly more weight than either self-monitoring or control Ss. At four-month follow-up, those Ss lost more weight. These findings are interpreted as providing a preliminary indication that self-reward strategies are superior to self-punitive and self-recording strategies in the modification of at least some habit patterns.

a

Despite the fact that self-control strategies have become increasingly popular in clinical and applied settings (Kanfer & Phillips, 1970), there has been an **appalling lack of research on the processes and parameters of these techniques** (Mahoney, 1974). In particular, there have been neither comparisons among the various techniques nor attempts to isolate their active components. **The present study addressed itself to an evaluative comparison of three of the more popular self-control techniques, self-reward, self-punishment, and self-monitoring.** Since the latter is also a component of the former, a partial component analysis was also provided.

SOURCE: Reprinted with permission from *Journal of Consulting and Clinical Psychology*, 1973, 40, 404–407.

ANALYSIS

REVIEW OF LITERATURE

In the first part of the experiment (a) Mahoney et al. point out the "appalling lack of research" on weight loss through self-control strategies and how the present experiment is addressed to a comparison of three popular techniques. In this section we note a very abbreviated statement

To provide a stringent test of the efficacy of the above techniques, they were each applied in the modification of a chronic and resistant habit pattern. The pattern chosen was overeating since, in addition to providing observable indices of treatment effectiveness, obesity has been one of the most resistant of maladaptive habit patterns (Stunkard, 1958). Moreover, previous studies have demonstrated that a core of behavioral techniques can—when consistently applied—dramatically improve human weight control (Harris, 1969; Penick, Filion, Fox, & Stunkard, 1971; Stuart, 1967, 1971; Wollersheim, 1970). **In general, behavior modification approaches to obesity have emphasized the alteration of everyday eating habits by manipulation of eating-related environmental cues (i.e., stimulus control). To date, these techniques have been combined with a variety of auxiliary maintenance strategies such as group support, covert sensitization, and therapist approval.** The present study focused on the effectiveness of various self-control strategies in motivating and maintaining individual applications of cue-altering weight control techniques. Specifically, **Ss were instructed to employ self-reward, self-punishment, or self-monitoring techniques in their modification of both their body weight and their eating habits.** A control group was provided with identical information on stimulus control techniques, but they did not receive self-regulatory instructions.

b

c

METHOD

Subjects

The Ss were invited by a newspaper ad to participate in a program for self-managed weight control. Eligibility criteria included **(a) a minimum age of 17 years, (b) nonpregnant status, (c) physician's consent, and (d) a minimum of 10% overweight.** Applicants who were concurrently enrolled in a reducing club and/or undergoing other treatments for weight loss were ineligible. Also, individuals reporting recent dramatic changes in body weight were excluded. A total of 53/ Ss (48 females, 5 males) enrolled and completed the program. Their average age was 37.9 years. Individual body weights varied from 107 to 270 with a mean of 166.3 pounds.

d

of the problem and research plan, which is augmented in the next paragraph and in the methods section. Despite the obvious redundancy of the literary style, it is useful in "setting the stage" for the rest of the article. Good scientific writing is characterized by its ability to convey a message in the most concise form possible while not being so lean that the reader is required to reread each passage for its content. The examples of good experimental design in this book also embody the balance between redundancy and parsimony of the language.

Previous studies that attempted to determine the relative efficacy of these techniques combined them with other supportive techniques, such

Prior to group assignment, the degree of obesity for each S was computed. This was done by dividing an individual's current weight by his ideal weight (Stillman & Baker, 1967), thereby controlling for height factors. The degree of obesity for all Ss in the study ranged from 13% with a mean of 48.6%. **The Ss were ranked according to degree of obesity and randomly assigned to experimental groups.**

e

Procedures

All Ss were required to place a refundable deposit of $10 with the Es for the duration of treatment (four weeks). A commercially available bathroom scale was employed. At their initial weigh in, all Ss were given a small booklet describing stimulus control approaches to weight loss. Treatment procedures varied as follows:

Self-Reward: Group 1 ($n = 12$)

The Ss in this group were asked to deposit an additional $11 with E for purposes of self-reward. They were asked to weigh in biweekly for 7 weigh-ins and to keep a daily graph of their weight at home. The Ss were provided with a weight chart and a behavioral diary in which they were to record the daily frequency of (a) "fat thoughts" (discouraging self-verbalizations), (b) "thin thoughts" (encouraging self-verbalizations), (c) instances of indulgence (eating a fattening food or excessive

as group support, therapist approval, and covert sensitization, thus failing to produce clear-cut results. The purpose of the present study (**b**), and a brief version of the research plan (**c**) including a control group, is stated. Specifically, the independent variable—the various treatment groups—is stated.

METHOD

In the method section, subjects are identified, including the criteria for selection (**d**), and the procedures of each treatment are described in great enough detail to allow replication. The subjects, for example, were required to be over 17 years of age, to be nonpregnant, to be at least 10 percent overweight, and to secure a physician's approval. Experimenters were not allowed to express approval or disapproval at weight change during the weigh-in sessions; the experimenter contact with the subjects was kept to a minimum of 5 to 10 minutes per weigh-in. A postquestionnaire was used to assess the subject's use of any method not specified in his or her treatment group.

In research in which the principal effect is susceptible to wide individual differences, it is essential that variation caused by those differences be tightly controlled. In the present experiment the assignment of subjects to treatment groups (**e**) was on the basis of a matched-subject design, by which, for example, one very obese subject in the self-reward group

quantity), and (d) instances of restraint (refusing a fattening food or reducing food intake). Self-reward was to take place at weigh-ins. Each S's $21 was transformed into 21 shares whose value increased when other Ss forfeited deposits or self-punished. A type of bank account was set up for each individual. The Ss began with an empty account and could self-reward by requesting that a deposit be made to their account. Each S had access to three shares per weigh-in. It was recommended that Ss self-reward two shares for a weight loss of one pound or more since their last weigh-in. The remaining share was to be self-awarded if adaptive behaviors (thin thoughts and restraint) had outnumbered nonadaptive behaviors since the last weigh-in. Beyond these recommendations, no external constraints were placed on S's standards or execution of self-reward. When an S chose not to self-reward at a weigh-in, his/her shares were placed in a community pool and divided among all other shareholders (thereby increasing the value of a share). The Ss were allowed three absences but were thereafter fined three shares for each absence. At their final weigh-in, they received the amount they had self-rewarded ($1/share) and were later mailed a dividend check covering increases in share value.

Self-Punishment: Group 2 ($n = 12$)

Procedures in this group were exactly parallel to those of Group 1 except that Ss began with a full 21-share account and were instructed to fine themselves shares for lack of weight loss and/or lack of behavior improvement. Self-punished shares were placed in the community pool and divided among all remaining shares. When Ss did not self-punish at a weigh-in, their shares remained in their account and were refunded to them at the final weigh-in. The share amounts and goals were identical to those in Group 1.

Self-Reward and Self-Punishment: Group 3 ($n = 8$)

Conditions in this group combined those of Groups 1 and 2. The Ss began with an empty account and could either deposit to it (self-reward) or fine themselves (self-punish) up to three shares per weigh-in.

would have a counterpart in the other groups. After ranking the subjects, the authors chose to randomly assign the subjects to experimental and control groups. This procedure is based on the assumption that the results of random selection (each subject having an equal opportunity to be selected and assigned to any specific group) will homogeneously distribute subjects throughout the treatment groups. It might have proved embarrassing to the researchers if, by some quirk of probability, all of the very obese subjects had fallen into one group and all the mildly obese into another. If you look at the Results section, you will notice that Mahoney et al. tested for both the pretreatment degree of obesity for each group (g) and the pretreatment number of pounds overweight (h) and found the differences among the groups to yield a very low F ratio, which suggested that subjects in one group did not significantly differ in the pretreatment weigh-in.

Self-Monitoring: Group 4 ($n = 5$)

The Ss in this group were asked to weigh in biweekly and to record their weight and adaptive and nonadaptive eating habits. The standard weight loss and behavior improvement goals used in Groups 1–3 were also suggested for these Ss. In short, conditions in this group duplicated those in the first three groups with the exception that no additional deposit was required and the self-reward and self-punishment strategies were not discussed.

Information Control: Group 5 ($n = 16$)

The Ss in this group received stimulus control booklets but did not participate in a second weigh-in until the four-week treatment period had ended. No self-monitoring materials were provided, and it was recommended that Ss refrain from any weigh-ins at home during that time.

f

Efforts were made to avoid any E approval or disapproval for weight change, and to minimize E contact (weigh-ins took 5–10 minutes). **After four weeks all Ss were weighed and instructed to continue self-application of their respective techniques. A postquestionnaire inquired about the use of any extraneous methods. Four months after their initial appointment, a follow-up weigh-in was conducted.**

RESULTS

g

Data analyses on the pretreatment degree of obesity for each group revealed no significant differences ($F = .31$, $df = 4/48$). Likewise, the groups did not initially differ on number of pounds overweight ($F = .44$). **A posttreat-

h

ment analysis of number of pounds lost yielded an overall F of 4.49 ($p <$ .005).** Newman-Keuls comparisons of treatment means showed that the self-reward Ss had lost significantly more pounds than either the self-mon-

i

itoring ($p < .025$) or the control group ($p < .025$). The self-punishment group did not differ significantly from any other. A difference approaching

---

After the treatment procedures are described, the authors note the principal dependent variables (f). It is important to notice that after the conclusion of the formal part of the experiment, a follow-up weigh-in was conducted. Many times weight-loss programs are temporarily effective, sometimes even spectacularly, but the long-range effects that are important to most people are not effective. The follow-up procedure is an important one that is so often lacking in psychological research.

RESULTS

The authors present the results in terse, almost blunt, language (i) "the self-reward Ss had lost significantly more pounds than either the self-significance at the .05 level was obtained when Group 3 Ss (self-reward and self-punishment) were compared with those in the self-monitoring group. **Average number of pounds lost per individual was 6.4, 3.7, 5.2, .8, and 1.4 for the five groups, respectively.**

An analysis of percentage of body weight lost also revealed significant group differences ($F = 3.44$, $p < .025$). Newman-Kuels comparisons indicated that the self-reward group lost significantly more in percentage of body weight than either the self-monitoring ($p < .05$) or the control group ($p < .05$).

An analysis of "follow-through" was performed for Groups 1–3 in order to assess any differential tendencies toward self-reward or self-punishment. Since the weigh-ins constituted the most accurately measured and observable behavior of Ss, analyses were done on the consistency with which Ss self-administered rewards or fines for weight change. Using the recommended standard (loss of one pound per weigh-in), Ss were scored on whether they appropriately transacted two shares for success or failure at the above criterion. At first glance the data indicate that follow-through was higher for self-reward. There were no instances where an individual made his weight criterion and failed to self-reward, whereas there were numerous instances where individuals failed to make the criterion but failed to self-punish. However, a further analysis revealed quite a few instances where individuals self-rewarded even though they had not met the weight criterion. (It should be recalled that if a self-reward S had not met the criterion and appropriately chose not to self-reward, he was inadvertently *punished* by the automatic transfer of his shares to the community pool.) A follow-through analysis using both halves of the dichotomy revealed no significant intergroup differences ($F = .53$, $df = 3/33$). The rate of follow-through was 58.1% for self-reward, 57.6% for self-punishment, and 67.1% for the combined group.

monitoring . . . or the control group. . . ." The average weight loss per individual by group is given in (j). A graphic representation in which immediate loss and long-term loss are depicted would have been very useful (see Figure 24.1).

Another way of analyzing the data is in percent weight loss, which is given in (k).

Throughout the remainder of the section, the authors demonstrate careful experimental analysis by looking at all suspected relevant factors. In the four-month follow-up, it should be observed that some subjects were lost, for a variety of reasons. So many were lost from one treatment group (see l) that the group had to be dropped from subsequent analysis. A difficult experimental question is posed by subject loss. On the one hand, the data from the follow-up are very important—some may say critical—to the experiment; on the other hand, it may be that the remaining subjects are not representative of the entire treatment group.

In order to control for nonspecific factors associated with weigh-ins, Groups 1–4 Ss who began the study but completed fewer than four (out of seven) weigh-ins were excluded from all data analyses. This restriction affected Group 4 (self-monitoring) more than the others. Only five (out of nine original) Group 4 Ss met the attendance criterion. In the first three groups, attendance was enhanced by levying fines after three absences. An analysis of attendance variations among Ss who did meet the criterion revealed no significant differences (F = .86).

1   Thirty-one Ss appeared for the four-month follow-up weigh-in. Of these, seven had to be excluded because of intervening or attendance factors (e.g., hormone shots, health farms, etc.). Because data were available from only

For example, if some subjects in one treatment group went "off the wagon" and ate savagely for four months, ballooning to leviathan corpulence, they may have been so ashamed of themselves that they would not submit to a follow-up. You will note that the authors struggle with this technical issue (1) and resolve it as best they can. There is no easy answer to the above dilemma, and data gathered from treatments in which subject loss is a factor should be interpreted with great caution.

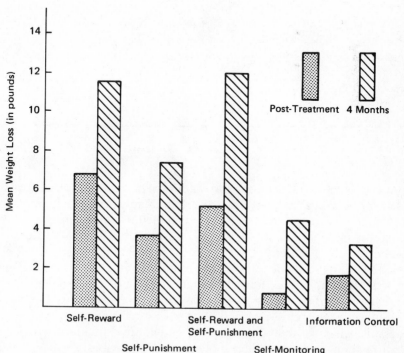

Figure 24.1   Weight loss by treatment groups.

two self-monitoring Ss, this group had to be excluded from follow-up analyses. An analysis of the four-month data revealed that the self-reward group and the combined group (self-reward and self-punishment) had both lost greater percentages of their body weight than information control Ss ($p <$ .05). Self-punishment Ss did not differ significantly from controls. Because follow-up Ss differed significantly in initial weight, analysis of number of pounds lost had to be converted to a proportion (number of pounds lost divided by number of pounds overweight). After an overall F of 5.03 ($p <$ .02), individual Newman-Keuls comparisons showed that both the self-reward and the self-reward and self-punishment group had lost significantly more weight than either the self-punishment ($p < .05$) or the control group ($p < .05$). No other group comparisons were significant. Over the entire four months, the average number of pounds lost per individual was 11.5 for self-reward, 7.3 for self-punishment, 12.0 for self-reward and self-punishment, 4.5 for self-monitoring, and 3.2 for controls.

DISCUSSION

m

The foregoing results provide some preliminary information on the relative effectiveness of several self-control strategies. In general, **it would appear that self-reward strategies may provide an effective incentive component in weight loss attempts.** Only those groups given self-reward opportunities differed significantly from any others in successful weight loss. Data from information control Ss indicate that the simple provision of relevant information on stimulus control techniques causes only minimal change in body weight. Moreover, variables associated with self-recording, frequent weigh-ins, and (minimal) E contact failed to produce substantial weight loss. This finding is in contrast with recent evidence on the reactive effects of self-

DISCUSSION

In the final section, Mahoney et al. cautiously state their main finding (**m**) and the negative findings. This is a preliminary answer to the question that initiated this study: Which self-control techniques are most efficient in a habit-control problem, specifically overeating? The experimenters comment on the problem of subject attrition (**n**) and suggest a deposit scheme similar to the one mentioned in the results section to counteract this attrition rate. The authors then state the implications of this study (**o**): (1) self-reward strategies seem to be superior to self-punishment and/ or self-monitoring techniques and (2) the parameters and components of successful self-reward strategies need further investigation in order to determine their applicability to problems other than obesity, to determine the effects of magnitude and scheduling of rewards, and so on.

monitoring (e.g., Broden, Hall, & Mitts, 1971; McFall, 1970; McFall & Hammen, 1971). However, subsequent research (Mahoney, 1974) has suggested that specific self-monitoring applications may vary considerably in their reactivity and in the permanence of their effects.

It should be noted that empirical comparison of self-reward and self-punishment strategies is complicated by such factors as control for frequency of application (which is, in turn, altered by their relative effectiveness). Moreover, **it would be desirable for subsequent studies to motivate *all* Ss' appearance at weigh-ins and follow-up by some form of deposit. Although there were no group differences in dropout rate, equal deposits are also advisable to insure against pretreatment motivational variations.** Finally, care should be taken to avoid *E* punishment of appropriate self-regulation. In the present study, follow-through rates for self-reward Ss may have been decreased by the fact that these Ss were adventitiously fined for not self-rewarding (irrespective of goal attainment). A methodologically improved study (Mahoney, 1974) has revealed average self-reward follow-through rates in excess of 90%.

The implications of the present findings are twofold. First, it would appear that self-reward strategies may be superior to self-punishment and/or self-recording techniques in the modification of at least some habit patterns. Second, the parameters and components of successful self-reward strategies need to be investigated. For example, how important are the factors of magnitude, scheduling, and focus in self-reward systems? Can the pattern of the present findings be generalized to behavior problems other than obesity? These and other issues must await clarification by further research in self-control.

QUESTIONS

1. What techniques would you suggest to prevent subject dropout?
2. Suggest another dependent variable that might be useful in obesity research.
3. What "practical" application could be made of these results?
4. Why were subjects required to deposit $10 at the beginning of the experiment?
5. What effects, if any, would the sex of the experimenter have on the results? The age of the experimenter?
6. Would magnitude and type of reward have an effect on the results? Design an experiment to test these potential effects.
7. Identify at least three other weight-loss strategies that could be experimentally tested.

REFERENCES

Broden, M., Hall, R. V., & Mitts, B. (1971). The effect of self-recording on the classroom behavior of two eighth-grade students. *Journal of Applied Behavior Analysis, 4,* 191–199.

Harris, M. B. (1969). Self-directed program for weight control: A pilot study. *Journal of Abnormal Psychology, 74,* 263–270.

Kanfer, F. H., & Phillips, J. S. (1970). *Learning foundations of behavior therapy.* New York: Wiley.

Mahoney, M. J. (1974). Self-reward and self-monitoring techniques for weight loss. *Behavior Therapy, 5,* 48–57.

McFall, R. M. (1970). The effects of self-monitoring on normal smoking behavior. *Journal of Consulting and Clinical Psychology, 35,* 135–142.

McFall, R. M., & Hammen, C. L. (1971). Motivation, structure, and self-monitoring: The role of nonspecific factors in smoking reduction. *Journal of Consulting and Clinical Psychology, 37,* 80–86.

Penick, S. B., Filion, R., Fox, S., & Stunkard, A. J. (1971). Behavior modification in the treatment of obesity. *Psychosomatic Medicine, 33,* 49–55.

Stillman, I. M., & Baker, S. S. (1967). *The doctor's quick weight loss diet.* New York: Dell.

Stuart, R. B. (1967). Behavioral control over eating. *Behavior Research and Therapy, 5,* 357–365.

Stuart, R. B. (1981). A three-dimensional program for the treatment of obesity. *Behavior Research and Therapy, 9,* 177–186.

Stunkard, A. J. (1958). The management of obesity. *New York State Journal of Medicine, 58,* 79–87.

Wollersheim, J. P. (1970). The effectiveness of group therapy based upon learning principles in the treatment of overweight women. *Journal of Abnormal Psychology, 76,* 462–474.

The authors would like to thank Albert Bandura for his assistance and suggestions.

Parents' Views

INTRODUCTION

Differences between the sexes have been a topic of interest to experimental psychologists as well as a lively social issue. Some assert that the behavior of males differs from that of females because of biological differences, while others advocate a position that attributes behavioral differences between the sexes to environmental factors. Both groups cite an impressive series of empirical results in support of their assertion, and yet many psychologists recognize that human behavior is the result of both inherited biological factors and environmental factors.

An important factor in the determination of behavior is the role of social expectation. Traditionally, in our society, boys are "expected" to be engineers, doctors, lawyers, aviators, and mechanics, while girls are "expected" to be nurses, homemakers, secretaries, garment workers, and models—to name but a few of the stereotypic sexual expectancies. Although these role types are being challenged in today's world, the influence of subtle forms of expectancy on behavior is still unknown.

Of particular interest to Rubin, Provenzano, and Luria is the question of how early in an infant's life behavioral expectancies are formed. In order to answer the question, the researchers evaluated parents' views of the characteristics of newborn children.

We think the article contains some surprising findings that may be interesting to you, as well as demonstrate some important design issues.

The article is reprinted from the *American Journal of Orthopsychiatry,* which frequently publishes articles of interest to psychologists but is a publication of the American Orthopsychiatric Association. The style is very similar to the APA style, but there are some minor changes. As psychologists occasionally read and report their results in publications outside of the journals published by the APA, it is important to be acquainted with these sources and their style.

The article is to be analyzed by the student. As you read the article, consider some of the following questions:

1. Notice how the researchers develop their argument in the Review of Literature section. What differences do you see in this article from previous articles?
2. What is the general conclusion reached in this paper?
3. What measures were made?
4. How did the authors treat their data?

The Eye of the Beholder: Parents' Views on Sex of Newborns

JEFFREY Z. RUBIN, FRANK J. PROVENZANO, and ZELLA LURIA

Thirty pairs of primiparous parents, fifteen with sons and fifteen with daughters, were interviewed within the first 24 hours postpartum. Although male and female infants did not differ in birth length, weight, or Apgar scores, daughters were significantly more likely than sons to be described as little, beautiful, pretty, and cute, and as resembling their mothers. Fathers made more extreme and stereotyped rating judgments of their newborns than did mothers. Findings suggest that *sex-typing* and *sex-role socialization* have already begun at birth.

As Schaffer (1971) observed, the infant at birth is essentially an asocial, largely undifferentiated creature. It appears to be little more than a tiny ball of hair, fingers, toes, cries, gasps, and gurgles. However, while it may seem that "if you've seen one, you've seen them all," babies are *not* all alike—a fact that is of special importance to their parents, who want, and appear to need, to view their newborn child as a creature that is special. Hence, much of early parental interaction with the infant may be focused on a search for distinctive features. Once the fact that the baby is normal has been established, questions such as, "Who does the baby look like?" and "How much does it weigh?" are asked.

Of all the questions parents ask themselves and each other about their infant, one seems to have priority: "Is it a boy or a girl?" The reasons for and consequences of posing this simple question are by no means trivial. The answer, "boy" or "girl," may result in the parents' organizing their perception of the infant with respect to a wide variety of attributes—ranging from its size to its activity, attractiveness, even its future potential. It is the purpose of the present study to examine the kind of verbal picture parents form of the newborn infant, as a function both of their own and their infant's gender.

The study reported here is addressed to parental perceptions of their infants at the point when these infants first emerge into the world. If it can be demonstrated that parental sex-typing has already begun its course at this earliest of moments in the life of the child, it may be possible to understand better one of the important antecedents of the complex process by which the growing child comes to view itself as boy-ish or girl-ish.

SOURCE: "The Eye of the Beholder: Parents Views on Sex of Newborns" by Jeffrey Z. Rubin, Frank J. Provenzano, and Zella Luria. *American Journal of Orthopsychiatry*, 1974, 44(4).

Based on our review of the literature, two forms of parental sex-typing may be expected to occur at the time of the infant's birth. First, it appears likely that parents will view and label their newborn child differentially, as a simple function of the infant's gender. Aberle and Naegele (1952) and Tasch (1952), using only fathers as subjects, found that they had different expectations for sons and daughters: sons were expected to be aggressive and athletic, daughters were expected to be pretty, sweet, fragile, and delicate. Rebelsky and Hanks (1971) found that fathers spent more time talking to their daughters than their sons during the first three months of life. While the sample size was too small for the finding to be significant, they suggest that the role of father-of-daughter may be perceived as requiring greater nurturance. Similarly, Pedersen and Robson (1969) reported that the fathers of infant daughters exhibited more behavior labeled (by the authors) as "apprehension over well-being" than did the fathers of sons.

The second form of parental sex-typing we expect to occur at birth is a function both of the infant's gender *and* the parent's own gender. Goodenough (1957) interviewed the parents of nursery school children, and found that mothers were less concerned with sex-typing their child's behavior than were fathers. More recently, Meyer and Sobieszek (1972) presented adults with videotapes of two seventeen-month-old children (each of whom was sometimes described as a boy and sometimes as a girl), and asked their subjects to describe and interpret the children's behavior. They found that male subjects as well as those having little contact with small children, were more likely (although not always significantly so) to rate the children in sex-stereotypic fashion—attributing "male qualities" such as independence, aggressiveness, activity, and alertness to the child presented as a boy, and qualities such as cuddliness, passivity, and delicacy of the "girl." We expect, therefore, that sex of infant and sex of parent will interact, such that it is fathers, rather than mothers, who emerge as the greater sex-typers of their newborn.

In order to investigate parental sex-typing of their newborn infants, and in order, more specifically, to test the predictions that sex-typing is a function of the infant's gender, as well as the gender of both infant and parent, parents of newborn boys and girls were studied in the maternity ward of a hospital, within the first 24 hours postpartum, to uncover their perceptions of the characteristics of their newborn infants.

METHOD

Subjects

The subjects consisted of 30 pairs of primiparous parents, 15 of whom had sons, and 15 of whom had daughters. The subjects were drawn from the available population of expecting parents at a suburban Boston hospital serving local, predominantly lower-middle-class families. Using a list of primiparous expectant mothers obtained from the hospital, the experimenter made contact with families by mail several months prior to delivery, and requested the subjects' assistance in "a study of social relations among parents and their first child." Approximately one week after the initial contact by mail, the experimenter telephoned each family, in order to answer any questions the prospective parents might have about the study, and to obtain their consent. Of the 43 families reached by phone, 11 refused to take part in the study. In addition, one consenting mother subsequently gave birth to a low birth weight infant (a 74-ounce girl), while another delivered an unusually large son (166 ounces). Because these two infants were at the two ends of the distribution of birth weights, and because they might have biased the data in support of our hypotheses, the responses of their parents were eliminated from the sample.

All subjects participated in the study within the first 24 hours postpartum—the fathers almost immediately after delivery, and the mothers (who were often under sedation at the time of delivery) up to but not later than 24 hours later. The mothers typically had spoken with their husbands at least once during this 24-hour period.

There were no reports of medical problems during any of the pregnancies or deliveries, and all infants in the sample were full-term at time of birth. Deliveries were made under general anesthesia, and the fathers were not allowed in the delivery room. The fathers were not permitted to handle their babies during the first 24 hours, but could view them through display windows in the hospital nursery. The mothers, on the other hand, were allowed to hold and feed their infants. The subject participated individually in the study. The fathers were met in a small, quiet waiting room used exclusively by the maternity ward, while the mothers were met in their hospital rooms. Every precaution was taken not to upset the parents or interfere with hospital procedure.

Procedure

After introducing himself to the subjects, and their congratulatory amenities, the experimenter (FJP) asked the parents: "Describe your baby as you would to a close friend or relative." The responses were tape-recorded and subsequently coded.

The experimenter then asked the subjects to take a few minutes to complete a short questionnaire. The instructions for completion of the questionnaire were as follows:

> On the following page there are 18 pairs of opposite words. You are asked to rate your baby in relation to these words, placing an "x" or a checkmark in the space that best describes your baby. The more a word describes your baby, the closer your "x" should be to that word.
>
> Example: Imagine you were asked to rate Trees.

Good   :__:__:__:__:__:__:__:__:__:__:__:__:__:__:__:__:__: Bad
Strong :__:__:__:__:__:__:__:__:__:__:__:__:__:__:__:__:__: Weak

If you cannot decide or your feelings are mixed, place your "x" in the center space. Remember, the more you think a word is a good description of your baby, the closer you should place your "x" to that word. If there are no questions, please begin. Remember, you are rating your baby. Don't spend too much time thinking about your answers. First impressions are usually the best.

Having been presented with these instructions, the subjects then proceeded to rate their baby on each of the following eighteen 11-point, bipolar adjective scales: firm–soft; large featured–fine featured; big–little; relaxed–nervous; cuddly–not cuddly; easygoing–fussy; cheerful–cranky; good eater–poor eater; excitable–calm; active–inactive; beautiful–plain; sociable–unsociable; well coordinated–awkward; noisy–quiet; alert–inattentive; strong–weak; friendly–unfriendly; hardy–delicate.

Upon completion of the questionnaire, the subjects were thanked individually, and when both parents of an infant had completed their participation, the underlying purposes of the study were fully explained.

Hospital Data

In order to acquire a more objective picture of the infants whose characteristics were being judged by the subjects, data were obtained from hospital records concerning each infant's birth weight, birth length, and Apgar scores. Apgar scores are typically assigned at five and ten minutes postpartum, and represent the physician's ratings of the infant's color, muscle tonicity, reflex irritability, and heart and respiratory rates. No significant differences between the male and female infants were found for birth weight, birth length, or Apgar scores at five and ten minutes postpartum.*

RESULTS

In Table 25.1, the subjects' mean ratings of their infant, by condition, for each of the eighteen bipolar adjective scales, are presented. The right-extreme column of Table 25.1 shows means for each scale, which have been averaged across conditions. Infant stimuli, overall, were characterized closer to the scale anchors of soft, fine featured, little, relaxed, cuddly, easygoing, cheerful, good eater, calm, active, beautiful, sociable, well coordinated, quiet, alert, strong, friendly, and hardy. Our parent-subjects, in other words, appear to have felt on Day 1 of their babies' lives that their newborn infants represented delightful, competent new additions to the world!

*Birth weight ($\overline{X}_{Sons}$ = 114.43 ounces, $\overline{X}_{Daughters}$ = 11.00, $t(28)$ = 1.04); Birth length ($\overline{X}_{Sons}$ = 19.80 inches, $\overline{X}_{Daughters}$ = 19.96, $t(28)$ = 0.52); 5 minute Apgar score ($\overline{X}_{Sons}$ = 9.07, $\overline{X}_{Daughters}$ = 9.33, $t(28)$ = 0.69); and 10 minute Apgar score ($\overline{X}_{Sons}$ = 10.00, $\overline{X}_{Daughters}$ = 10.00).

Analysis of variance of the subjects' questionnaire responses (1 and 56 degrees of freedom) yielded a number of interesting findings. There were *no* rating differences on the eighteen scales as a simple function of Sex of Parent:

Table 25.1   MEAN RATING ON THE 18 ADJECTIVE SCALES AS A FUNCTION OF SEX OF PARENT *(MOTHER VS. FATHER)* AND SEX OF INFANT *(SON VS. DAUGHTER)*[a]

SCALE	EXPERIMENTAL CONDITION				
(I)–(II)	M–S	M–D	F–S	F–D	$\overline{X}$
Firm–Soft	7.47	7.40	3.60	8.93	6.85
Large featured–Fine featured	7.20	7.53	4.93	9.20	7.22
Big–Little	4.73	8.40	4.13	8.53	6.45
Relaxed–Nervous	3.20	4.07	3.80	4.47	3.88
Cuddly–Not cuddly	1.40	2.20	2.20	1.47	1.82
Easygoing–Fussy	3.20	4.13	3.73	4.60	3.92
Cheerful–Cranky	3.93	3.73	4.27	3.60	3.88
Good eater–Poor eater	3.73	3.80	4.60	4.53	4.16
Excitable–Calm	6.20	6.53	5.47	6.40	6.15
Active–Inactive	2.80	2.73	3.33	4.60	3.36
Beautiful–Plain	2.13	2.93	1.87	2.87	2.45
Sociable–Unsociable	4.80	3.80	3.73	4.07	4.10
Well coordinated–Awkward	3.27	2.27	2.07	4.27	2.97
Noisy–Quiet	6.87	7.00	5.67	7.73	6.82
Alert–Inattentive	2.47	2.40	1.47	3.40	2.44
Strong–Weak	3.13	2.20	1.73	4.20	2.82
Friendly–Unfriendly	3.33	3.40	3.67	3.73	3.53
Hardy–Delicate	5.20	4.67	3.27	6.93	5.02

[a]The larger the mean, the greater the rated presence of the attribute denoted by the second (right-hand) adjective in each pair.

Parents appear to agree with one another, on the average. As a function of Sex of Infant, however, several significant effects emerged: Daughters, in contrast to sons, were rated as significantly softer ($F = 10.67$, $p < .005$), finer featured ($F = 9.27$, $p < .005$), littler ($F = 28.83$, $p < .001$), and more inattentive ($F = 4.44$, $p < .05$). In addition, significant interaction effects emerged for seven of the eighteen scales: firm–soft ($F = 11.22$, $p < .005$), large featured–fine featured ($F = 6.78$, $p < .025$), cuddly–not cuddly ($F = 4.18$, $p < .05$), well coordinated–awkward ($F = 12.52$, $p < .001$), alert–inattentive ($F = 5.10$, $p < .05$), strong–weak ($F = 10.67$, $p < .005$), and hardy–delicate ($F = 5.32$, $p < .025$).

The meaning of these interactions becomes clear in Table 25.1, in which it can be seen that six of these significant interactions display a comparable pattern: fathers were more extreme in their ratings of *both* sons and daughters than were mothers. Thus, sons were rated as firmer, larger featured, better coordinated, more alert, stronger, and hardier—and daughters as softer, fine featured, more awkward, more inattentive, weaker, and more delicate—by their fathers than by their mothers. Finally, with respect to the other significant interaction effect (cuddly–not cuddly), a rather different pattern was found. In this case, mothers rated sons as cuddlier than daughters, while fathers rated daughters as cuddlier than sons—a finding we have dubbed the "oedipal" effect.

Responses to the interview question were coded in terms of adjectives used and references to resemblance. Given the open-ended nature of the question, many adjectives were used—healthy, for example, being a high frequency response cutting across sex of babies and parents. Parental responses were pooled, and recurrent adjectives were analyzed by Ψ^2 analysis for sex of child. Sons were described as big more frequently than were daughters ($\Psi^2(1)$ = 4.26, p < .05); daughters were called little more often than were sons ($\Psi^2(1)$ = 4.28, p < .05). The "feminine" cluster—beautiful, pretty, and cute—was used significantly more often to describe daughters than sons ($\Psi^2(1)$ = 5.40, p < .05). Finally, daughters were said to resemble mothers more frequently than were sons ($\Psi^2(1)$ = 3.87, p < .05).

DISCUSSION

The data indicate that parents—especially fathers—differently label their infants, as a function of the infant's gender. These results are particularly striking in light of the fact that our sample of male and female infants did *not* differ in birth length, weight, or Apgar scores. Thus, the results appear to be a pure case of parental labeling—what a colleague has described as "nature's first projective test" (personal communication, Leon Eisenberg). Given the importance parents attach to the birth of their first child, it is not surprising that such ascriptions are made.

The results of the present study are, of course, not unequivocal. Although it was found, as expected, that the sex-typing of infants varied as a function of the infant's gender, as well as the gender of both infant and parent, significant differences did not emerge for all eighteen of the adjective scales employed. Two explanations for this suggest themselves. First, it may simply be that we have overestimated the importance of sex-typing at birth. A second possibility, however, is that sex-typing is more likely to emerge with respect to certain classes of attributes—namely, those which denote physical or constitutional, rather than "internal," dispositional, factors. Of the eight different adjective pairs for which significant main or interaction effects emerged, six (75%) clearly refer to external attributes of the infant. Conversely, of the 10 adjective pairs for which no significant differences were found, only three (30%) clearly denote external attributes. This suggests that it is physical and constitutional factors that especially lend themselves to sex-typing at birth, at least in our culture.

Another finding of interest is the lack of significant effects, as a simple function of sex of parent. Although we predicted no such effects, and were therefore not particularly surprised by the emergence of "non-findings," the implication of these results is by no means trivial. If we had omitted the sex of the infant as a factor in the present study, we might have been led to conclude (on the basis of simply varying the sex of the parent) that *no* differences exist in parental descriptions of newborn infants—a patently erroneous conclusion! It is only when the infant's and the parent's gender are considered together, in interaction, that the lack of differences between overall parental mean ratings can be seen to reflect the true differences between the parents. Mothers rate both sexes closer together on the adjective pairs than do fathers (who are the stronger sex-typers), but *both* parents agree on the direction of sex differences.

The central implication of the study, then, is that sex-typing and sex-role socialization appear to have already begun their course at the time of the infant's birth, when information about the infant is minimal. The Gestalt parents develop, and the labels they ascribe to their newborn infant, may well affect subsequent expectations about the manner in which their infant ought to behave, as well as parental behavior itself. This parental behavior, moreover, when considered in conjunction with the rapid unfolding of the infant's own behavioral repertoire, may well lead to a modification of the very labeling that affected parental behavior in the first place. What began as a one-way street now bears traffic in two directions. In order to understand the full importance and implications of our findings, therefore, research clearly needs to be conducted in which delivery room stereotypes are traced in the family during the first several months after birth, and their impact upon parental behavior is considered. In addition, further research is clearly in order if we are to understand fully the importance of early parental sex-typing in the socialization of sex roles.

REFERENCES

Aberle, D., and Naegele, K. (1952). Middleclass fathers' occupational role and attitudes toward children. *American Journal of Orthopsychiatry, 22*(2), 366–378.

Goodenough, E. (1957). Interest in persons as an aspect of sex differences in the early years. *Genetic Psychology Monograph, 55,* 287–323.

Harari, H., and McDavid, J. Name stereotypes and teachers' expectations. *Journal of Educational Psychology.* (In press).

Meyer, J., and Sobieszek, B. (1972). Effect of a child's sex on adult interpretations of its behavior. *Developmental Psychology, 6,* 42–48.

Pedersen, F. and Robson, K. (1969). Father participation in infancy. *American Journal of Orthopsychiatry, 39,* 466–472.

Rebelsky, F., and Hanks, C. (1971). Fathers' verbal interaction with infants in the first three months of life. *Child Development, 42,* 63–68.

Schaffer, H. (1971). The Growth of Sociability. Baltimore: Penguin Books.

Tasch, R. (1952). The role of the father in the family. *Journal of Experimental Education,*
20, 319–361.

Zajonc, R. (1968). Attitudinal effects of mere exposure. *Journal of Personality and Social*
Psychology Monograph, Supplement 9, 1–27.

ADDITIONAL QUESTIONS

1. What is the problem?
2. What is the hypothesis?
3. What control issues are raised in this experiment? How did the authors deal with these issues?
4. Were some subjects lost? If so, what justification was presented for loss of subjects?
5. What type of scale was used in the experiment?
6. How would you classify this research?
7. Present the data from Table 25.1 in a graphic form.
8. The results are predicated on the assumption that the infants were essentially identical shortly after birth. Discuss this assumption and suggest another design that would have eliminated the need for this assumption.
9. Briefly summarize the authors' discussion section.
10. Does this research alter your concept of sex-role typing? If so, in what way?

APPENDIX

A

Computational Procedures for Basic Statistics

In this appendix the mathematical procedures for a few of the statistical tests mentioned in this book are discussed. Only the rudiments are described.

CENTRAL TENDENCY

A score representative of all scores in a distribution is called *central tendency*. One measure is the *mean* $(\overline{X})$, or average, which is calculated by adding all scores and dividing by the number of scores:

$$\overline{X} = \frac{\Sigma X}{N}$$

Example: Five students took an examination with a possible high score of 10 and low score of 1. The results were:

Paul	7			
Mary	3			
Jill	9	$\overline{X} = \dfrac{\Sigma X}{N} = \dfrac{30}{5} = 6$
Bob	6	
Curt	5	
$\Sigma X =$	30	
$N =$	5	

MEASURE OF VARIABILITY

The *standard deviation*, the most frequently used measure of the distribution of scores, is an index of how much diversion there is among the scores. The formula for calculating a standard deviation is

$$\text{SD} = \sigma = \sqrt{\frac{\Sigma X^2 - N\bar{X}^2}{N - 1}}$$

The calculation of standard deviation (SD) for the examination scores mentioned above is

	X		X^2
Paul	7		49
Mary	3		9
Jill	9		81
Bob	6		36
Curt	5		25
$\Sigma X = 30$			$\Sigma X^2 = 200$
$N = 5$			

$$\sigma = \sqrt{\frac{200 - (5 \cdot 6^2)}{5 - 1}} = 2.24$$

$$\bar{X} = \frac{30}{5} = 6$$

MEASURE OF ASSOCIATION

The strength between two sets of data can be determined by the *correlation coefficient*. Two methods are described below.

Spearman Rank-Order Correlation

The first is called the Spearman rank-order correlation coefficient (*rho*) and is used with *ordinal* data, or data that can be ranked such as the order in which horses might finish in a race and the ranked popularity of their jockeys.

The formula for *rho* is

$$rho = 1 - \frac{6\Sigma d^2}{N^3 - 1}$$

Example: In addition to test scores, the students (see above) also were given a physical examination and ordered for health with a high score indicating good health. The two sets of rank order scores were as follows:

	TEST SCORES (RANK ORDER) X_t	HEALTH SCORES (RANK ORDER) X_h	d DIFFERENCE $(X_t - X_h)$	d^2
Paul	4	5	-1	1
Mary	1	1	0	0
Jill	5	2	3	9
Bob	3	3	0	0
Curt	2	4	-2	4
$N = 5$				$\Sigma d^2 = 14$

$$rho = 1 - \frac{6 \cdot 14}{5^3 - 1} = +.32$$

The significance of *rho* can be determined by finding the *critical values* (a probability level set by the experimenter that reaches *statistical significance*) in Table B.1 in Appendix B.

The critical value for *rho* is found by finding N in Table B.1, which represents the number of pairs. In the case of our example *rho* = .32 with $N = 5$. The value of *rho* is far short of "significance," which, given the limited number of observations, is not surprising.

Pearson Product-Moment Correlation

A second correlation coefficient, called the Pearson product-moment correlation coefficient (r), is used to measure the relationship between two interval scales, such as temperature and test scores.

The formula for r is

$$r = \frac{N\Sigma XY - \Sigma X\Sigma Y}{\sqrt{N\Sigma X^2 - (\Sigma X)^2} \sqrt{N\Sigma Y^2 - (\Sigma Y)^2}}$$

Example: In addition to test scores, the students also were given a test of intelligence. The test had a high score of 20 and a low score of 0. The two sets of scores were the following:

TEST SCORES		INTELLIGENCE SCORES			
	X_t	X_t^2	Y_i	Y_i	X_iY_i
Paul	7	49	18	324	126
Mary	3	9	5	25	15
Jill	9	81	15	225	135
Bob	6	36	12	144	72
Curt	5	25	10	100	50
ΣX = 30		ΣX^2 = 200	ΣY = 60	ΣY^2 = 818	ΣXY = 398

$N = 5$
$df = N - 2$
$df = 5 - 2 = 3$

$$r = \frac{5(398) - (30)(60)}{\sqrt{5(200) - 30^2} \sqrt{5(818) - 60^2}}$$

$$= \frac{1990 - 1800}{\sqrt{1000 - 900} \sqrt{4090 - 3600}}$$

$$= \frac{190}{\sqrt{100} \sqrt{490}}$$

$$= \frac{190}{10 \cdot 22.14}$$

$$= +.86$$

To find the level of significance for r, the *degrees of freedom*, or *df*, must be found. Then consult Table B.2 and determine the level of significance. In our example $r = +.86$ with $df = 3$, which falls between .10 and .05.

CHI-SQUARE

The chi-square (χ^2) test is used to determine if the observed frequency of scores differs from the expected frequency of scores. The formula for chi-square is

$$\chi^2 = \Sigma \frac{(O - E)^2}{E}$$

where O is the observed frequency and E is the expected frequency.

Example: An experimenter is interested in the question of gender and television programs. A sample of 100 subjects is asked which of three television programs they would prefer watching (*Dating Game, 60 Minutes,* or *Wheel of Fortune*). The data (fictitious) were the following:

	DG	60	WF	ROW TOTALS
Males ($N = 50$)	12	20	18	50
Females ($N = 50$)	22	10	18	50
Column Totals	34	30	36	$N = 100$

Find the expected frequencies for each of the cells by the following formula:

$$E = \frac{\text{row total} \times \text{column total}}{N}$$

In the first cell (male–*Dating Game*) the row total is 50, the column total is 34, and $N = 100$. Thus

$$E = \frac{50 \times 34}{100} = 17$$

In the second cell (male–*60 Minutes*) the row total is 50, the column total is 30, and $N = 100$. Thus

$$E = \frac{50 \times 30}{100} = 15$$

Continue to find all expected frequencies and place them in a table:

	DG/E	60/E	WF/E	ROW TOTALS
Males ($N = 50$)	12/17	20/15	18/18	50
Females ($N = 50$)	22/17	10/15	18/18	50
Totals	34	30	36	$N = 100$

Then calculate the squared values for each of the observed expected frequencies: $(O - E)^2 / E$ for each cell. These values are summed as follows:

$$\chi^2 = \frac{(12 - 17)^2}{17} + \frac{(20 - 15)^2}{15} + \frac{(18 - 18)^2}{18} +$$

$$\frac{(22 - 17)^2}{17} + \frac{(10 - 15)^2}{15} + \frac{(18 - 18)^2}{18} = 6.27$$

The degrees of freedom (df) is found by the formula

$$df = (R - 1)(C - 1)$$

where R = rows and C = columns. In our example, $R = 2$, $C = 3$. To find the probability level for the value of 6.27 (in our example) consult Table B.3. The value with 2 df is slightly more than the .05 level.

t-TEST

Here the computational procedures for the t-test are shown. The t-test is used to determine if the probability of the differences between two independent groups occurs by chance. Another form of the t-test is available for correlated measures. For further details and assumptions about the t-test and other statistical procedures mentioned here consult a standard statistics text.

The formula for the t-test for independent groups is

$$t = \frac{\overline{X}_1 - \overline{X}_2}{\sqrt{\left(\frac{SS_1 + SS_2}{(n_1 - 1) + (n_2 - 1)}\right)\left(\frac{1}{n_1} + \frac{1}{n_2}\right)}}$$

where $\overline{X}_1$ is the mean for the first group, $\overline{X}_2$ is the mean for the second group, n is equal to the number of subjects in each group, and SS can be found by the following formula:

$$SS = \Sigma X^2 - \frac{(\Sigma X)^2}{n}$$

Example: A group of 10 randomly selected sample of subjects is randomly assigned to two equal groups. One group (control) is asked to copy notes from a score of music as originally composed and a second group is asked to copy notes from the same score, except that the subjects hear the music being played at the same time as they are copying the notes. The independent variable is the playing of the music; the dependent variable is the time taken to copy the score.

The (hypothetical) results were

TIME (IN SECONDS) TO COPY MUSIC

X_1 CONTROL	X_1^2	X_2 EXPERIMENTAL	X_2^2
7	49	5	25
12	144	8	64
10	100	7	49
11	121	10	100
14	196	9	81
$\Sigma X_1 = 54$	$\Sigma X_1^2 = 610$	$\Sigma X_2 = 39$	$\Sigma X_2^2 = 319$

Find components of formula and substitute values for computation of t.

$$SS_1 = \Sigma X_1^2 - \frac{(\Sigma X_1)^2}{n_1} = 610 - \frac{(54)^2}{5} = 26.8$$

$$SS_2 = \Sigma X_2^2 - \frac{(\Sigma X_2)^2}{n_2} = 319 - \frac{(39)^2}{5} = 14.8$$

$$\overline{X}_1 = \frac{\Sigma X_1}{n_1} = \frac{54}{5} = 10.8$$

$$X_2 = \frac{\Sigma X_2}{n_2} = \frac{39}{5} = 7.8$$

$$t = \frac{10.8 - 7.8}{\sqrt{\left(\frac{26.8 + 14.8}{(5-1)+(5-1)}\right)\left(\frac{1}{5} + \frac{1}{5}\right)}}$$

$$= \frac{3}{\sqrt{\left(\frac{41.6}{8}\right)\left(\frac{2}{5}\right)}} = \frac{3}{\sqrt{2.08}} = \frac{3}{1.44} = 2.08$$

$$df = n_1 + n_2 - 2$$
$$df = 5 + 5 - 2 = 8$$

Consult Table B.4 to find the values of t. The value of t (2.08) with 8 df falls between p of .05 and .10 and would be considered significant at $p < .10$. Most experiments in psychology require at least a $p < .05$ to be considered statistically significant.

ANALYSIS OF VARIANCE

The *analysis of variance* (ANOVA) is one of the most useful and versatile of all statistics used in psychology today. It can be used with between subjects or within subjects designs and with experiments that have several levels of the independent variable. Even more complex designs may be treated by the ANOVA.

The main formula for F is

$$F = \frac{MS_{bg}}{MS_{wg}} \qquad\qquad df_{bg} = k - 1$$

$$MS_{bg} = \frac{SS_{bg}}{df_{bg}} \qquad\qquad df_{wg} = N - k$$

$$MS_{wg} = \frac{SS_{wg}}{df_{wg}}$$

$$SS_{bg} = \frac{(\Sigma X_1)^2}{n_1} + \frac{(\Sigma X_2)^2}{n_2} + \frac{(\Sigma X_3)^2}{n_3} - \frac{(\Sigma X_t)^2}{n_t}$$

$$SS_{wg} = SS_{tot} - SS_{bg}$$

$$SS_{tot} = \Sigma X_t^2 - \frac{(\Sigma X_1 + \Sigma X_2 + \Sigma X_3)^2}{N_t}$$

Example: An experimenter is interested in the effect of imagery on learning. She defines *imagery* as how well a word conjures up a picture in the head. For example, the words *arrow, salary,* and *context* were rated from high to low in imagery. A group of these three types of words, or levels of the independent variable, was given to three independent samples of subjects. The subjects studied the words and then were asked to recall them. The independent variable was the three levels of imagery. The dependent variable was the number of items recalled.

The results were as follows:

NUMBER OF WORDS RECALLED AS A FUNCTION OF IMAGERY

GROUP 1 LOW IMAGERY		GROUP 2 MODERATE IMAGERY		GROUP 3 HIGH IMAGERY	
X_1	X_1^2	X_2	X_2^2	X_3	X_3^2
3	9	9	81	12	144
5	25	4	16	14	196
2	4	8	64	12	144
4	16	9	81	10	100
6	36	10	100	15	225
$\Sigma X_1 = 20$	$\Sigma X_1^2 = 90$	$\Sigma X_1 = 40$	$\Sigma X_2^2 = 342$	$\Sigma X_3 = 63$	$\Sigma X_3^2 = 839$
$n_1 = 5$		$n_2 = 5$		$n_3 = 5$	

The calculation of F follows:

$$\Sigma X_t^2 = 90 + 342 + 839 = 1271$$

$$\Sigma X_t = 20 + 40 + 63 = 123$$

$$SS_{tot} = 1271 - \frac{(123)^2}{15} = 262.4$$

$$SS_{bg} = \frac{(20)^2}{5} + \frac{(40)^2}{5} + \frac{(63)^2}{5} - \frac{(123)^2}{15} = 185.2$$

$$SS_{wg} = 262.4 - 185.2 = 77.2$$

$$df_{bg} = k - 1 = 3 - 1 = 2$$

$$df_{wg} = N - k = 15 - 3 = 12$$

$$MS_{bg} = \frac{185.2}{2} = 92.60$$

$$MS_{wg} = \frac{77.2}{12} = 6.43$$

$$F = \frac{92.60}{6.43} = 14.40$$

To determine if the obtained F value is statistically significant consult Table B.5. In order to find the appropriate F number two degrees of freedom must be used: the df for the numerator (the df associated with the between group data) and the df for the denominator (the df associated with the within group data). These are found by the following formulas:

$$df_{bg} = k - 1$$
$$df_{wg} = N - k$$

Where k is the number of cases of experimental conditions. In the above example, $k = 3$. And N is the total number of observations. In our example $N = 15$. In the calculation of F we found df_{bg} and df_{wg} to be 2 and 12 respectively. At the intersection of 2 and 12 in Table B.5, we find that our value of 14.40 is greater than the value for the .01 level of significance. This is expressed as: $F(2,12) = 14.40$, $p < .01$.

APPENDIX
B

Statistical Tables

Table B.1 CRITICAL VALUES OF *rho*
(SPEARMAN RANK-ORDER
CORRELATION COEFFICIENT)

N	$p = .0500$	$p = .0100$
5	1.000	—
6	.886	1.000
7	.786	.929
8	.738	.881
9	.683	.833
10	.648	.794
12	.591	.777
14	.544	.715
16	.506	.665
18	.475	.625
20	.450	.591
22	.428	.562
24	.409	.537
26	.392	.515
28	.377	.496
30	.364	.478

Computed from Olds, E. G., Distribution
of the sum of squares of rank differences
for small numbers of individuals, *Annals
of Mathematical Statistics*, 1938, IX,
133–148, and the 5% significance levels
for sums of squares of rank differences
and a correction, *Annals of
Mathematical Statistics*, 1949, XX, 117–
118, by permission of the Insitute of
Mathematical Statistics.

Table B.2   CRITICAL VALUES OF *r* (PEARSON PRODUCT-MOMENT CORRELATION COEFFICIENT)

df	LEVEL OF SIGNIFICANCE FOR TWO-TAILED TEST		
	.10	.05	.01
1	.988	.997	.9999
2	.900	.950	.990
3	.805	.878	.959
4	.729	.811	.917
5	.669	.754	.874
6	.622	.707	.834
7	.582	.666	.798
8	.549	.632	.765
9	.521	.602	.735
10	.497	.576	.708
11	.476	.553	.684
12	.458	.532	.661
13	.441	.514	.641
14	.426	.497	.623
15	.412	.482	.606
16	.400	.468	.590
17	.389	.456	.575
18	.378	.444	.561
19	.369	.433	.549
20	.360	.423	.537
25	.323	.381	.487
30	.296	.349	.449
35	.275	.325	.418
40	.257	.304	.393
45	.243	.288	.372
50	.231	.273	.354
60	.211	.250	.325
70	.195	.232	.303
80	.183	.217	.283
90	.173	.205	.267
100	.164	.195	.254

Adapted from R. A. Fisher, *Statistical Methods for Research Workers*. 14th Edition. Copyright 1973. Hafner Press.

Table B.3   CRITICAL VALUES OF CHI-SQUARE

df	$p = .05$	$p = .01$
1	3.84	6.64
2	5.99	9.21
3	7.82	11.34
4	9.49	13.28
5	11.07	15.09
6	12.59	16.81
7	14.07	18.48
8	15.51	20.09
9	16.92	21.67
10	18.31	23.21
11	19.68	24.72
12	21.03	26.22
13	22.36	27.69
14	23.68	29.14
15	25.00	30.58
16	26.30	32.00
17	27.59	33.41
18	28.87	34.80
19	30.14	36.19
20	31.41	37.57
21	32.67	38.93
22	33.92	40.29
23	35.17	41.64
24	36.42	42.98
25	37.65	44.31
26	38.88	45.64
27	40.11	46.96
28	41.34	48.28
29	42.56	49.59
30	43.77	50.89

Table B.3 is taken from Table 4 of Fisher & Yates, *Statistical Tables for Biological, Agricultural and Medical Research*, published by Longman Group Ltd., London (previously published by Oliver and Boyd Ltd., Edinburgh). By permission of the authors and publishers.

Table B.4   CRITICAL VALUES OF t

df	$p = .10$	$p = .05$	$p = .02$	$p = .01$
1	6.314	12.706	31.821	63.657
2	2.920	4.303	6.965	9.925
3	2.353	3.182	4.541	5.841
4	2.132	2.776	3.747	4.604
5	2.015	2.571	3.365	4.032
6	1.943	2.447	3.143	3.707
7	1.895	2.365	2.998	3.499
8	1.860	2.306	2.896	3.355
9	1.833	2.262	2.821	3.250
10	1.812	2.228	2.764	3.169
11	1.796	2.201	2.718	3.106
12	1.782	2.179	2.681	3.055
13	1.771	2.160	2.650	3.012
14	1.761	2.145	2.624	2.977
15	1.753	2.131	2.602	2.947
16	1.746	2.120	2.583	2.921
17	1.740	2.110	2.567	2.898
18	1.734	2.101	2.552	2.878
19	1.729	2.093	2.539	2.861
20	1.725	2.086	2.528	2.845
21	1.721	2.080	2.518	2.831
22	1.717	2.074	2.508	2.819
23	1.714	2.069	2.500	2.807
24	1.711	2.064	2.492	2.797
25	1.708	2.060	2.485	2.787
26	1.706	2.056	2.479	2.779
27	1.703	2.052	2.473	2.771
28	1.701	2.048	2.467	2.763
29	1.699	2.045	2.462	2.756
30	1.697	2.042	2.457	2.750
60	1.671	2.000	2.390	2.660
∞	1.645	1.960	2.326	2.576

Table B.4 is taken from Table 3 of Fisher & Yates, *Statistical Tables for Biological, Agricultural and Medical Research*, published by Longman Group Ltd., London (previously published by Oliver & Boyd Ltd., Edinburgh). By permission of the authors and publishers.

Table B.5 CRITICAL VALUES OF F (TOP NUMBER IN EACH CELL IS FOR TESTING AT .05 LEVEL; BOTTOM NUMBER FOR TESTING AT .01 LEVEL.)

DEGREES OF FREEDOM FOR NUMERATOR—df_{bg}

df_{wg}	1	2	3	4	5	6	8	12	24	∞
1	161.45	199.50	215.72	224.57	230.17	233.97	238.89	243.91	249.04	254.32
	4032.10	4999.03	5403.49	5625.14	5764.08	5859.39	5981.34	6105.83	6234.16	6366.48
2	18.51	19.00	19.16	19.25	19.30	19.33	19.37	19.41	19.45	19.50
	98.49	99.01	99.17	99.25	99.30	99.33	99.36	99.42	99.46	99.50
3	10.13	9.55	9.28	9.12	9.01	8.94	8.84	8.74	8.64	8.53
	34.12	30.81	29.46	28.71	28.24	27.91	27.49	27.05	26.60	26.12
4	7.71	6.94	6.59	6.39	6.26	6.16	6.04	5.91	5.77	5.63
	21.20	18.00	16.69	15.98	15.52	15.21	14.80	14.37	13.93	13.46
5	6.61	5.79	5.41	5.19	5.05	4.95	4.82	4.68	4.53	4.36
	16.26	13.27	12.06	11.39	10.97	10.67	10.27	9.89	9.47	9.02
6	5.99	5.14	4.76	4.53	4.39	4.28	4.15	4.00	3.84	3.67
	13.74	10.92	9.78	9.15	8.75	8.47	8.10	7.72	7.31	6.88
7	5.59	4.74	4.35	4.12	3.97	3.87	3.73	3.57	3.41	3.23
	12.25	9.55	8.45	7.85	7.46	7.19	6.84	6.47	6.07	5.65
8	5.32	4.46	4.07	3.84	3.69	3.58	3.44	3.28	3.12	2.93
	11.26	8.65	7.59	7.01	6.63	6.37	6.03	5.67	5.28	4.86
9	5.12	4.26	3.86	3.63	3.48	3.37	3.23	3.07	2.90	2.71
	10.56	8.02	6.99	6.42	6.06	5.80	5.47	5.11	4.73	4.31
10	4.96	4.10	3.71	3.48	3.33	3.22	3.07	2.91	2.74	2.54
	10.04	7.56	6.55	5.99	5.64	5.39	5.06	4.71	4.33	3.91
11	4.84	3.98	3.59	3.36	3.20	3.09	2.95	2.79	2.61	2.40
	9.65	7.20	6.22	5.67	5.32	5.07	4.74	4.40	4.02	3.60
12	4.75	3.88	3.49	3.26	3.11	3.00	2.85	2.69	2.50	2.30
	9.33	6.93	5.93	5.41	5.06	4.82	4.50	4.16	3.78	3.36
14	4.60	3.74	3.34	3.11	2.96	2.85	2.70	2.53	2.35	2.13
	8.86	6.51	5.56	5.03	4.69	4.46	4.14	3.80	3.43	3.00

DEGREES OF FREEDOM FOR DENOMINATOR—df_{wg}

df										
16	4.49 / 8.53	3.63 / 6.23	3.24 / 5.29	3.01 / 4.77	2.85 / 4.44	2.74 / 4.20	2.59 / 3.89	2.42 / 3.55	2.24 / 3.18	2.01 / 2.75
18	4.41 / 8.28	3.55 / 6.01	3.16 / 5.09	2.93 / 4.58	2.77 / 4.25	2.66 / 4.01	2.51 / 3.71	2.34 / 3.37	2.15 / 3.01	1.92 / 2.57
20	4.35 / 8.10	3.49 / 5.85	3.10 / 4.94	2.87 / 4.43	2.71 / 4.10	2.60 / 3.87	2.45 / 3.56	2.28 / 3.23	2.08 / 2.86	1.84 / 2.42
25	4.24 / 7.77	3.38 / 5.57	2.99 / 4.68	2.76 / 4.18	2.60 / 3.86	2.49 / 3.63	2.34 / 3.32	2.16 / 2.99	1.96 / 2.62	1.71 / 2.17
30	4.17 / 7.56	3.32 / 5.39	2.92 / 4.51	2.69 / 4.02	2.53 / 3.70	2.42 / 3.47	2.27 / 3.17	2.09 / 2.84	1.89 / 2.47	1.62 / 2.01
40	4.08 / 7.31	3.23 / 5.18	2.84 / 4.31	2.61 / 3.83	2.45 / 3.51	2.34 / 3.29	2.18 / 2.99	2.00 / 2.66	1.79 / 2.29	1.52 / 1.82
50	4.03 / 7.17	3.18 / 5.06	2.79 / 4.20	2.56 / 3.72	2.40 / 3.41	2.29 / 3.19	2.13 / 2.89	1.95 / 2.56	1.74 / 2.18	1.44 / 1.68
60	4.00 / 7.08	3.15 / 4.98	2.76 / 4.13	2.52 / 3.65	2.37 / 3.34	2.25 / 3.12	2.10 / 2.82	1.92 / 2.50	1.70 / 2.12	1.39 / 1.60
70	3.98 / 7.01	3.13 / 4.92	2.74 / 4.07	2.50 / 3.60	2.35 / 3.29	2.23 / 3.07	2.07 / 2.78	1.89 / 2.45	1.67 / 2.07	1.35 / 1.53
80	3.96 / 6.98	3.11 / 4.88	2.72 / 4.04	2.49 / 3.56	2.33 / 3.26	2.21 / 3.04	2.06 / 2.74	1.88 / 2.42	1.65 / 2.03	1.32 / 1.49
90	3.95 / 6.92	3.10 / 4.85	2.71 / 4.01	2.47 / 3.53	2.32 / 3.23	2.20 / 3.01	2.04 / 2.72	1.86 / 2.39	1.64 / 2.00	1.30 / 1.46
100	3.94 / 6.90	3.09 / 4.82	2.70 / 3.98	2.46 / 3.51	2.30 / 3.21	2.19 / 2.99	2.03 / 2.69	1.85 / 2.37	1.63 / 1.98	1.28 / 1.43
200	3.89 / 6.76	3.04 / 4.71	2.65 / 3.88	2.42 / 3.41	2.26 / 3.11	2.14 / 2.89	1.98 / 2.60	1.80 / 2.28	1.57 / 1.88	1.19 / 1.28
∞	3.84 / 6.64	2.99 / 4.60	2.60 / 3.78	2.37 / 3.32	2.21 / 3.02	2.09 / 2.80	1.94 / 2.51	1.75 / 2.18	1.52 / 1.79	1.00 / 1.00

DEGREES OF FREEDOM FOR DENOMINATOR (CONTINUED)

Adapted from Table F of H. E. Garrett, *Statistics in Psychology and Education*, 5th Edition, Copyright 1958, David McKay Co., Inc.

References

American Psychological Association. (1983). *Ethical standards of psychologists* (rev. ed.). Washington, DC: Author.

American Psychological Association. (1983). *Standards for providers of psychological services* (rev. ed.). Washington, DC: Author.

American Psychological Association. (1981). *Directory of the American Psychological Association.* Washington, DC.

Asch, S. (1952). *Social psychology.* Englewood Cliffs, NJ: Prentice-Hall.

Atkinson, R.C., & Raugh, M. R. (1975). An application of the mnemonic keyword method to the acquisition of a Russian vocabulary. *Journal of Experimental Psychology: Human Learning and Memory, 104,* 126–133.

Ayllon, T. (1963). Intensive treatment of psychotic behavior by stimulus satiation and food reinforcement. *Behavior Research and Therapy, 1,* 53–61.

Ayllon, T., & Azrin, N. H. (1968). *The token economy: A motivational system for therapy and rehabilitation.* Englewood Cliffs, NJ: Prentice-Hall.

Bayoff, A. G. (1940). The experimental social behavior of animals: II. The effect of early isolation of white rats on their competition in swimming. *Journal of Comparative Psychology, 29,* 293–306.

Benbow, C., & Stanley, J. C. (1980). Sex differences in mathematical ability: fact or artifact? *Science 210 (4475):* 1262–1264.

Bitterman, M. E. (1969). Thorndike and the problem of animal intelligence. *American Psychologist, 24,* 444–453.

Bower, G. H., Karlin, M. B., & Dueck, A. (1975). Comprehension and memory for pictures. *Memory and Cognition, 3,* 216–220.

Bryson, J. B., & Hamblin, K. (1988). Reporting Infidelity: The MUM effect and the double standard. Paper presented at WPA meeting: April 29, 1988, Burlingame, CA.

Campbell, D. T., & (1969). Reforms as experiments. *American Psychologist, 24,* 409–429.

Campbell, D. T., & Stanley, J. C. (1966). *Experimental and quasi-experimental designs for research.* Chicago: Rand McNally.

Chi, M. T. (1978). Knowledge structures and memory development. In R. S. Siegler (Ed.), *Children's thinking: What develops?* Hillsdale, NJ: Erlbaum.

Conant, J. B. (1951). *Science and common sense:* New Haven, CT: Yale University Press.

Cook, T. D., & Campbell, D. T. (1979). *Quasi-experimentation: Design & analysis issues for field settings.* Chicago: Rand McNally.

Crespi, L. (1942). Quantitative variation of incentive and performance in the white rat. *American Journal of Psychology, 55,* 467–517.

Czeisler, C. A., Moore, C., Ede, M. C., & Coleman, R. M. (1982). Rotating shift work schedules that disrupt sleep are improved by applying circadian principles. *Science, 217,* 460.

Dempster, F. N. (1981). Memory span: Sources of individual and developmental differences. *Psychological Bulletin, 89,* 66.

Ehrenfreund, D., & Badia, P. (1962). Response strength as a function of drive level and pre- and postshift incentive magnitude. *Journal of Experimental Psychology, 63,* 468–471.

Elliott, M. H. (1928). The effect of change of reward on the maze performance of rats. *University of California Publications in Psychology, 4,* 19–30.

Ferster, C. B., & Perrott, M. C. (1968). *Behavior principles.* Englewood Cliffs, NJ: Prentice-Hall.

Festinger, L. (1957). *A theory of cognitive dissonance.* New York: Harper & Row.

Fisher, J. L., & Harris, M. B. (1973). Effect of note taking and review on recall. *Journal of Educational Psychology, 65,* 321–325.

Goodenough, D. R., Shapiro, A., Holden, M., & Steinschriber, L. (1959). A comparison of "dreamers" and "nondreamers": Eye movements, electroencephalograms, and the recall of dreams. *Journal of Abnormal and Social Psychology, 59,* 295–302.

Hayes, S. (1981). Single case experimental design and empirical clinical practice, *Journal of Consulting and Clinical Psychology, 49,* 193–211.

Heidmets, M., & Niit, T. (1984). An activity analysis of office environments: Reality and preferences. *Problems of Perception and Social Interaction.* Publication of Tartu University, Tartu, Estonia.

Hilgard, E. R. (1969). Pain as a puzzle for psychology and physiology. *American Psychologist, 24,* 103–113.

Hirsch, I. J. Audition. (1966). In J. B. Sidowski (Ed.), *Experimental methods and instrumentation in psychology.* New York: McGraw-Hill.

Howes, D. H., & Solomon, R. L. (1950). A note on McGuinnes' "Emotionality and perceptual defense." *Psychological Review, 57,* 229–234.

Howes, D. H., & Solomon, R. L. (1951). Visual duration threshold as a function of word-probability. *Journal of Experimental Psychology, 41,* 401–410.

Johnson, H. H., & Scileppi, J. A. (1969). Effects of ego-involvement conditions on attitude change to high and low credibility communicators. *Journal of Personality and Social Psychology, 13*, 31–36.

Lambert, W. W., & Solomon, R. L. (1952). Extinction of a running response as a function of block point from the goal. *Journal of Comparative and Physiological Psychology, 45*, 269–279.

Larson, C. C. (1982). Animal research: Striking a balance. *APA Monitor, 13* (Jan.), 1 ff.

Linder, D. E., Cooper, J., & Jones, E. E. (1967). Decision freedom as a determinant of the role of incentive magnitude in attitude change. *Journal of Personality and Social Psychology, 6*, 245–254.

Lorge, I. (1930). Influence of regularly interpolated time intervals upon subsequent learning. *Teachers College, Columbia University Contributions to Education* (Whole No. 438).

Loftus, E. F., & Palmer J. C. (1974). Reconstruction of automobile destruction: An example of the interaction between language and memory. *Journal of Verbal Learning and Verbal Behavior, 13*, 585–589.

Mahoney, M. J., Moura, N. G. M., & Wade, T. C. (1973). Relative efficacy of self-reward, self-punishment, and self-monitoring techniques for weight loss. *Journal of Consulting and Clinical Psychology, 40*, 404–407.

Marshall, J. (1969). *Law and Psychology in Conflict.* New York: Anchor Books.

More, A. J. (1969). Delay of feedback and the acquisition and retention of verbal materials in the classroom. *Journal of Educational Psychology, 60*, 339–342.

Paul, G. L. (1966). *Insight versus desensitization in psychotherapy.* Stanford, CA: Stanford University Press.

Posner, M. I. (1969). Abstraction and the process of recognition. In J. T. Spence & G. H. Bower (Eds.), *The psychology of learning and motivation: Advances in learning and motivation* (Vol. 3). New York: Academic Press.

Posner, M. I., Boies, S. J., Eichelman, W., & Taylor, R. L. (1969). Retention of visual and name codes of single letters. *Journal of Experimental Psychology, 73*, 28–38.

Posner, M. I., & Keele, S. W. (1968). On the genesis of abstract ideas. *Journal of Experimental Psychology, 77*, 353–363.

Postman, L., Bronson, W. C., & Gropper, G. L. (1952). Is there a mechanism of perceptual defense? *Journal of Abnormal and Social Psychology, 48*, 215–224.

Pryor, K. W., Haag, R., & O'Reilly, J. (1969). The creative porpoise: Training for novel behavior. *Journal of the Experimental Analysis of Behavior, 12*, 653–661.

Rosenthal, R. (1966). *Experimenter effects in behavioral research.* New York: Appleton-Century-Crofts.

Rosenthal, R. (1969). Interpersonal expectations: Effects of the experimenter's hypothesis. In R. Rosenthal & R. L. Rosnow (Eds.), *Artifact in behavioral research.* New York: Academic Press.

Rosenthal, R., & Fode, K. (1963). The effects of experimenter bias on the performance of the albino rat. *Behavioral Science, 8*, 183–189.

Rosenthal, R., & Rosnow, R. L. (Eds.). (1969). *Artifact in behavioral research.* New York: Academic Press.

Sands, S. F., Lincoln, C. E., & Wright, A. A. (1982). Pictorial similarity judgments and the organization of visual memory in the Rhesus monkey. *Journal of Experimental Psychology: General, 3(4),* 369.

Simon, C. W., & Emmons, W. H. (1956). Responses to material presented during various levels of sleep. *Journal of Experimental Psychology, 51,* 89–97.

Smith, A. P., Tyrrell, D. A. J., Coyle, K., & Willman, J. S. (1987). Selective effects of minor illnesses on human performance. *British Journal of Psychology, 78,* 183–188.

Solso, R. L. (1987a). The social-political consequences of the organization and dissemination of knowledge. *American Psychologist, 42,* 824–825.

Solso, R. L. (1987b). Recommended readings in psychology over the past 33 years. *American Psychologist, 42,* 1130–1132.

Solso, R. L., & McCarthy, J. E. (1981). Prototype formation of faces: A case of pseudomemory. *British Journal of Psychology, 72,* 499–503.

Solso, R. L., & Short, B. A. (1979). Color recognition. *Bulletin of the Psychonomic Society, 14,* 275–277.

Snodgrass, J. G., Levy-Berger, G., & Haydon, M. (1985). *Human experimental psychology.* New York: Oxford.

Sprafkin, J. N., Liebert, R. M., & Poulos, R. W. (1975). Effects of a prosocial televised example on children's helping. *Journal of Experimental Child Psychology, 20,* 119–126.

Stevens, S. S., & Davis, H. (1938). *Hearing, its psychology and physiology.* New York: Wiley.

Stevenson, H. W., & Odom, R. D. (1962). The effectiveness of social reinforcement following two conditions of social deprivation. *Journal of Abnormal and Social Psychology, 65,* 429–431.

Suls, J. M., & Miller, R. L. (1976). Humor as an attributional index. *Personality and Social Psychology Bulletin, 2,* 256–259.

Supa, M., Cotzin, M., & Dallenbach, K. M. (1944). "Facial vision": The perception of obstacles by the blind. *American Journal of Psychology, 57,* 133–183.

Terkel, J., & Rosenblatt, J. S. (1968). Maternal behavior induced by maternal blood plasma injected into virgin rats. *Journal of Comparative and Physiological Psychology, 65,* 479–482.

Thumin, F. J. (1962). Identification of cola beverages. *Journal of Applied Psychology, 46,* 358–360.

Underwood, B. J. (1983). *Attributes of memory.* Glenview, IL: Scott-Foresman.

Underwood, B. J. (1975). *Psychological research.* New York: Appleton-Century-Crofts.

Underwood, B. J., Rehula, R., & Keppel, G. (1962). Item selection in paired-associate learning. *American Journal of Psychology, 75,* 353–371.

Walk, R. D. (1969). Two types of depth discrimination by the human infant with five inches of visual depth. *Psychonomic Science, 14,* 253–254.

Walter, R. H., & Parke, R. D. (1964). Emotional arousal, isolation, and discrimination learning in children. *Journal of Experimental Child Psychology, 1,* 163–173.

Webb, E. J., Campbell, D. T., Schwartz, R. D., & Sechrest, L. (1966). *Unobtrusive measures: Nonreactive research in the social sciences.* Chicago: Rand McNally.

Wilson, G. D., Nias, D. K. B., & Brazendale, A. H. (1975). Vital statistics, perceived sexual attractiveness, and response to risqué humor. *The Journal of Social Psychology, 95,* 201–255.

Wynder, E. L., & Stellman, S. D. (1977). Comparative epidemiology of tobacco-related cancer. *Cancer Research, 37,* 4608–4622.

Name Index

Subject Index

The experimental effects of review on recall are more definitive. Howe (1970a) found that subjects who were allowed to review notes that they had taken had significantly higher mean recall scores than subjects who attended to an interfering task. DiVesta and Gray (1972) found that rehearsal immediately after listening increased the number of words recalled.

The functions of note taking have been discussed in various publications. Notes have been described as serving as an "external memory device" (Miller, Galanter, & Pribram, 1960) for storing data for later retrieval and study. Note taking has also been described as a kind of "encoding" process, in which the learner reorganizes the input data, and by putting it in his own words, in a sense, transforms the data and "makes it his own" (DiVesta & Gray, 1972; Howe, 1970b).

DiVesta and Gray (1972) as well as Howe (1970a) see the second or "encoding" function of note taking as the more important of the two functions in terms of learning. DiVesta and Gray in fact suggest that the external memory function of note taking may possibly lead the learner to feel that having a "good set of notes" is the equivalent of studying and influence him to bypass any form of transformational coding. Howe (1970a) also discusses the importance of the second function of note taking and points out that if the external memory function were the only reason for taking notes, "It would be simpler to provide mimeographed outlines leaving the student free to attend and react to other things. . . . (p. 100)."

b   **The present study sought to discover if note taking and review of notes do aid students in short-term and long-term recall of material presented in a typical lecture setting.** It further investigated whether note taking serves as an encoding device, an external memory device, or a combination of the two, by manipulating the various review conditions of the study. It also attempted to assess whether there was a relationship between students' opinions about taking notes (liking, disliking, etc.) and their subsequent recall of material and if there were any significant correlations between quality of notes taken and subsequent recall, as well as efficiency of notes taken and subsequent recall.

c   **It was predicted that subjects who took notes, regardless of review condition, would perform better than subjects who did not take notes (the "encoding" function) and that those who had notes to review (either their**

the relationship between a student's attitude toward note taking and recall. In (b) the fundamental plan of the experiment is clearly identified. In most of the experiments reviewed in this book the authors wisely chose to add a sentence or two in the introduction that sharply delineate the experimental plan. The use of that technique is very effective in enhancing the understanding of the experiment.

In (c) the hypothesis of the experiment is stated in the form of a pre-